Proceedings of the 2nd International Conference on LASERS in manufacturing

26-28 March 1985
Birmingham U.K.

Edited by Dr. M.F. Kimmitt

LIM-2

An international event organised and sponsored by IFS (Conferences) Ltd, Kempston, Bedford, UK

Co-sponsors: Laser Institute of America
Lasers and Optics International

Co-published by: IFS (Publications) Ltd and North-Holland
(a division of Elsevier Science Publishers BV)

Proceedings of the 2nd International Conference on LASERS IN MANUFACTURING

An international event organised and sponsored by:

IFS (Conferences) Ltd
Kempston, Bedford

CO-SPONSORS

Laser Institute of America
Lasers and Optics International

PROGRAMME COMMITTEE

Chairman: Dr M. F. Kimmitt, University of Essex
Mr P. Anthony, Rofin Sinar Laser UK Ltd
Mr R. P. Ashcroft, Control Laser Ltd
Mr A. J. Bishop, Lasercut Products Ltd
Mr M. J. Bragg, Laser Scientific Services Ltd
Dr R. C. Crafer, The Welding Institute
Dr R. B. Dennis, Edinburgh Instruments Ltd
Mrs S. Gardner, Lasers and Optics International
Mrs K. Gibbs, IFS (Conferences) Ltd
Mr G. W. Hamilton, Barr & Stroud Ltd
Mr J. Leece, Ferranti plc
Miss A. R. Pollock, IFS (Conferences) Ltd
Dr I. J. Spalding, UKAEA – Culham Laboratory
Dr W. M. Steen, Imperial College of Science and Technology
Dr T. M. Weedon, JK Lasers Ltd

© March 1985, IFS (Conferences) Ltd and authors

Jointly published by:
IFS (Publications) Ltd, UK
35-39 High Street, Kempston, Bedford MK42 7BT, England
ISBN 0-903608-85-5

North-Holland (a division of Elsevier Science Publishers BV)
PO Box 1991, 1000 BZ Amsterdam, The Netherlands
ISBN 0-444-877339

In the USA and Canada:
Elsevier Science Publishing Company Inc
52 Vanderbilt Avenue, New York, NY 10017

Printed by: Cotswold Press, Oxford, UK

CONTENTS

Lasers in Perspective

Keynote address: Why choose a laser for materials processing?
J. T. Luxon, GMI Engineering & Management Institute, USA 1

What choice for high integrity joints: electron beam or laser beam welding?
G. Sayegh, Sciaky SA, France 11

Energy efficient laser machining
L. J. Li, Hunan University, People's Republic of China and J. Mazumder, University of Illinois, USA 23

Laser Processes – Cutting

Cut edge quality improvement by laser pulsing
J. Powell, T. G. King and I. A. Menzies, Loughborough University of Technology, UK 37

Laser cutting of sheet metal in modern manufacturing
J. M. Weick and R. Wollermann-Windgasse, Trumpf GmbH & Co, West Germany 47

Laser cutting of Al 7075 sheets
A. Di Ilio, Universita dell'Aquila, G. Dionoro, Universita di Cagliari, and F. Memola Capece Minutolo and V. Tagliaferri, Universita di Napoli, Italy 57

Prevention of dross attachment during laser cutting
F. N. Birkett, D. P. Herbert and J. Powell, Loughborough University of Technology, UK 63

Excimer lasers in photolithography
M. C. Gower, SERC Rutherford Appleton Laboratory, UK 67

Laser Processes – Surface Treatment

Laser surface alloyed Fe-Cr-C
S. Das, I. Dumler and J. Mazumder, University of Illinois, USA 73

In situ clad alloy formation by laser cladding
T. Takeda, W. M. Steen and D. R. F. West, Imperial College of Science and Technology, UK 85

Laser surface alloying
E. F. Semiletova and T. H. Dumbadze, Georgian Polytechnic Institute, USSR 97

Properties of laser melted SG iron
H. W. Bergmann, IWW, Universität Clausthal, West Germany and M. Young, University of Birmingham, UK 109

Laser hardening of a 12%-Cr steel
M. Roth, Brown Boveri Research Center, Switzerland and M. Cantello, RTM Institute, Italy 119

Non-Destructive Testing and Optical Sensors

Process monitoring of high power CO_2-lasers in manufacturing
W. König. F. U. Meis, H. Willerscheid and Cl. Schmitz-Justen, Fraunhofer-Institut für Produktionstechnologie (IPT), West Germany 129

The use of electronic speckle pattern interferometry (ESPI) as an inspection tool
P. C. Montgomery and J. Tyrer, Loughborough University of Technology, UK 141

Non-destructive testing of adhesive joints
B. P. Holownia, Loughborough University of Technology, UK 151

Application of immersion technique on the measurement of some engineering products
M. M. El Sayed, Helwan University and M. M. Koura, Ain Shams University, Egypt 159

Optical polar profilometer: a new method for analysis of surfaces with circular symmetry
G. Laufer and E. Lenz, Israel Institute of Technology, Israel and Y. Fainman, University of California, USA 169

Optical fibre sensors in industrial process control
N. Macfadyen, Barr & Stroud Ltd, UK 173

Laser diagnostics of combustion devices and chemical reactors using coherent Anti-Stokes Raman spectroscopy
D. A. Greenhalgh, Harwell Laboratories, UK 183

Equipment

New CO_2-lasers of 1-4 kilowatt power with fast axial gas flow
L. Bakowsky, Messer Griesheim GmbH, West Germany 195

The start of a new generation of CO_2-lasers for industry
P. Hoffmann, Laser Innovation GmbH & Co KG, West Germany 201

The CO_2 waveguide laser (a laser with designs on industry)
I. E. Ross, Ferranti plc, UK 209

A laser robot for cutting and trimming deeply stamped metal sheets
A. Delle Piane, Prima Progetti SpA, Italy 219

Laser safety in perspective
E. A. Cox, Consultant, UK 225

Laser Systems

The use of lasers in manufacturing – relevant research at the Welding Institute
P. J. Oakley, M. N. Watson, C. J. Dawes and N. R. Stockham, The Welding Institute, UK 237

A flexible laser manufacturing system based on the composition of several laser beams
V. Fantini, L. Garifo and G. Incerti, CISE SpA, and I. Franchetti and L. Grisoni, Alfa Romeo Auto SpA, Italy 249

Laser-cutting and welding in a flexible manufacturing system
H. Uetz, G. Hardock and H.-J. Warnecke, Fraunhofer-Institute for Manufacturing Engineering and Automation (IPA), West Germany 261

Programmable laser character generation
D. I. Greenwood, A. McNeish and J. J. Harris, Isomet Laser Systems Ltd, UK 279

The management of industrial lasers
B. G. Green and M. J. Bragg, Laser Scientific Services Ltd, UK 285

Supplementary Paper

Manufacturing WO_3 and Fe_2O_3 films by laser chemical vapour deposition
M. S. Chiu, C. C. Chou and G. P. Shen, Shanghai Institute of Laser Technology, People's Republic of China 295

Late Papers

Heat treatment of steels using different high power CO_2 laser beam intensity distributions
W. Cerri and A. Vendramini, CISE SpA and E. Ramous, Università di Padova, Italy 301

Optical problems of beam delivery 309
N. Forbes, Ferranti Industrial Electronics Ltd, UK

AUTHOR INDEX

L. Bakowsky 195
  Messer Griesheim GmbH, West Germany
H. W. Bergmann 109
  IWW, Universität Clausthal, West Germany
F. N. Birkett 63
  Loughborough University of Technology, UK
M. J. Bragg 285
  Laser Scientific Services Ltd, UK
M. Cantello 119
  RTM Institute, Italy
M. S. Chiu 295
  Shanghai Institute of Laser Technology, People's Republic of China
C. C. Chou 295
  Shanghai Institute of Laser Technology, People's Republic of China
E. A. Cox 225
  Consultant, UK
S. Das 73
  University of Illinois, USA
C. J. Dawes 237
  The Welding Institute, UK
A. Delle Piane 219
  Prima Progetti SpA, Italy
A. Di Ilio 57
  Universita dell'Aquila, Italy
G. Dionoro 57
  Universita di Cagliari, Italy
T. H. Dumbadze 97
  Georgian Polytechnic Institute, USSR
I. Dumler 73
  University of Illinois, USA
M. M. El Sayed 159
  Helwan University, Egypt
Y. Fainman 169
  University of California, USA
V. Fantini 249
  CISE SpA, Italy
I. Franchetti 249
  Alfa Romeo Auto SpA, Italy
L. Garifo 249
  CISE SpA, Italy
M. C. Gower 67
  SERC Rutherford Appleton Laboratory, UK
B. G. Green 285
  Laser Scientific Services Ltd, UK
D. A. Greenhalgh 183
  Harwell Laboratories, UK
D. I. Greenwood 279
  Isomet Laser Systems Ltd, UK
L. Grisoni 249
  Alfa Romeo Auto SpA, Italy
G. Hardock 261
  Fraunhofer-Institute for Manufacturing Engineering and Automation (IPA), West Germany
J. J. Harris 279
  Isomet Laser Systems Ltd, UK
D. P. Herbert 63
  Loughborough University of Technology, UK
P. Hoffmann 201
  Laser Innovation GmbH & Co KG, West Germany

B. P. Holownia 151
  Loughborough University of Technology, UK
G. Incerti 249
  CISE SpA, Italy
T. G. King 37
  Loughborough University of Technology, UK
W. König 129
  Fraunhofer-Institut für Produktionstechnologie (IPT), West Germany
M. M. Koura 159
  Ain Shams University, Egypt
G. Laufer 169
  Israel Institute of Technology, Israel
E. Lenz 169
  Israel Institute of Technology, Israel
L. J. Li 23
  Hunan University, People's Republic of China
J. T. Luxon 1
  GMI Engineering & Management Institute, USA
N. Macfadyen 173
  Barr & Stroud Ltd, UK
J. Mazumder 23, 73
  University of Illinois, USA
A. McNeish 279
  Isomet Laser Systems Ltd, UK
F. U. Meis 129
  Fraunhofer-Institut für Produktionstechnologie (IPT), West Germany
F. Memola Capece Minutolo 57
  Universita di Napoli, Italy
I. A. Menzies 37
  Loughborough University of Technology, UK
P. C. Montgomery 141
  Loughborough University of Technology, UK
P. J. Oakley 237
  The Welding Institute, UK
J. Powell 37, 63
  Loughborough University of Technology, UK
I. E. Ross 209
  Ferranti plc, UK
M. Roth 119
  Brown Boveri Research Center, Switzerland
G. Sayegh 11
  Sciaky SA, France
Cl. Schmitz-Justen 129
  Fraunhofer-Institut für Produktionstechnologie (IPT), West Germany
E. F. Semiletova 97
  Georgian Polytechnic Institute, USSR
G. P. Shen 295
  Shanghai Institute of Laser Technology, China
W. M. Steen 85
  Imperial College of Science and Technology, UK
N. R. Stockham 237
  The Welding Institute, UK
V. Tagliaferri 57
  Universita di Napoli, Italy
T. Takeda 85
  Imperial College of Science and Technology, UK
J. Tyrer 141
  Loughborough University of Technology, UK

H. Uetz 261
    Fraunhofer-Institute for Manufacturing Engineering and
    Automation (IPA), West Germany
H. J. Warnecke 261
    Fraunhofer-Institute for Manufacturing Engineering and
    Automation (IPA), West Germany
M. N. Watson 237
    The Welding Institute, UK
J. M. Weick 47
    Trumpf GmbH & Co, West Germany
D. R. F. West 85
    Imperial College of Science and Technology, UK
H. Willerscheid 129
    Fraunhofer-Institut für Produktionstechnologie (IPT),
    West Germany
R. Wollermann-Windgasse 47
    Trumpf GmbH & Co, West Germany

M. Young 109
    University of Birmingham, UK

Late Papers

W. Cerri 301
    CISE SpA, Italy
N. Forbes 309
    Ferranti Industrial Electronics Ltd, UK
E. Ramous 301
    Università di Padova, Italy
A. Vendramini 301
    CISE SpA, Italy

FOREWORD

Conference Chairman:
Dr M. F. Kimmitt
University of Essex, UK

IT IS exactly 25 years since laser light was seen for the first time. Furthermore, it is just 20 years since the arrival of the carbon dioxide laser, arguably the most important of these new sources for use in industry. While some will be surprised to realize that the laser has been around so long, others may ask why it has taken many years for them to take a significant place in manufacturing industry. Indeed, the question becomes more interesting when one remembers that ruby lasers were being used for piercing holes in diamond dies for wire-drawing in the early 1960s. Before the end of that decade lasers were also being used for such diverse applications as balancing armatures, cutting out men's suits and micromachining of photolithographic masks, as well as for welding and cutting purposes.

While the answer is not straightforward, it is worth considering the various reasons for this relatively slow take-off, after such a promising start. The early lasers were somewhat cumbersome, particularly the carbon dioxide ones. They were reliable, but perhaps not quite reliable enough. Above all, it was difficult to integrate them into an industrial production process. Added to this, there were fears – mostly unjustified – about their safety. It is interesting that in a recent survey of a large number of industrial laser users the reasons given for purchase included reduction of rejects, less wastage of material, elimination of tool wear, lower operating and maintenance costs and the ability to increase throughput in a fully-automated production system. All of them agreed that lasers saved them money.

It is significant that these answers show lasers as replacements for conventional machine tools and it is important to realize this as we hear of some more exotic applications in this conference. Of course conferences are only of use if they do look at the state-of-the-art and into the future, but it is worth reminding ourselves that we are already in the middle of the decade where lasers are assuming a vital role in manufacturing.

What has produced this change from interest in lasers in the early 1970s to enthusiasm in the 1980s? Lasers have certainly become very reliable, they are more compact and cheaper to run but, perhaps surprisingly, it is carbon dioxide and Nd:YAG, the old favourites, which still lead the field. However, what has advanced much more is the ability to control the beam and apply it to the target. The output of the laser is now uniform and stable, and, with microprocessor control and robotics, it is much easier to integrate the laser into a complete production system. There are still problems but in many applications it is now easier to use a laser than conventional methods.

Finally, we must realize that the next generation of lasers is moving from research to production. Alexandrite is challenging Nd:YAG and Nd:glass. The exotic ultraviolet excimer lasers show high promise for submicron-sized processing and on the horizon are the tunable 'free electron' laser and even x-ray lasers. In this conference we have the privilege of hearing the experts; in surface processing, for example, we will be told of processes which only very high-power lasers can perform. At the same time we must not forget that a small 25 Watt sealed-off carbon dioxide laser, scarcely larger than a helium neon one, can routinely cut out plastic parts with great accuracy, and for weeks on end, with no maintenance. Lasers, from the very big to the very small, are an accepted alternative in a wide range of manufacturing applications. I am sure that this conference will give us all an increased awareness of their possibilities.

LASERS IN PERSPECTIVE

Keynote address: Why choose a laser for materials processing?

J. T. Luxon
GMI Engineering & Management Institute, USA

The potential reasons for using lasers in materials processing are presented in this paper. The major materials processing lasers, Nd-YAG, Nd-Glass and carbon dioxide are reviewed and new designs discussed. The advantages and disadvantages of using lasers in such applications as joining, material removal and surface modification are discussed. Incorporation of lasers into robotics and other automated systems and recent and future developments in the areas of fiber optics and excimer lasers are described. Approximate costs and expected up-time are presented for laser systems.

INTRODUCTION

Materials processing applications for lasers are growing at a rapid rate as evidenced by worldwide sales of 103 million dollars in 1984. A 35% growth in sales is projected for 1985. Improvements in design have occurred for both YAG and carbon dioxide lasers leading to more compact lasers with higher-power output (for their size) and improved beam quality.[1]

Interest is growing in the use of lasers in welding, soldering, surface modification, marking, cutting, hole drilling and scribing. For example, in the automotive industry in the U.S., carbon dioxide lasers in the 5 to 9 kW power range have been placed in production for welding and flexible laser cutting systems are being purchased for sheet metal and softgoods cutting.

Many product and manufacturing engineers are acutely aware of the unique capabilities that lasers offer. Consequently, products and manufacturing systems are frequently designed to take advantage of those capabilities to reduce cost, increase productivity and/or quality. The days of the James Bond syndrome are pretty much a thing of the past. Engineers and managers now recognize lasers as a useful tool to add to their arsenal of sophisticated manufacturing systems.

Lasers are not used instead of conventional tools unless they offer a unique advantage. They do provide supplementary techniques to conventional and exotic techniques such as E-beam, EDM, ECM and water jet cutting. Lasers provide a broader range of design possibilities and manufacturing methods.

MATERIALS PROCESSING LASERS

The dominant lasers in materials processing are the carbon dioxide, Nd-YAG* and Nd-Glass lasers with carbon dioxide lasers accounting for the largest percentage of sales. In this section these three types of lasers and their output characteristics will be briefly described.

Nd-YAG

This laser employs neodymium (Nd) as the lasant material doped in a YAG crystal. Crystal rod sizes for the typical Nd-YAG laser range from 5-10 mm diameter to 6-15 cm in length. Excitation is by one or two krypton or xenon lamps. The wavelength is 1.06 μm, which is close enough to the visible spectrum that conventional optics can be used for lasers and windows. However, cavity mirrors must be of the dielectric enhanced reflection type to achieve sufficient reflectance.

Nd-YAG lasers can be continuously operated from a few watts to several hundred watts, but in most applications pulsed operation is preferred. For low power applications, such as marking or scribing (e.g. thick or thin film resistor trimming), Q-switching is used to achieve peak powers of kilowatts at kilohertz rates. For high power applications, such as cutting or hermetic seam welding, electronic pulsing of the lamps is utilized to produce 5 to 10 pulses per second at tens of joules of energy per pulse and pulse lengths of 0.1 to 10 mS, depending on the application. Figure 1 is a photograph of a 10W average power marking laser. The laser head is in the box on top of the cabinet, the power supply, controls and computer are in the right-hand cabinet, the garro-mirrors and marking area are in the left-hand cabinet which is partially shown.

Nd-Glass

The Nd-Glass laser utilizes neodymium as the lasant, but the host material is glass. Due to the poor thermal characteristics of glass this laser is used chiefly for spot welding and hole piercing with pulse rates of one pulse per second and 30-50 joules of energy per pulse. However, more than one laser head can be operated from the same power supply, typically two, in an alternating fashion. Glass laser rods may be circular or rectangular cylinders. Rectangular cross-section rods are frequently used for spot welding applications. Figure 2 contains a photograph of a Nd-Glass laser head.

Carbon Dioxide

Carbon dioxide lasers come in a variety of power ranges, sizes and designs. All use the molecular vibrations of carbon dioxide as the lasing mechanism. Generally, a mixture of carbon dioxide, nitrogen and helium are employed, the nitrogen is active in the excitation process and helium acts as an internal heat sink. Sometimes a small amount of oxygen is added to reduce contamination from carbon monoxide and carbon.

Carbon dioxide lasers emit radiation at 10.6 μm, which is quite far into the infra-red. This causes problems with respect to reflectance in processing metals such as copper, silver and gold, but alternatively these metals can be used as mirror materials internally or externally.

Some small sealed off carbon dioxide lasers, such as the waveguide types, use rf excitation, but most use dc electrical discharge excitation. The three major designs used in industrial processing applications are illustrated in Figure 3. These are the slow axial gas flow with axial discharge, the fast axial gas flow with axial discharge and the fast transverse gas flow with transverse discharge. The slow flow design relies on thermal conduction for cooling of the gases and consequently the cross-sectional area of the discharge region is limited; relatively narrow bore tubes must be used. Roughly 50-70 watts of power per meter of tube length can be obtained from this type of design. Beam

*Nd is the abbreviation for neodymium and YAG is the acronym for yttrium-aluminum-garnet.

quality is generally good with near Gaussian output being attainable because of the long-narrow bore tube.

The fast flow (about 60 m/s) designs use convection cooling so that much larger discharge cross sections can be achieved. Hence, the power per unit length is much higher, 600 W/m for fast axial flow and 2500 W/m for transverse fast flow. In fast transverse flow designs the beam is folded back and forth through the discharge region several times. Unstable resonator configurations are frequently employed for multikilowatt carbon dioxide lasers to eliminate transmissive optics.

The efficiency for these carbon dioxide lasers is approximately 10% (total input power divided into useful output power). This is about three times the efficiency of the other two industrial lasers described above. Figure 4 is a photograph of a 2.5 kW transverse fast flow carbon dioxide laser and laboratory workstation.

CONSIDERATIONS FOR LASER APPLICATIONS

In this section some of the advantages and disadvantages that must be weighed when considering the use of a laser in various applications will be presented.

General Considerations

The following is a list of advantages that could apply to any application:

- No tool contact
- Work in normally difficult to reach areas
- Work in a variety of atmospheres
- Minimal HAZ
- High speed
- Easily automated
- High up-time
- Minimal operator skill required

Lack of tool contact eliminates tool wear, but is also important in applications where tool force on the part is a problem. Since a laser beam propagates over long distances with little spreading, the beam can be piped around a work area using mirrors and conduit and focused inside of cylinders or other hollow parts. With Nd-YAG or Nd-Glass lasers the beam can be transmitted through transparent media such as glass or plastic before focusing on the part.

The laser beams used in materials processing are not markedly attenuated in air and virtually any gas can be used as a shield gas, examples are nitrogen, helium, argon, carbon dioxide and air.

A very high percentage of the energy absorbed by a part from a laser beam goes to perform the function intended, very little appears as waste heat. Consequently very small heat-affected zones (HAZ) are produced. This implies minimal induced chemical change and reduced damage due to thermal cracking and distortion.

In many applications the laser can perform the operation faster than conventional or other exotic techniques. Because of the lack of tool contact and ease with which a laser beam can be delivered to the workpiece, the laser is one of the most easily automated machine tools.

Industrial processing lasers, in spite of the relatively short time they have existed, are highly reliable. Up-time for carbon dioxide lasers in heat treating and welding applications of 90 to 95% are routinely reported and results for other types of lasers should be comparable. Failures are usually not directly laser related, but are the same types of problems encountered with more traditional machinery.

The skill and training required for laser operators is minimal. Lasers are really very simple to operate once a system has been properly set-up.

The following is a list of disadvantages that will apply in most cases:

- High initial cost
- Operating cost
- Routine maintenance

- Skilled maintenance people
- Safety

Lasers and the optical delivery systems that go with them are expensive. The cost (in U.S. dollars in the U.S.) for a carbon dioxide laser will run from over $150.00 per watt for a 500 watt laser to less than $100.00 per watt for multikilowatt lasers. Operating costs, including maintenance and consumables will run from $3.00 to $10.00 per hour or higher. Costs given here do not include operator wages or shield or cutting gas costs. A 20 watt Nd-YAG laser scribing system could cost $100,000.00 and cost about $2.00 per hour to operate. A 400 watt Nd-YAG welder will cost over $80,000.00, exclusive of beam delivery system, and will cost about $4.00 an hour to operate.

Lasers do require routine maintenance such as lamp replacement, cleaning of optics or the electrodes in some carbon dioxide lasers. This maintenance must be done carefully by properly trained and well supervised technicians.

There are special safety considerations associated with lasers. Aside from the serious danger of electrocution from the high voltage power supplies used in these lasers there is a severe risk of injury (particularly to the eyes) as a result of exposure to the direct, reflected or scattered radiation. Also, some high power carbon dioxide lasers produce x-rays which can be hazardous if not properly shielded. A laser safety officer should be properly trained and have authority to specify and approve safe operating procedures in all industrial installations.

Specific Considerations

There are some advantages and disadvantages which are specific to particular applications, some of those will be listed and discussed briefly here. Some advantages are:
- Weld dissimilar materials
- Weld magnetic materials
- Make deep penetration welds
- Skid welding
- Drill high aspect ratio holes
- Drill holes at large angles to the surface
- Cut complex patterns
- Narrow kerf for cut parts
- Use oxygen assist for cutting and drilling

The laser frequently does a superior job of welding dissimilar metals such as Inconel to copper or tungsten to stainless steel. Magnetic fields do not affect laser beams so demagnetization is unnecessary for magnetic materials. High aspect ratio welds can be made as a result of the phenomenon of "keyholing" when using multikilowatt power levels. The technique of "skid welding" permits filet welding of a seam between right angle pieces of metal with the underbead being produced on the seam as a result of light tracking through the seam.

High aspect ratio holes can be laser drilled or pierced (10 to 1 routinely with 20 to 1 being achieved in special cases) with precise location and repeatability. Holes can be pierced at large angles to the surface, $60°$ is not uncommon with $15°$ being routine in aircraft engine turbine blades. Low aspect ratio holes (large diameter) can be cut by a process called "trepanning" which simply involves bringing the beam through a focusing lens off center and rotating the lens.

In cutting applications an extremely narrow kerf can be achieved (0.1 mm or less) with extremely smooth edges free from dross and recast material. Oxygen or air can be used in gas assist cutting and drilling to increase speed and quality when working on oxidizable metals.

Some of the disadvantages specific to particular types of applications are:
- Cracking and/or embrittlement due to high carbon content during welding
- Hot and/or cold cracks in welding

- Stress cracking due to thermal shock
- Spatter on optics from welding
- Toxic vapors

The extremely rapid heating and cooling that occurs during laser processing does lead to problems on occasion. In welding this can lead to extremely brittle welds in high carbon steel which are subject to cracking and porosity. Also, steels containing low vapor point alloy materials such as lead or sulfur can result in cracking and porosity.

When cutting or drilling brittle materials such as ceramics, thermal stress cracking can be a problem. This is minimized by using rapid short pulses to reduce the heat input required for the job. This is usually accomplished by Q-switching.

Spatter damage to optics is a serious problem, particularly during welding of certain types of metal alloys such as high carbon steel or aluminum. With Nd-YAG or Glass lasers a glass shield can be placed in front of the focusing optic. With carbon dioxide lasers a strong crossflow of air may be required and use of the longest possible focal length lens or focusing mirror is recommended.

Toxic vapors may be produced, particularly during material removal operations. Since little is known about the type or toxicity of most of these vapors adequate ventilation and filtering must be provided to protect workers and others who might be exposed to these vapors.

AUTOMATION

Lasers are readily adapted to existing automated parts manipulation systems. A number of machine tool companies have incorporated lasers into existing machines designed for cutting (nibbling) and punching operations. The laser basically replaces the nibbling process because it is faster and is much less wasteful. Specially designed machines for cutting large sheet material frequently employ one axis to move the sheet and a second axis to translate the beam. The z-axis, the one that controls focal point location and nozzle position, usually floats freely on a ball bearing arrangement that rides on the sheet. When contoured parts are to be cut some sort of position sensor is used to feedback position data to the computer to numerically control the z-axis. Sensors may be capacitive, inductive, air flow or mechanical contact.

More complex automated systems may use six axes of motion or more. Frequently these systems are hybrids, utilizing part motion for some of the translational movements and beam manipulation for high speed motion and angular and/or rotational motions.

There is a tremendous interest in the marriage of lasers and robots. Again, such systems may utilize part manipulation by the robot or in more flexible systems the robot is used to manipulate the beam. Figure 5 is a schematic of a robot system for a carbon dioxide laser which uses an articulated mirror system to direct the beam to a focusing assembly at the end of the robot's wrist. The beam is directed down a conduit which has two or three knuckles to give it flexibility. Figure 6 is a schematic of a beam-guide. Each knuckle contains three mirrors which maintain the direction of the beam along the axis of the conduit regardless of their orientation. Such robot systems may have six or seven degrees of freedom and show great promise for laser welding and cutting of complex shapes. Examples are automotive welding and trimming of stamped metal parts.

The use of fiber optics to deliver Nd-YAG or Nd-Glass laser beams to the workpiece for welding, drilling or cutting shows great promise. A system is already available for soldering which delivers the beam in this manner. Powers of several hundred watts have been transmitted through 1 mm diameter quartz fibers over distances in excess of 25 m and prototype laser robot systems have been built and tested.[2] This approach provides a high degree of flexibility since the laser can be quite remote from the robot and work area and losses are extremely low for the length of optical fiber required. You simply have to design a system so that the fiber will not be twisted or too severely bent and an optical system for launching the laser beam into the fiber and properly focusing it after it exits. The robot manipulates the focusing package which is placed at the end of its wrist.

Unique safety considerations are required in the installation of robotic laser beam manipulation systems because of the potential for uncontrollable movements of the robot while the laser is on, particularly if the focusing element is broken allowing the raw, undiverged beam, to become exposed.

FUTURE DEVELOPMENTS

Possible future developments are too numerous to describe in detail. Only a few of those which at present appear to have a likelihood of making a significant impact in the near future will be briefly mentioned.

Substantial improvements in existing laser types such as the carbon dioxide, Nd-YAG and Nd-Glass will be made. The next generation of carbon dioxide lasers will extend the power range to the 20-30 kW range for routine industrial applications. These lasers will be of the transverse flow types. Continued beam quality improvements will be made for lower power carbon dioxide lasers. The development of coated potassium chloride optics (KCl) may result in stable resonator designs up to 20 kW or higher with good beam quality. Very compact carbon dioxide lasers capable of 1000 W power output may become available for mounting directly on to robots. Zinc selinide (ZnSe) or other types of optical waveguides may become available for carbon dioxide laser radiation for remote location of the laser and robot manipulation.

Nd-YAG and Nd-Glass lasers using slabs instead of rods and employing total internal reflection will extend the power range and improve beam quality for the lasers making them more attractive for flexible manufacturing systems.

Excimer laser power levels and reliability will improve to the point that they become practical for many materials processing applications. An excimer laser utilizes a compound of an inert gas in an excited state, and a halogen as the lasant, e.g. the xenon chloride (XeCl) laser. When the inert gas atom returns to the good state, the excimer "molecule" dissociates and the energy is given off as a photon which leads to laser action under the proper conditions. Such lasers have extremely high gain and operate in the ultraviolet (uv) part of the spectrum. The absorbtion of metals in the uv is nearly 100%, which suggests many interesting materials processing possibilities. Additional safety problems are posed because of the use of toxic gasses such as chlorine and flourine and the potential for x-ray generation in the workpiece.

References

1. Levitt, M., "The Prognosis for 1985: Technical Advances and Policy Issues". Laser Focus, Vol.21, No.1, pp.10-12 (January 1985).
2. Marshall, G. J. and G. Georgalas, "Flexible Beam Delivery for Material Processing Laser Power Through a Fiber Optic Cable". Laser News, Vol.6, No.4, pp.11-12 (July 1984).

Fig.1. Nd-YAG marking system.

Fig.2. Nd-Glass laser head.

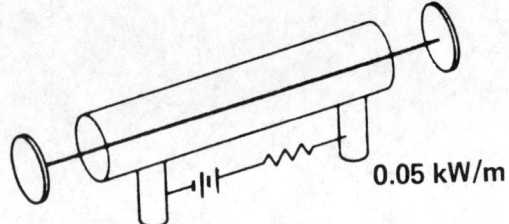

Fig.3a. Slow axial flow.

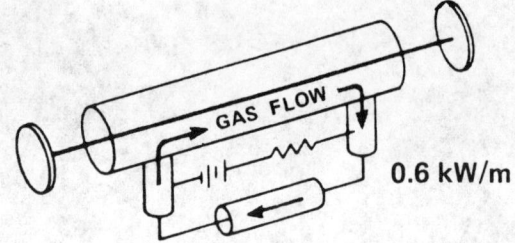

Fig.3b. Fast axial flow.

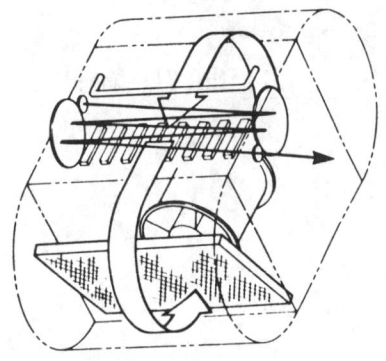

Fig.3c. Transverse electrical discharge with fast transverse gas flow.

Fig.4. 2.5kW fast transverse flow CO_2 laser and work station.

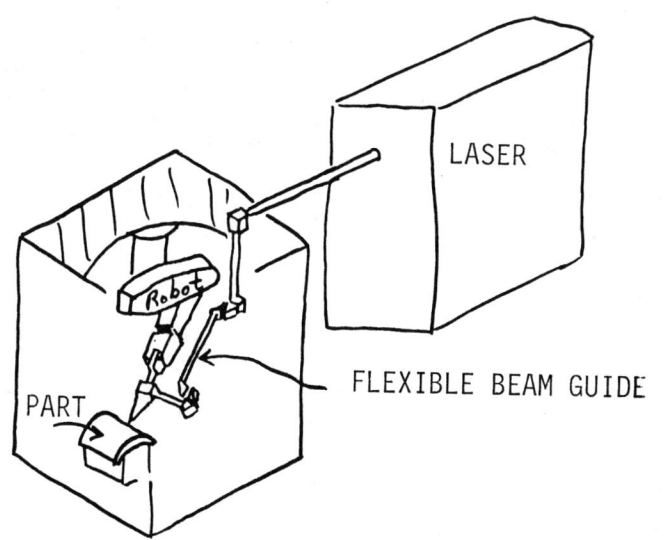

Fig.5. Schematic of robot-laser system.

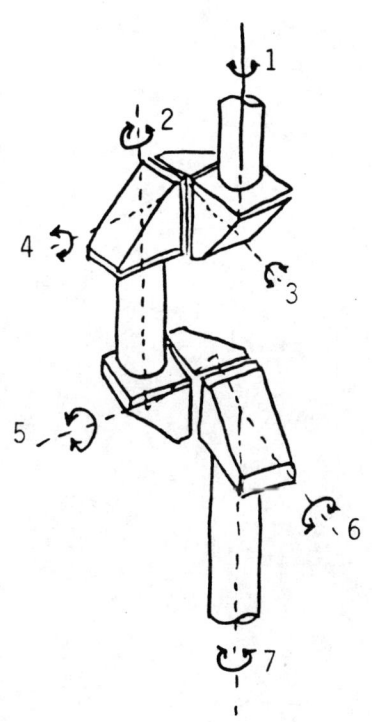

Fig.6. Schematic of mirror beam-guide system for a CO_2 laser.

What choice for high integrity joints: electron beam or laser beam welding?
G. Sayegh
Sciaky SA, France

Through few industrial application examples, specific characteristics and advantages of high energy density beams (EB and LB) will be illustrated ; thus justifying the use of the processes in production. Technical performance comparison of the two processes and economical considerations will permit to define three types of applications : thin gauge material for which LB is more appropriate ; thick gauge material for which EB is more appropriate ; and medium gauge material for which effective and strong competition exist between LB and EB.

INTRODUCTION

Electron Beams (EB) and Laser Beams (LB) can be concentrated to produce very high energy density heat sources. When used in welding they can achieve high integrity joints with specific characteristics.

High energy density beams are regularly employed in metal working production and specially in welding. Several thousands of welding systems are used in industrial production in the various fields: aeronautics, nuclear, automobile, electrical appliances etc.

EBW was introduced in industry about 25 years ago thanks to its many advantages, among which was the fact that it produced welds impossible to achieve with other welding processes.

Multikilowatt LBW was used about 6 years ago in production. It can produce welds very similar to EBW and do have some additional advantages in some specific conditions. One can note today a strong competition between the two processes to resolve the same type of problems with the same high performances.

In the first part of the paper we shall present the general characteristics of high energy density beam welds. The present technical situation of the process will be analysed through a number of existing industrial applications.

By comparing the technical performances and the economical positions of the two processes, the paper will define the domains and the type of application where each process is most suitable. Three types of applications will be recognized : those most suitable for LBW, those most suitable for EBW and those where a strong competition exists between the two processes.

1 - SPECIFIC FEATURES OF EB AND LB WELDING -

I.1. High power density.

Power density obtained by EB and LB is about 1000 to 10000 times that of conventional welding heat sources. This high power density level engenders beam/material interaction that differs significantly from conventional processes. During welding, the metal under the beam spot sublimates, creating a capillary which is instantaneously filled up with plasma type metallic vapors. Heat transfer from the beam into the joint occurs along the entire thickness of the joint instead of being diffused by conduction from the surface as in the case of conventional processes. This produces joints having the following characteristics :

- high depth to width ratio (10 to 50)
  commercially available beam powers exceed 100kW for EB and 10kW for LB. The majority of applications concern thicknesses less than 30mm for EBW achieved by about 20kW and less than 4 to 5mm for LBW achieved by about 3kW (figure 1 gives some macrographies of welds),

- minimal heat input and heat affected zone (HAZ). Owing to the nature of heat transfer between the beam and the joint, the total heat input in EB/LB is much lower than in conventional processes (a factor of 10 and more for heavy sections),

- minimal distorsion and residual stress. EB/LB welded assemblies are usable as welded without intermediate heat treating (if metallurgy permits) or machining operations. This production advantage, fully exploited by the automotive industry for the assembly of gears and other mechanical components, is a consequence of the low heat input. Over 60% of EB/LB industrial installations produce components for the motor industry.

1.2. Flexibility and automation in production.

- Use as a flexible tool.

The electrons transported in the beam have almost insignificant masses ($m_e = 9.1 \times 10^{-31}$ kg, therefore practically inertialess) and travel at a velocity between one-half and two-thirds that of light when accelerated by a voltage on the order of 60 to 150kV. They can be focussed or deflected in various patterns by the application of electromagnetic fields. The action is instantaneous and similar to what is obtained on a television screen. It is thus possible to rapidly displace the beam according to pre-defined heat distributions for better control the thermo-mechanical phenomena. Additionally, the characteristics inherent to electrons allows their use for auxiliary operations, such as automatic seam tracking.

The laser beam is composed of photons that travel at the speed of light ; it can be focussed and transmitted by ordinary optical lenses and mirrors. This permits manipulation and transmission from one point to another, and confers high mobility and flexibility to the process. Their flexibility is probably the most fascinating feature of lasers in welding and metal working. Flexible manufacturing centers utilizing a central laser beam supplying several work stations are being implemented in the US and Japan.

. Automation and productivity of the processes.

Because of the flexibility of the processes and their automation which could be CNC control ; high productivity could be achieved with EB/LB techniques. It is mainly because of the high productivity that industrial use of such processes can be justified economically even though the initial investment is much larger than conventional processes.

1.3. Pressure environment in EB/LG.

Electron beam technology requires a high vacuum level (10^{-4} to 10^{-5} mbar) surrounding the cathodic emitting surface and the high voltage electrodes.

Laser beam technology requires an accurate control of the pressure (30 to 70mbars) in the lasing cavity and constant characteristics of gaz (mixture, temperature, flow, speed, etc.).

To prevent beam dispersion caused by the electrons colliding with gaz molecules, the pressure in the zone through which the beam travels should be less than 10^{-2} mbar. This means that the workpiece should be entirely put in a vacuum chambre, or eventually only the seam area should be brought to the low vacuum (local vacuum systems). The vacuum in EB can bring an advantage by protecting the molten zone against oxidation and hydrogen susceptibility. High quality welds can be made in reactive materials such as Zircalloy, Ti and refractory metal .

Laser beam can travel at atmospheric pressure without significant dispersion, thus it does not need a vacuum chambre. Nevertheless a protective gaz is needed around the molten metal to avoid its oxidation. On the other hand laser beam welding in vacuum (some torrs) produce deeper welds in heavy sections ($>$ 6mm) than in atmosphere (2 to 3 times more) by reducing the dispersion of laser beam by the plasma created at the point of impact.

2 - RECENT DEVELOPMENT AND INDUSTRIAL EXAMPLES OF EB/LB WELDING -

Our analysis will concern welding applications which employ multikilowatt power beam. Consequently, we shall not consider cutting or micro welding applications which employ laser sources not exceeding few hundred of watts.

The majority of the applications will be drawn from the motor industry where the following advantages of EB/LB welding can be seen :

. complex geometries are reduced to simple to manufacture components which are then joined,

. each component in an assembly can be made from the best suited material,

. EB/LB welding processes lend themselves to automation and high productivity,

Here are some examples which can be achieved indifferently by EB or LB welding.

Pignon gears shown on figure 2 are welded on a diameter of 60mm, with a beam power of 2kW and a welding speed of 2.2m/mn.

Deformations of the gear in two perpendicular directions were 0,05mm and 0,08mm. Consequently these parts can be used as welded without any further machining.

High energy density beam welding of gear pignon have the following advantages :

- machining of the gears is highly facilitated,
- smaller volume of the gears,
- weight of the gear box will be reduced, vibrations and noise also.

In EBW several types of machines can be used to satisfy the industrial production:

- a machine with one station bottom delivery for low production 50 parts/hour,
- a rotating table with 3, 4, 6 or 8 stations équiped with 1 or 2 guns for high production rates 200 to 500 parts/hour,
- a sliding seal system in which the pumping time is completely masked for very high production rates of 1000 parts/h.

In LBW machines, different systems are used according the industrial productivity:

- one working station with one welding head for medium production,
- two working stations supplied by the same laser ; while welding is realised on one station ; loading and unloading is achieved on the second ; thus reducing the idle time. This solution, used currently in welding gears is adapted for high production , about 500-600 parts /hour.

Figure 3 shows a planetary carrier. The part was originally machined from a forging. With EB/LB welding the two 1010 steel cages are welded to the 1045 disk with the split beam technique : a square wave signal applied to the deflection coil in EB time-shares the beam and thus achieves simultaneous joining of the two seams. The laser beam can be splitted by an appropriate optical mirror into two seperate beams before they are concentrated on the joint. A principal beam of about 4,5kW can achieve the two welds at a speed of 2m/minute in about 5mm thick components.

EB/LB weldings can present a very big advantage in welding dissimilar metals by precise control of the molten volume of each part and thus the final composition of the weld. As a matter of fact beams can be directed with high precision (better than 0.1mm) on a given impact point. This principle was used successfully for welding sintered powder steel (type 20NiMoCr6E) to mild steel as shown on figure 4. In order to achieve good metallurgical weld qualities two conditions should be respected :

- realise a very narrow weld ($\simeq$ 0.5mm wide) in order to reduce as much as possible the molten zone,
- direct the beam on the joint about 0.1mm in the direction of mild steel in order to reduce the rate of molten sintered part in the weld and thus avoid porosity in the molten zone and improve its metallurgical characteristics.

Complete automation of EB/LB welding processes contributes to increased product quality. Recent computer control systems integrated into production have significantly improved :

- product quality,
- productivity,
- maintenance conditions.

Monitoring of all welding parameters and operating conditions ensure optimal finished product quality. High and low operating limits are established for each parameter in addition to the setpoints. If the measured parameter is out of these preset limits, the operator is alerted.

All parameters and measurements for each part are stored in memory and can be printed out on a "follow up" document that can be used in product quality control procedures.

EB/LB applications are not limited to the automobile industry. They cover many other industrial fields from aeronautic to the heavy industries. As an illustration we shall present two examples :

A. A multifunction EB equipment for heavy industry (fig.5).
A $265m^3$ chamber is equipped with a 45 (or 100kW) electron gun that slides along the 5.5 meter long vertical axis. The parts are placed on the 3.2 meter diameter rotating table that supports a load of 30 tonnes. The computer numerical control handles all machines functions including the sequence and the optimized part programmes.

Although the machine does not automatically assemble all shape and sizes that arrive in random, it represents a flexibility increase for production of parts of varied dimensions and geometries.

The possibilities of the machine include :

a) Circular butt welding of 3 meter diameter, 5.5 meter high shells. The shells, placed on the turntable, are rotated as the gun remains stationary.

b) Longitudinal welding of semicircular shells up to 5.5 meters in height. The gun moves vertically along the Z axis.

c) Nozzle welds are realized by combining the circular movement of the turntable along with the vertical gun movement.

This application shows the degree of flexibility which can be obtained by judicious use of EB welding, computer controls, and specialized software packages.

In spite of the relatively high initial investment, such equipment can be amortized in less than 18 months.

B. Spacer assembly of insulating windows.
This very spectacular application uses a dozen of 1kW laser sources in a roll forming mills line to produce an interrupted seam weld in thin gauge aluminium spacer. Welding at a rate of 100m/mn is achieved on one line.

Although aluminium alloy is a very bad absorber of CO_2 laser beam, it was used successfully in this application because the weld qualities required from the joint were not very restrictive. As a matter of fact the role of the weld is only to provide the necessary mechanical rigidity to the spacer during manipulation.

One of the major advantage given by the user was the non-contact aspect of the process which allows to reduce filler material resulting in a substantial material cost reduction.

3 - COMPARISON BETWEEN EB AND LB WELDING : PROSPECTS -

3.1. Technical performances.

The main advantages of high energy density beams when used in welding are :

a) Small distorsion eliminating post machining operation.

b) No filler material and no protective gaz for EB.

c) Flexibility and reproducibility of results because of the high automation.

d) Energy saving processes.

e) Cleanliness and nice environment.

The main draw back is the relatively high investment cost of an industrial equipment ; but in spite of the investment the high productivity of the machine could justify economically its acquirements. Table 1 compares the high energy density beams techniques to PLASMA and TIG processes when welding 4mm of low carbon steel. It can be seen that when high production is considered, EB and LB could compete economically with TIG or plasma welding.

TABLE 1 : Comparison of EB, LB, PLASMA and TIG welding of 4mm thick low alloy steel.

PROCESS	LB	EB	PLASMA	TIG
Source power (kW)	5	5	4	2
Installed power (kW)	60	6	6	3
Welding speed mm/sec.	33	33	10	3
Energy/mm (J/mm)	150	150	400	660
Shrinkage (mm)	$\simeq 0$	$\simeq 0$	$\simeq 1$	$\simeq 1$
Angular distorsion	parallel	parallel	5°	6°
Investment of the source (relative values)	2000	900	120	100
Investment/welded mm/sec.(relative values)	180	82	36	100

TABLE 2 : advantages and limits of EB/LB techniques.

	ELECTRON BEAM	LASER BEAM
ADVANTAGES	1) High efficiency 2) High flexibility for modifying focal point, beam power 3) High power are available 4) Easy to automatise 5) High power beams (up to 100kW) are available commercially.	1) No vacuum 2) Transmission of beam on long distance 3) No effect of magnetic field 4) Can be directed to hidden surface by mirrors 5) No X rays are produced during processing 6) Could be used in time sharing several stations.
DRAWBACKS	1) Vacuum is needed 2) Should be protected from magnetic fiel 3) Protection against X rays is needed 4) Less flexible for time sharing of the beam.	1) Low efficiency 2) High cost of consummable 3) Difficult to modify focal point 4) Protection of operaters from hazardeous light 5) Beam power limited to 10 to 15kW commercially.

Table n°2 compares EB/LB techniques showing these advantages and limits .

Figure 6 showsthe penetration obtained in a SS304 steel when welded at a speed of 1m/mn with different beam powers of EB and LB. Focussing conditions were arranged so that to operate at the same power density spots. One can draw the following conclusions.

a) For penetration up to 4 to 5mm, EB and LB produce the same performances ; indeed the fundamental parameters which affect the penetration in metal (vapor pressure, surface tension..) are of the same order of magnitude for both processes.

b) When the penetration exceeds 5mm, EB is more performant for the welding conditions of figure 6 executed at atmospheric pressure with a protective gaz. The lack of penetration in laser welding is attributed to the reflection of the beam by the plasma gazes located in the capillary.

In Nov.84, Prof Arata reported some laser welding results realised in a vacuum chamber where the pressure varied between 10^{-4} and few torr. He obtained laser welds in about 20 to 25mm steel with the same laser beam power as with EB power when the pressure in the chambre is 1torr. He thus concluded that plasma effect in laser welding is completely eliminated when operating in vacuum. Curiously it can be said that the best conditions and the highest efficiency of laser beam welding are obtained in vacuum !

but then one of the main advantages of laser beam compared to EB disappears. Nevertheless, this is a direction to follow in the future.

3.2. Economical comparison

The choice of a new production technique for an industrial application is dictated by one or several of the following considerations :

a) The application can be realized only by the new technique.

b) The new technique is cost effective (lower cost/unit).

c) For equal cost/unit, better technical characteristics of the product.

d) For equal cost/unit, better operating conditions.

These rules are valid when considering the use of high energy density beams in production. The evaluation of the cost/unit is difficult to analyse as it depends on factors that the users keep confidential. Here are some indications which should be taken into consideration when establishing the cost effectiveness of the process.

. Investment for the process and the toolings.

. Cost of the preparation pre and post processing.

. Consummables.

. Direct and indirect labors.

. Maintenance and repair.

Table 3 compares EBW and LBW equipment cost to produce gears as shown on fig.2. Welded diameter is 50mm with a depth of 3.8mm achieved at a speed of 1m/mn. Production rate is about 240 parts/minutes for LBW and 220 parts/minute for EBW which uses a 3.5kW beam power at a welding speed of 1.25m/mn.

TABLE 3 : Economical analysis for EB and LB equipment for welding gears.

	LB	EB
	2kW	5kW
Working conditions . hours/year . Nb of operators	4000 1	4000 1
Investment for equipment	1.500.000	1.900.000
Operating cost/hour . capital amortization Fr/h . interest on capital Fr/h (12%) . maintenance . consumables (electricity, gaz, water..) TOTAL	125 45 18,75 74 262,75	158,35 57 14,25 9,70 239,30
Labor and overhead	165	165
Total operating cost/hour Production rate part/hour	427,75 240	404,30 220
Cost per part (Fr/part)	1,782	1,84

The mechanical tooling is a simple rotating table with three working stations ; its cost was evaluated to 300.000 Frs for the LB and 750.000 Frs for EBW because of vacuum system.

Laser source of 2kW estimated to 1.200.000 Frs including the focussing system ; where as a 5kW EBW is used at a price of 1.150.000 Frs . Turn back is supposed to be in 3 years for two shifts operations (12000h).

An important factor is the "beam availability for welding" which can be defined as the ratio between the welding time and total cycle time. We have considered a factor of 75% for EBW (25% is allocated for puming and transfer) and 95% for LBW (5% for transfer). It is obvious that this parameter is an important factor in comparing EBW and LBW ; as a matter of fact if an EBW application uses small "beam availability factor" LBW is more appropriate for the application.

Figure 7 compares in relative values the cost investments for EBW and LBW equiped with various powers for welding of small components as those encountered in automobile industry. This simplified presentation assumes four hypotheses :

. "Beam availability" factor for the application remains around 75% for EBW and 95% for LBW.

. Welding speed is not a critical factor for metalurgical qualities thus alloying operation at highest beam power level.

. Production is sufficient to justify the use of high power beams.

. Laser price is almost directly proportional to beam power, where as EB sources is almost the same for powers between 3kW and 15kW.

From this figure one can draw the following conclusion :

a) For relatively thin gauge applications, smaller than 2 to 3mm which needs about 1 to 1.5kW power, laser beams are most cost effective. Lower is the power level needed for the application more cost effective the laser will appear. It is in this range of power that lasers are mostly used in metalworking. Indeed the total number of laser systems employed in metalworking is evaluated to about 2000 systems in the Western countries ; among which 1500 are of the subkilowatt power level type.

Laser Manufacturers are developing miniaturised sources of this power level which can be integrated to robots for metalworking applications. This will present a big advantage of LB in metalworking applications.

b) For relatively thick gauges applications ; higher than 6 to 8mm, which needs more than 5kW beam power, electron beams are more cost effective. Higher is the power level needed for the application, more cost effective is EB. Very few industrial applications use laser beams higher than 5kW because of cost investment involved and technical reliability which should be improved to ensure productivity. Laser sources manufacturers are working on new generation of sources with some technological innovations in order to improve their reliability.

Flexibility of laser beam represents a very important advantage for its use in "Flexible Manufacturing Centers" for multifunctions operations in metalworking. The success of such centers necessitates the resolution of many technical problems which are under consideration by the manufactueres of very high power laser sources.

c) For medium thick gauges applications (between 3 to 6mm) which needs a power level of 2 to 5kW ; EB and LB systems can be very competitive. It is necessary to conduct a detailed economical study similar to the one of table 3, for the actual part before drawing conclusion for the most appropriate process."Beam availability" factor plays a very important role in the choice of one or other process for this range of beam power. When this factor is small ($<$ 30 to 40%) laser beam will be more cost effective ; but when it is high enough ($\simeq$ 80 to 90%) EB could be more cost effective because the investment per watt will be much lower for EB as the power increases. Present competition between the two processes is very strong in the automobile industry for welding parts as indicated on figures 2,3,4. Advantage seems to lean towards laser welding which benifits from its "à la mode" situation.

CONCLUSIONS

High energy density beams (EB & LB) are currently used in industrial welding applications to achieve high integrity joints with specific characteristics. They are integrated naturally in machine tools yard and are manipulated by qualified workers.

The technical performances of EB & LB expressed as penetration/beam power are very similar, for thicknesses under 4 to 5mm.

When laser welding is achieved at atmospheric pressure, EB is more performant for thicker gauges (> 6 to 8mm) ; but when LB is achieved in vacuum, performances are the same.

Economical considerations allow to distinguish three types of potential applications for EB & LB weldings :

- thin gauge applications which can be achieved with beam power smaller than 1 to 1,5kW ; LBW is more advantageous,
- thick gauge applications which can be achieved with beam power higher than 5kW, EBW is more advantageous,
- medium gauge applications which can be achieved with beam power range of 2 to 5kW, EB & LB techniques are very competitive and a detailed analysis is needed before concluding.

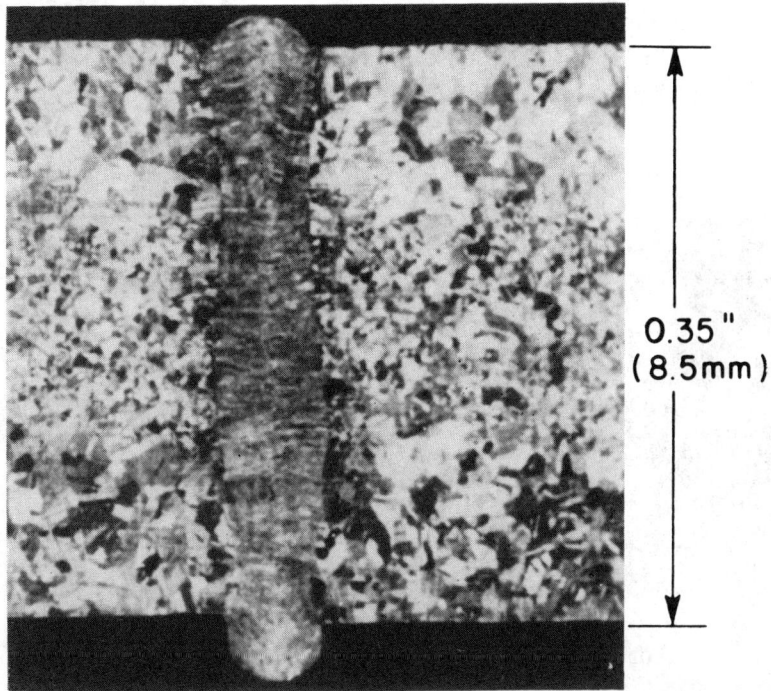

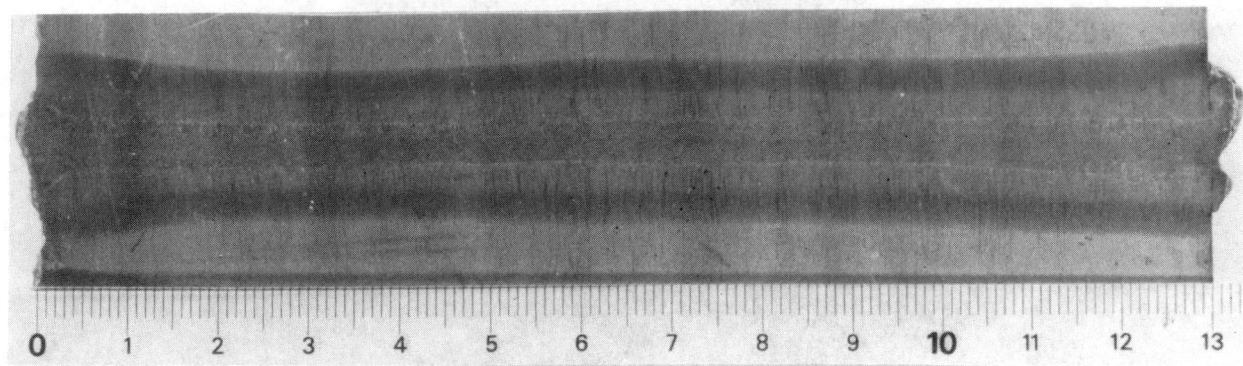

Fig 1 : Macrographies of LB & EB welds :

Up. SS 304 ; Laser welded at 10kW and 180cm/mn
Bottom 533B nuclear steel EB welded at 50kW and 15cm/mn.

Fig. 2 : Typical parts for gear box EB/LB welded.

Fig. 3 : Simultaneous welding of 2 joints in a planetary carrier shaft.

Fig. 4 : Sintered powdered - mild steel component.

Fig. 5 : Large EB welder for the heavy industry

    Up : General view of vacuum chambre

    Bottom : Inside view of the chambre showing gun on its support and rotating table for 30^t load.

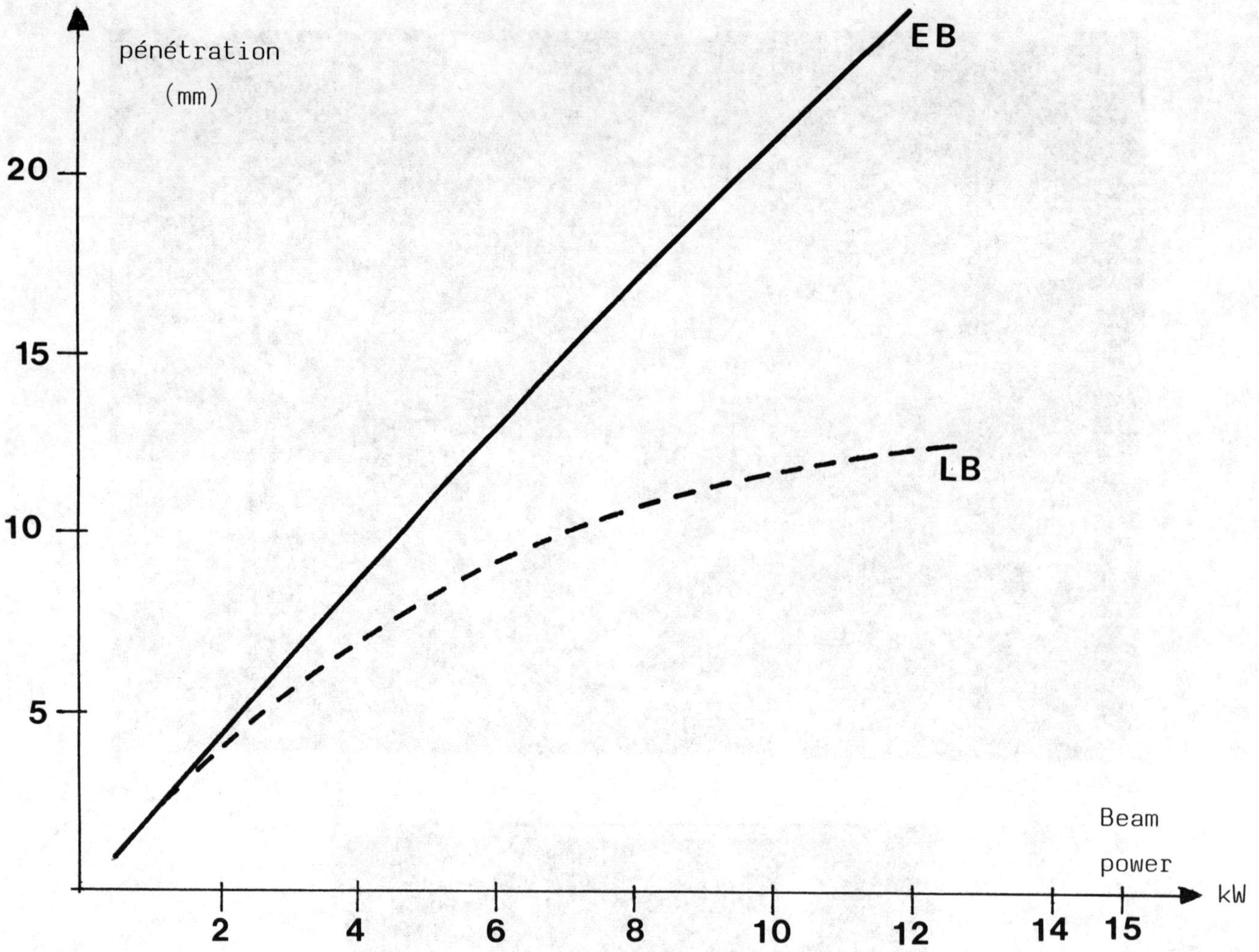

Fig. 6 : penetration of EB and LB as function of beam power in SS304 at 1m/mn.

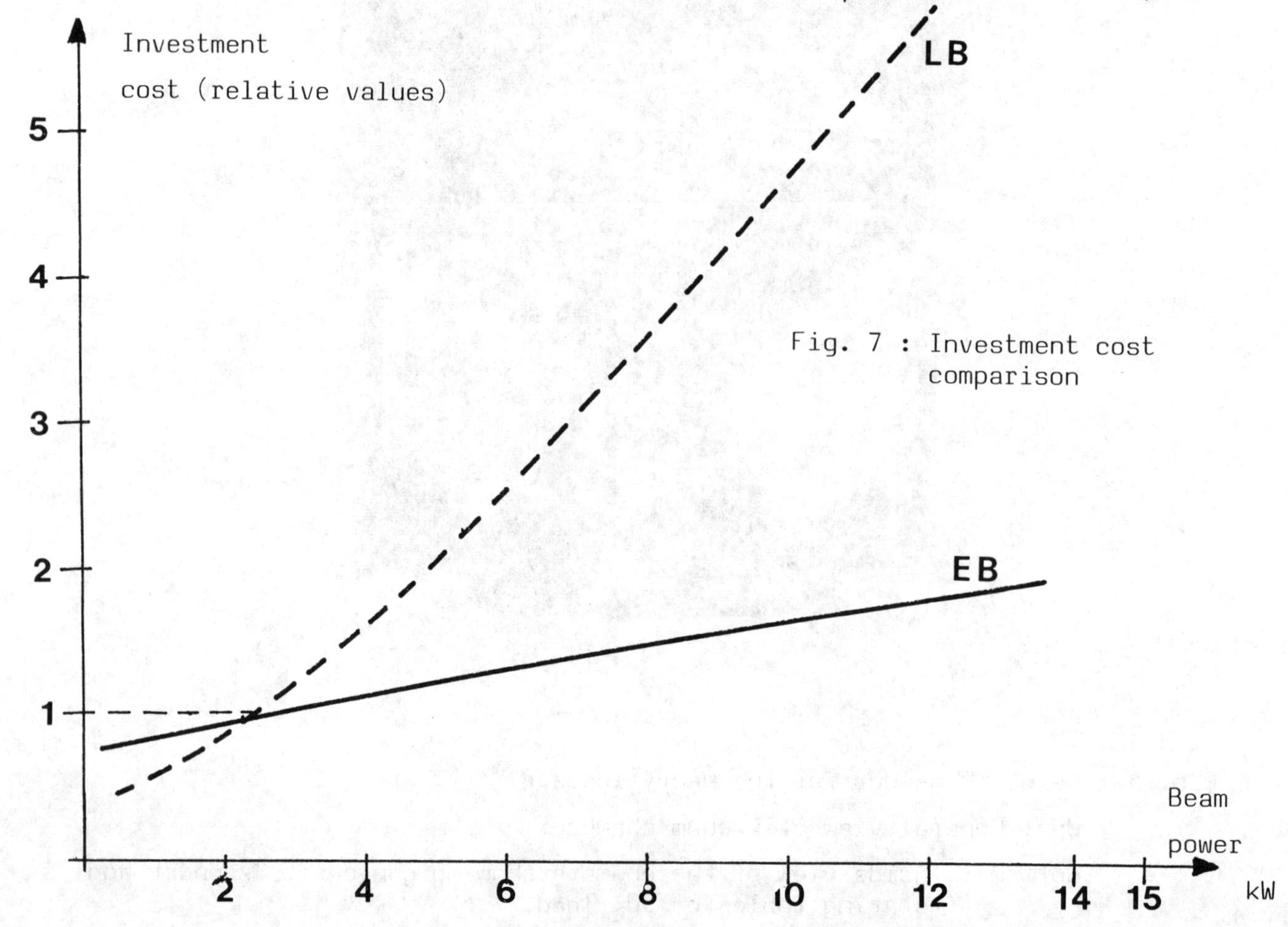

Fig. 7 : Investment cost comparison

Energy efficient laser machining

L. J. Li
Hunan University, People's Republic of China
and
J. Mazumder
University of Illinois, USA

ABSTRACT

In this paper, a new machining process using a laser beam as the tool is proposed and preliminary experiments are carried out to prove the concept. In this process, a laser beam, which is focused by cylindrical lenses to a light strip (band source), is used to melt the micro-area at the interface between the material to be removed and the main body. The chip separated due to melting is removed by a stripping tool during relative motion between the tool and the workpiece.

The cutting force is small since the material is removed by micro-area melting as opposed to plastic deformation by a tool for traditional cutting process. The "laser cutting tool" is free from wear and inertia-related problems. The process is particularly suitable for machining difficult-to-cut metals such as nickel base alloys. Initial data indicates that the finer the focused beam, the better the cutting performance.

INTRODUCTION

The traditional machining process has its inherent drawbacks. It is at first a mechanical process where a tool is used to produce severe local plastic deformation on the workpiece material to strip part of the material as a chip. Great force is developed in this process and a large amount of energy is needed. Specific energies of major machining processes are as follows[1]:

Process	Specific Energy, (J/mm^3)
Grinding	10-200
Single Point Tool Cutting	1-10

---

* Former Visiting Scholar at the University of Illinois at Urbana-Champaign.

Considering that the energy needed to melt one cubic millimeter of steel from room temperature is about 10 J, the specific energy consumed in the traditional machining process is fairly high.

The large cutting force associated with the traditional machining process is responsible for the deformation and chatter of the machining system. This results in poor surface finish and frequent tool failure due to wear. To withstand the high applied cutting force, the steady and dynamic stiffness of machine tools has been a major problem.

The work done by the cutting force becomes almost entirely heat energy and leads to a high temperature in the cutting region. Cutting tools wear and fail rather rapidly due to the large forces exerted on them at the high temperature.

It is attractive to introduce inertialless tools such as lasers in the machining process to solve some of these problems.

Copley and Bass[2,3] studied two processes: laser assisted machining and laser machining. In the former case, a laser beam is focused onto the cut surface at a distance in front of the edge of the cutting tool to heat and soften the workpiece material for subsequent machining. In laser machining processes, the focused laser beam is used to form grooves on the surface of ceramic parts by vaporization. Such grooves are overlapped to form machined surfaces. The laser assisted machining helps to reduce the cutting force to some extent, whereas laser machining makes it possible to machine extremely hard materials. The problem in question with these two processes is the energy utilization efficiency. To heat or vaporize all the allowance by the laser beam needs a high power laser unit. In the laser assisted machining process, the low absorptivity of metal surface to the laser beam makes this problem even more serious, and therefore it becomes difficult to compete with the plasma torch assisted machining process with respect to energy utilization efficiency[4].

To use the laser more effectively in the machining process, a new laser machining process was proposed[5]. A mathematical model for the process using finite difference technique is developed and the predicted trends are experimentally verified using a 10 kW CO_2 laser at the University of Illinois at Urbana-Champaign.

THE CONCEPT

The purpose of a machining process is to separate the allowance from the required products. To do this, it is not necessary to plastically deform all the allowance as in traditional cutting processes or to heat all the allowance as in the laser assisted machining. If we can melt a micro-area at the interface between the allowance and the required product, we can easily separate them. The laser beam is capable of doing this. With the proper optical system like a cylindrical lens, a laser beam can be focused as a light strip to heat and melt the interfacial area. In order to expose the interfacial area between the workpiece and the chip to the laser beam, a stripping tool is needed to strip the separated chip during relative movement. Figure 1 illustrates this process with a single cutting edge.

Since it is a micro-area heating process, higher utilization of energy is expected. Moreover, in this case, there is a wedge angle included between the bottom surface of the chip and the machined surface. As a result, the laser beam would be reflected from both these surfaces leading to a concentrated beam at the tip of the angle due to the "waveguide" effect of internal reflections. Thus almost all the incident laser beam energy would be absorbed during the multi-reflection process which is very favorable to the utilization of laser beam.

It is possible to overcome the inherent drawbacks of the traditional machining process by this laser machining technique. The force required in the laser machining process is fairly small. The focused laser beam used in the machining process has a convergent angle. In order to clear the path for the laser beam to the tip, the chip should be bent large enough so that the slope of the chip at the tip would be larger than the convergent angle of the laser beam. This is realized by the deformation of the material around the root of the chip. The stripping tool is away from the tip as shown in Fig. 1. The force needed to bend the chip is much less than the force required to shear the workpiece material on the shear plane in the traditional cutting process. As only plastic materials can withstand strong bending deformation without chip breakage, this process is suitable for machining plastic materials.

The focused laser beam--the "laser cutting tool"--is free from wear. A laser machining process is essentially a thermal process. The effect of the process depends mainly on the thermophysical properties of the workpiece material. Laser machining may offer a new procedure for machining materials which are difficult-to-cut for traditional cutting processes.

MAJOR CONSIDERATIONS IN DESIGNING THE OPTICAL SYSTEM

The energy utilization efficiency and the quality of the machined surface are dependent on the fineness of the focused laser beam--the sharpness of the "laser cutting tool". Therefore, achieving the most narrow light strip on the focal plane is the most important factor in developing a laser machining process. The focused spot diameter of a TEM_{00} mode Gaussian beam focused by a aberration free objective is $D = 1.27 \lambda f/d$, where d is the diameter of the input laser beam, f is the focal distance of the objective and λ is the wavelength[6].

D expresses approximately the diffraction limit of the minimum width of the light strip focused from a TEM_{00} laser beam.

In a laser machining process, only a micro-area is melted; therefore, the objective is not subjected to contamination by material vapor and ejection of molten material or radiation from vast amounts of heated material. In this process, the objective of lower focal distance can be used. The width of the light strip is inversely proportional to d/f. The larger the d/f, the narrower the light strip, but d/f equals roughly the convergence angle of the focused laser beam expressed in radians. The smaller the convergence angle, the smaller is the required chip bending and the angle included between the beam axis and cutting movement direction. If we assume that $d/f = 1/3$, then the convergence angle would be 19 degrees and $D = 3.8\lambda$.

D is directly proportional to the wavelength. For the commonly used CO_2 laser and the YAG laser, $D = 0.04$ mm and 0.004 mm, respectively. As aberration is unavoidable in a practical optical system, the actual width of light strip must be somewhat larger than its physical limit.

Another important requirement for the light strip is its power distribution. For an orthogonal machining process, a nearly uniform power distribution along the longitudinal direction of the light strip of hundreds of Watts per millimeter is needed. To meet this condition, an objective which only converges a laser beam in one plane, e.g. a single cylindrical lens, is usually insufficient. As described later, in this experiment two cylindrical lenses were used to converge part of a 10 kW CW CO_2 laser beam in two planes, respectively, which were perpendicular to each other.

MODELING OF THE PROCESS

A two-dimensional heat transfer model has been developed for a laser machining process using finite difference numerical techniques.

The model is based on the following assumptions:

1. An orthogonal machining process with one light strip (band source) perpendicular to the direction of traverse is considered. Therefore this process can be treated approximately as a two-dimensional case.
2. A quasi-steady state condition has been assumed for a continuous machining process.
3. The workpiece and the chip are infinite except for a wedge angle included between the bottom surface of the chip and the machined surface.
4. A wedge angle w = 20 degrees included between the bottom surface of the chip and the machined surface has been assumed. A focused laser beam with a convergence angle of less then 20 degrees is used to heat the tip area of the wedge angle. The symmetric lines of the two angles are coincident.
5. The focal plane of the laser beam passes through the tip of the wedge angle. The center of the Gaussian beam is at the tip.
6. The thermal conductivity, specific heat and thermal diffusivity of the workpiece material are assumed to be constants as their respective average values at the concerned temperature region.
7. The radiative and convective heat losses are assumed to be negligible.

In order to simplify the boundary condition, two Cartesian coordinate systems are used for the workpiece and the chip which are connected at the dotted line as shown in Fig. 2. For both coordinate systems, the same heat conduction equation has been used, i.e.,

$$V(\partial T/\partial x) - \alpha[(\partial^2 T/\partial x^2) + (\partial^2 T/\partial y^2)] = 0$$

where V is the constant machining speed and α is the thermal diffusivity of the workpiece material. This equation is then put in a finite difference form for the computer model. As the tip of the wedge angle, the most interesting part of the whole problem, is a singular point, an exponentially expanding grid from the tip is used to get the accurate solution for the tip and meet computer hardware constraints. Since the present problem involves a steady-state situation, a Gauss-Seidel method was chosen to solve the equation.

The temperature gradient at the surfaces, the boundary condition of the partial differential equation, is calculated by an application of Fourier's law of heat conduction, the surface heat flux being the absorbed laser beam power per unit area of the surface.

The principal ray parallel to the symmetric line of the laser beam would get reflected from the surfaces by 180°/w times towards the tip and back as shown in Fig. 2, where w is the wedge angle. Each time the laser beam impacts the surfaces, part of it is absorbed. The absorptivity of the surfaces to the laser beam is almost independent of the incident angle of the beam but strongly dependent on the surface temperature[7,8]. The temperature dependence of the surface absorptivity can be expressed as C(T) = A + BT.

The constants A and B can be determined by using the data of absorptivities at different temperatures in the above equation.

Knowing the beam power per unit length of the light strip and the width of the Gaussian beam, the total absorbed power distribution along the two surfaces can be calculated by integrating the absorbed power for each of the reflections of the laser beam on the surfaces as

$$AB(j) = \sum_{i=1}^{n} Q_i(j)[A + B \times T(j)]$$

where $AB(j)$ is the total absorbed power density at grid point j. $T(j)$ is the temperature at point j and $Q_i(j)$ is the power density at point j of the laser beam impacting the surfaces the ith time. $Q_1(j)$ is the power density at point j of the incident laser beam. $Q_2(j)$ is the power density at point j of the laser beam reflected from the incident beam and so on. n is the number of times the laser beam gets reflected from the surfaces and equals 180°/w.

For Inconel workpiece material, calculated isotherms at different powers per unit length of light strip p, widths of Gaussian beam D, and machining speeds V, are shown in Fig. 3.

The temperatures at the tips are all 1400°C, the melting point of the material. But the tip temperature is not the highest in the cutting region. The highest temperature point is on the surfaces away from the tip. Keeping the tip temperature at the melting point of the material, the narrower the light strip (i.e., smaller the beam width) and the lower the machining speed, the lower would be the power per unit length of light strip p needed and the lower would be the maximum temperature and the smaller would be the distance between the maximum temperature point and the tip.

These conclusions from the numerical calculation are coincident with Jaeger's analytical work[9]. The present heat transfer problem can be analyzed approximately by separating the workpiece, chip, and the incident beam into two symmetric parts, each of which can be treated as a moving heat band source problem with semi-infinite substrate which has been analyzed[9]. The highest temperature point is not at the front edge of the band source which is the tip of the wedge angle in our case, but somewhere near the rear edge of the source. The narrower the band source and the lower the moving speed, the smaller would be the distance between the maximum temperature point and the front edge.

If we decrease the width of the band source or moving speed but do not change the power per unit length of the band source, the temperature of the substrate would be higher. But if we decrease the width of the band source or moving speed and keep the temperature at the front edge unchanged as the melting point of the substrate, the power needed would be lower and as the difference between the maximum temperature and the temperature at the front edge decreases, the maximum temperature would be lower.

In the above stated condition, as soon as the tip temperature reaches the melting point of the material being cut, the laser machining can be carried out. From now on, by the power needed for laser machining, we mean the power needed for the tip temperature to reach the melting point of the material.

Stainless steel, Inconel, Ti-6Al-4V are taken into consideration to calculate the relationship between the main parameters.

The relationship between the beam power per unit length of the light strip P required for laser machining and the machining speed V with different beam widths D and workpiece materials are shown in Fig. 4.

The powers needed to cut these materials are very similar. This is due to the fact that their thermophysical properties (thermal diffusivities) are similar although their mechanical properties are quite different.

The beam energy required for machining one unit area is $R = P/V$. R may be called the specific energy of the laser machining process. The lower the R value, the better would be the efficiency of the laser machining process.

The values of specific energy with different machining speeds V, workpiece materials, and beam widths D are shown in Fig. 5. In the low speed region when speed increases, the specific energy decreases significantly but in the high speed region, the values of specific energy tend to stabilize and gradually become constant. This is due to the two conflicting factors that influence the process. Increasing the machining speed will reduce the heat conduction time and thus decrease the heat affected zone but will increase the power needed and the maximum temperature as mentioned before and in this way increase the energy needed.

In order to protect the workpiece from overheating and to avoid the interruption of the beam path due to molten material formed at the high temperature, the machining speed should not be too high. To improve the working efficiency, the best way is to increase the cutting width with a longer light strip.

The relationship between power per unit length of light strip P required and beam width D is shown in Fig. 6 with different machining speeds and workpiece materials. The specific energy R for V = 20 cm/s is also shown in Fig. 6. From Figs. 5 and 6 it can be seen that beam width has a very strong influence on the power and specific energy needed.

In order to make a laser machining process more efficient and maintain high quality of the machined surface, the beam width should be as small as possible or the light strip should be as narrow as possible.

In a traditional cutting process, specific energy means the energy needed for cutting one cubic unit of material. To compare the specific energy of a laser machining process with that of a traditional process, it is necessary to know the chip thickness of the laser machining process. Of course, in this process almost any chip thickness within reason can be machined. This will be function of laser power. Assuming that the chip thickness as thin as 1 mm, the values of specific energy obtained here will be equal to that defined for traditional cutting process. As long as the beam width is small enough, the specific energy of a laser machining process will be much smaller than that of a traditional cutting process.

EXPERIMENTAL RESULTS AND DISCUSSION

An experiment was conducted to verify the feasibility of the proposed process and to find out its proper working conditions. The laser available was an AVCO 10 kW CW CO_2 laser. It is a powerful laser but the annular output beam cannot be focused to a narrow light strip. Therefore, a mask was used to utilize only part of the output beam. Two KCl cylindrical lenses, one of which was a plano-convex and the other was a meniscus lens, were used. Each lens converged the beam passed through the mask in one plane. The two convergent planes of the two lenses were perpendicular to each other. Using this optical system, a light strip 1.4 mm in length and 0.25 mm in width was obtained. The optical system is shown in Fig. 7.

It was determined that only 5 percent of the laser power was utilized. When the total output was 10 kW, the power of the light strip was 500 W and the power per unit length of the light strip was 357 W/mm.

A Kyon 2000 Si-Al-O-N ceramic tool bit offered by Kennametal, Inc., was used as a stripping tool. Stainless steel 304 plate and Inconel 718 plate were used as the workpiece. The thickness of these plates was 0.8 mm which was the cutting width. The undeformed chip thickness was 1.5 to 2.0 mm. In order to start the experimental machining process, the workpieces were mechanically cut to form chips for the stripping tool to bend them. A motor drove a slide way with the workpiece moving against the stable stripping tool and the laser beam. A time delay relay was used to adjust the time delay between the laser beam output start and the workpiece movement. An encoder was used to measure the displacement of the workpiece and the machining speed. Figure 8 is a photograph of the experimental set up for the laser machining. Figure 9 shows the detail of the machining region. The machining speeds obtained for stainless steel 304 and Inconel 718 workpieces were 25 and 22 mm/s, respectively. According to the calculation, with a beam power per unit length of 357 W/mm and a beam width of 0.25 mm, the machining speeds for stainless steel and Inconel workpiece should be 74 and 92 mm/s, respectively. We believe the main reason for the difference between the theoretical and experimental speeds is that a molten pool is formed around the tip and the laser beam can only heat the tip area through the molten pool thus, the heat spreads to a larger area and the laser machining process becomes less efficient. The formation of a molten pool around the tip may be unavoidable in a laser machining process but the wider the light strip, the higher the power needed and the higher the maximum temperature on the surfaces and the greater the distance between the maximum temperature point and the tip and therefore the larger the molten pool. As the beam width in the experiment was as large as 0.25 mm, a rather large molten pool was formed and its influence on the machining process was significant.

A controlled force laser machining process was also performed in which a weight was used to pull the slide way and the workpiece instead of the motor and screw which drove the slide way and the workpiece with constant speed. Within the cutting parameters mentioned before (0.8 mm cutting width and 1.5 to 2.0 mm undeformed chip thickness), the cutting forces were less than 180 N which is negligible compared to that of the traditional cutting process.

For a laser machining process to work properly, it is important to synchronize the workpiece moving speed with the melting speed of the interfacial area by the laser beam. If the workpice speed is faster than the melting speed, the tip of the wedge angle will move forward and away from the focal plane of the laser beam and the process cannot be continued. If the workpiece moving speed is too slow, a bulk material will be melted and this will lead to low machining efficiency, a large heat affected zone, and a rough machined surface.

It is difficult to synchronize these two speeds at the beginning of the laser machining process, especially with our simple workpiece driving system. To solve this problem partially in our experiment, an inert gas jet was used from one side of the workpiece plate to blow the extra molten material away whenever it was formed.

Figure 10 is a macrograph of a machined workpiece. The accuracy, roughness of the machined surface, and the dimension of the heat affected zone are of the same order of the beam width.

In order to get better laser machining results with high working efficiency and accuracy, more work is needed with a much finer focused laser beam.

CONCLUSION

A new machining process, laser machining by localized melting, is proposed. A two-dimensional heat transfer model has been developed for the process. The feasibility of the process has been demonstrated experimentally.

The important working conditions derived by analysis, process modeling, and experiment are as follows:
1. The light strip used as a "laser cutting tool" should be as narrow as possible or the focused beam width should be as small as possible.
2. The chip should be bent large enough that the slope of the chip at

its root would be larger then the convergent angle of the focused laser beam. Therefore the convergent angle of the focused laser beam should not be too large.

3. The choice of the F number of the objective should be based on such consideration that the smaller the F number, the finer the light strip but the larger the convergent angle. An F number of 3 or larger is chosen in the experiment.
4. The power distribution along the light strip should be nearly uniform.
5. The machining speed of a laser machining process should not be too high. A suitable way to improve the working efficiency is to increase the cutting width with a longer light strip.
6. The workpiece moving speed and the melting speed of the interface area between the chip and the main body by the laser beam should be synchronized.

The force produced in a laser machining process is negligible comparing to that of the traditional cutting process. The "laser cutting tool" is free from wear.

There is not much difference between machining a stainless steel workpiece and an Inconel workpiece by laser machining as their thermophysical properties are similar. This process is particularly suitable for machining difficult-to-cut metals.

The accuracy, roughness of the machined surface, and the dimension of the heat affected zone are of the same order of the beam width.

In order to get better laser machining results with higher working efficiency and accuracy, more work is needed with a much finer focused laser beam.

ACKNOWLEDGMENT

The cooperation of Mr. Jon Culton and Mr. Thomas Casale in the laser experiments is appreciated. This work is partially funded by the Materials Processing Consortium of the University of Illinois at Urbana-Champaign. One of the authors, Lijun Li, acknowledges the financial support from the People's Republic of China.

REFERENCES

1. Nakayacun, K., "Grinding Wheel with Helical Grooves," Annals of CIRP, 1977.
2. Bass, M., S. M. Copley, and D. Beck, "Laser Assisted Machining," Proceedings of the Fourth European Electro-Optic Conf., Utrecht, The Netherlands, 1978.
3. Copley, S. M., M. Bass, and R. G. Wallace, "Shaping Silicon Compound Ceramics with a Continuous Wave Carbon Dioxide Laser," Proceedings of the Second International Symposium on Ceramic Machining and Finishing, Gaithersburg, Md., 1978.
4. Moore, A. I. W., "Hot Machining for Single-Point Turning--A Breakthrough," Tooling and Production, Nov. 1977.
5. Li, L. J., "An Idea About Laser Cutting Process," Journal of Hunan University, China, No. 4, 1982.
6. Duley, W. W., CO_2 Lasers--Effects and Applications New York, 1976.
7. Touloukian, Y. S. (ed.), The Thermophysical Properties of Matter, New York, 1970.
8. Goldman, L., Applications of the Laser, CRC Press, Inc., 1973.
9. Jaeger, J. C., "Moving Sources of Heat and Temperature at Sliding Contacts," Proceedings of the Royal Society of New South Wales, 1942.

LIST OF FIGURES

Figure 1   Schematic illustration of laser machining.

Figure 2   Laser beam alignment and its reflection path in the wedge angle and coordinate systems used in process modeling.

Figure 3   Temperature distributions at different powers per unit length of light strip P, widths of Gaussian beam D and machining speeds V for Inconel workpiece material.

Figure 4   Relation between the beam power per unit length P and the machining speed V with different beam widths D and workpiece materials

Figure 5   Relation between specific energy R and machining speed V with different beam widths D and workpiece materials.

Figure 6   Relation between power per unit length p, specific energy R (for V = 20 cm/s) and beam width D with different machining speeds V and workpiece materials.

Figure 7   Optical system.

Figure 8   Experimental set up for the laser machining.

Figure 9   Detail of the machining region.

Figure 10   A sample of laser machining.

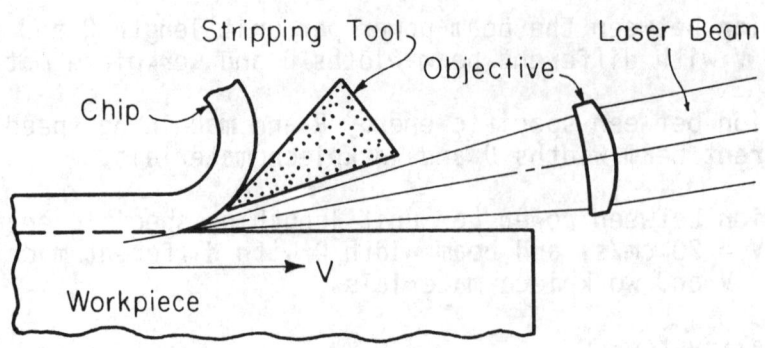

Figure 1   Schematic illustration of laser machining.

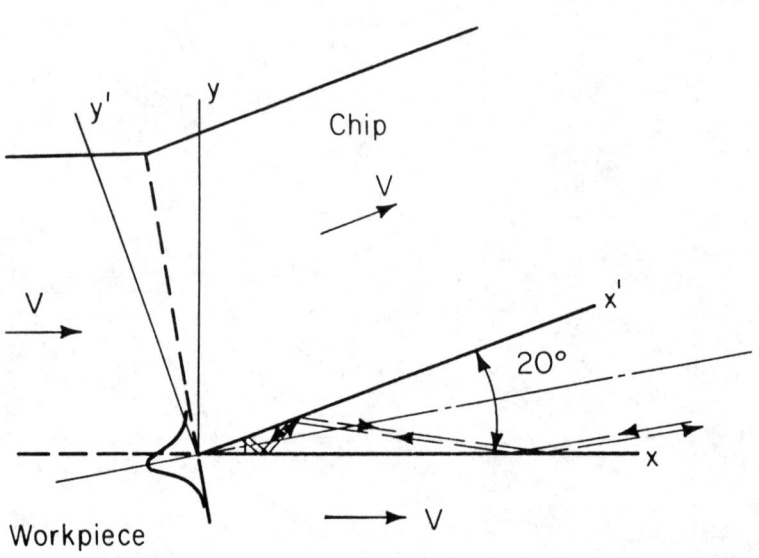

Figure 2   Laser beam alignment and its reflection path in the wedge angle and coordinate systems used in process modeling.

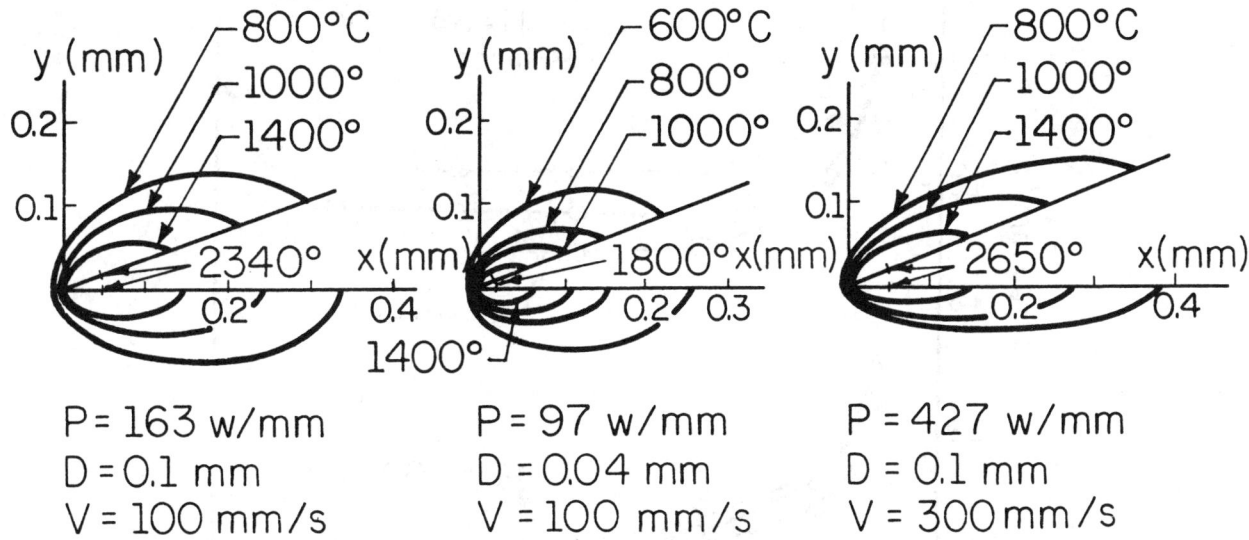

Figure 3 Temperature distributions at different powers per unit length of light strip P, widths of Gaussian beam D and machining speeds V for Inconel workpiece material.

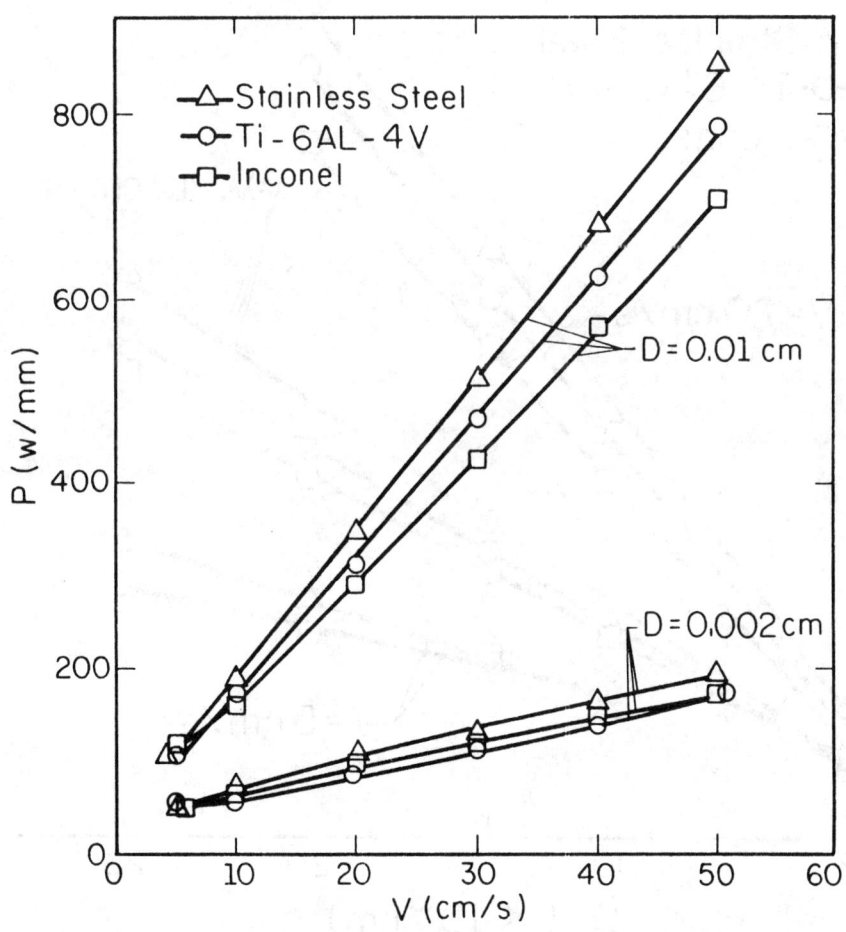

Figure 4 Relation between the beam power per unit length P and the machining speed V with different beam widths D and workpiece materials

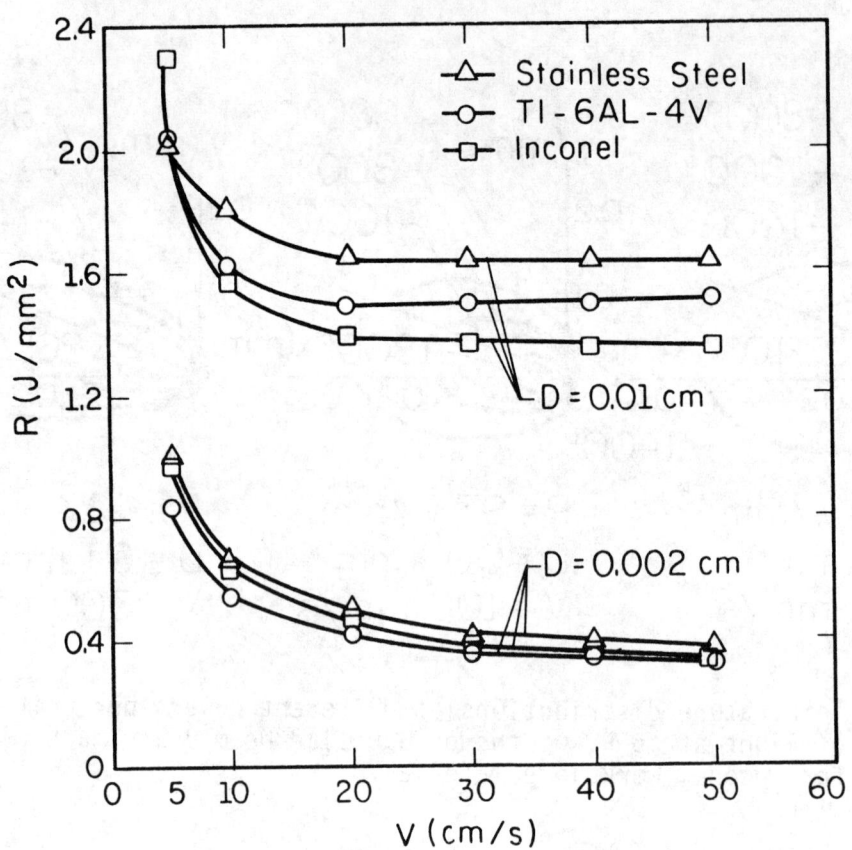

Figure 5 Relation between specific energy R and machining speed V with different beam widths D and workpiece materials.

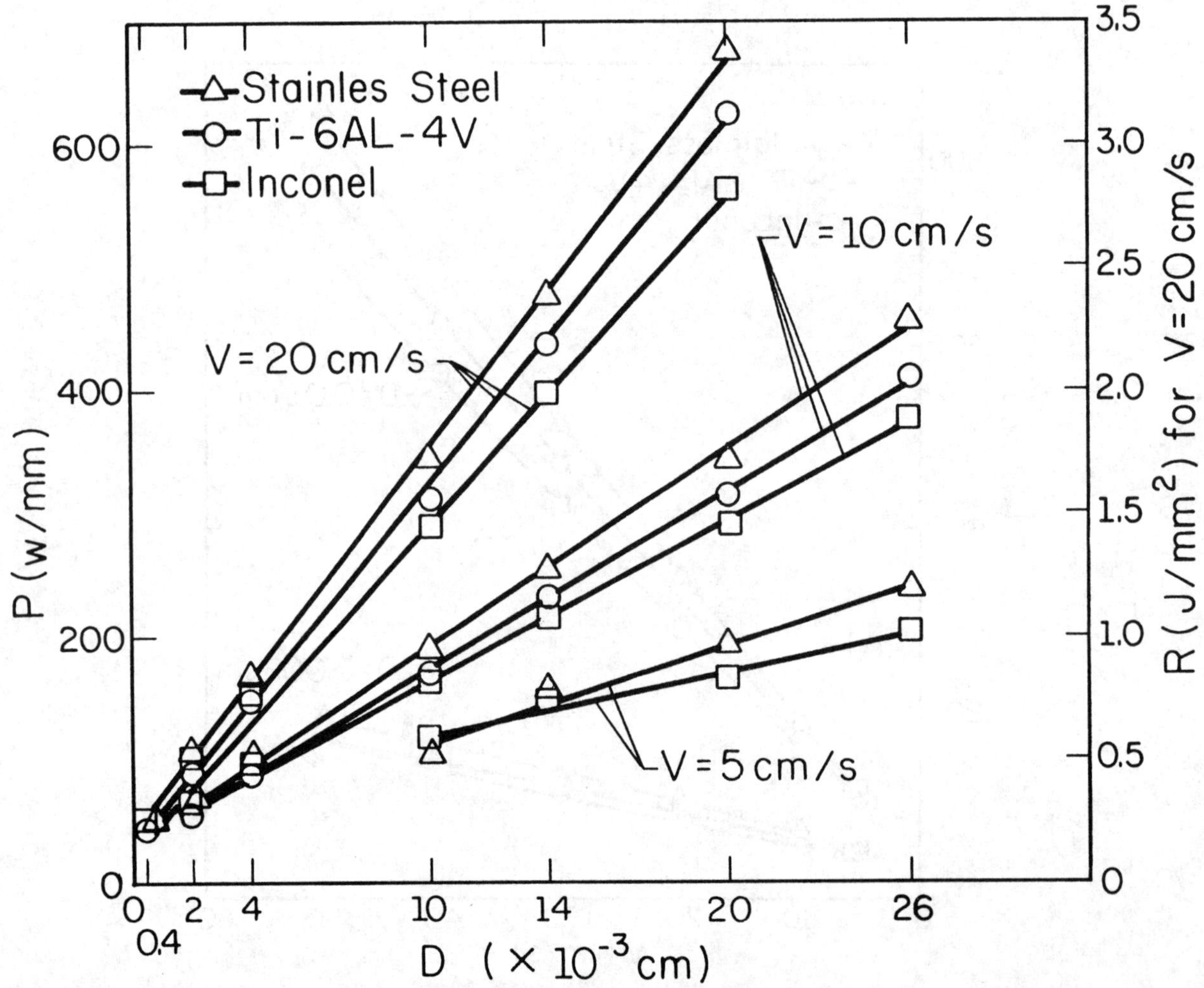

Figure 6 Relation between power per unit length p, specific energy R (for V = 20 cm/s) and beam width D with different machining speeds V and workpiece materials.

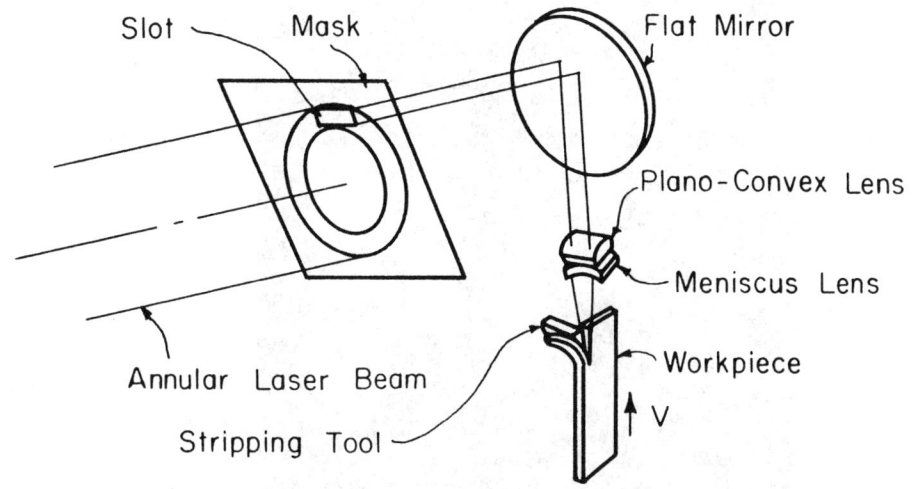

Figure 7 Optical system.

Figure 8 Experimental set up for the laser machining.

Figure 9 Experimental Setup for the Laser Machining

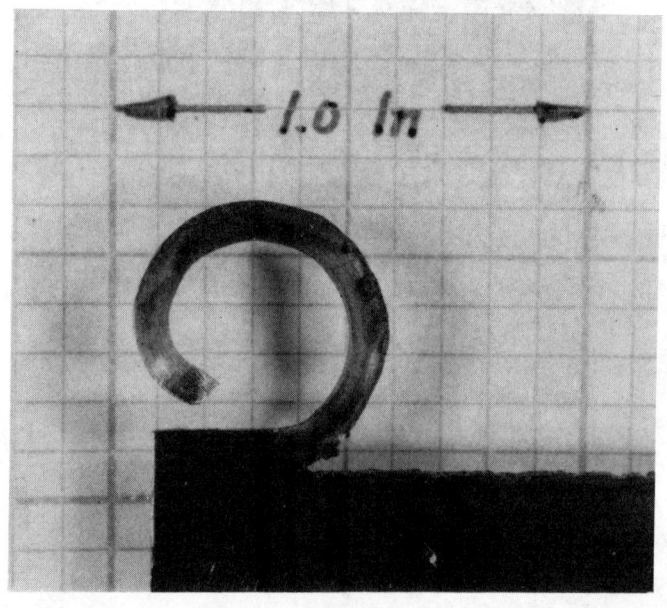

Figure 10 A Sample of Laser Machining

LASER PROCESSES – CUTTING

Cut edge quality improvement by laser pulsing
J. Powell, T. G. King
and
I. A. Menzies
Loughborough University of Technology, UK

ABSTRACT

Analysis of the CO_2 laser-oxygen jet cutting of mild steel has led to the development of a new cutting technique utilising a specific range of laser pulsing frequencies. Profilometry and S.E.M. analysis of the cut edges have shown remarkable improvements in quality, Ra values of the surface roughness have been reduced by a factor of three or more.

INTRODUCTION

The rapid growth in the use of high power lasers for profiling sheet materials has generated a great deal of scientific and commercial interest in the quality of the cut edges produced. Generally, when cutting non ferrous metals, plastics, wood etc., the material removal process is a consequence of the high energy density on the material surface resulting in either melting or evaporation. Fluids thus generated are removed from the cutting zone with the aid of pressurised gas jets. In the case of polymers (e.g. polymethyl methacrylate) this process can often result in a high quality fire polished cut edge. The cutting of ferrous materials, however, is achieved by a different type of laser material interaction. In this case the incident laser energy is associated with a co-axial pressurised oxygen jet. The steel is heated by the laser spot to a temperature at which it will ignite in the presence of the oxygen. The resulting exothermic oxidation reaction results in a combustion front which burns radially away from the laser hot spot at a rate exceeding the average cutting speed. This oxidation front comes to rest just out side the laser material interaction area and the subsequent encroachment of the laser beam on the new cutting front repeats the oxidation initiation process. This initiation-propogation event is repeated as the laser beam moves across the material surface resulting in a characteristic cut edge consisting of regularly spaced striations.

This cutting mechanism has been investigated in depth by Arata et al. (refs 1 and 2). A typical laser-O_2 cut mild steel edge is shown in figure 1a.

Although the edge quality of laser cut steel is generally superior to that achieved by mechanical sawing, the periodic striations place a limit on the applicability of laser cutting to the production of finished component edges.

This paper reports an experimental investigation into the improvement of cut-edge quality. The approach adopted in this experimental programme was to disturb the 'natural' frequency of the periodic striations associated with continuous wave laser cutting by modulating the incident laser power.

Thus the natural striation would be replaced by carefully controlled striations whose amplitude and frequency would be dependant on laser input characteristics rather than material properties.

EXPERIMENTAL

The carbon dioxide laser used for these experiments was a Coherent 'Everlase - 525.2' providing a maximum continuous wave (cw) otuput of 450 watts. This laser can also be used in a pulsed output mode in which higher peak powers can be generated repetitively. In pulsed mode, the laser output was monitored using a digital frequency meter to measure pulse rate, an oscilloscope for mark-space ratio and the Everlase power meter, which because of its long response time gives readings of integrated power output.

The material to be cut was moved under the stationary beam using an Aerotech dc servo motor driven X-Y table controlled by microcomputer. After completion of cutting the edge topography was examined using scanning electron microscopy and a Taylsurf surface profile measuring instrument. The Talysurf was interfaced to a microcomputer to facilitate profile analysis. A micrometer translation stage enabled accurate positioning of the specimen so that profile traces could be made at differing distances from the top edge of the cut. The kerf width of the tops and bottoms of the cuts were measured using the microscope measuring gates of a Vickers hardness tester which gives a direct reading in microns.

The specimen material used was mild steel with a thickness of 1.25mm which was guillotined into 30mm by 100mm blanks. These blanks were cut along 90% of their length to provide 'blind' cuts suitable for the kerf width investigation as well as the topographical analysis.

The process parameters used for the cuts are given in table 1. The average laser power was kept constant throughout the experiments at a value of 350W. The other important parameter held constant was the pressure of the oxygen supply to the cutting jet which was maintained at 30 p.s.i.

Pilot tests had shown that laser pulse frequencies of the order of twice the 'natural striation frequency' (calculated from the speed of cutting and the wavelength of striations produced during continous wave cutting) generated obvious improvements in the cut edge. For this particular laser-material interaction the natural striation frequency was approximately 250Hz and therefore it was chosen to investigate a range of pulsing frequencies between 100 and 1000Hz. As expected, the greatest improvements were observed in the 400 to 500Hz range and these frequencies were further investigated by altering the mark-space ratio of the laser output without altering the average power.

RESULTS AND DISCUSSION

The effects of laser pulse frequency:

Figure 2 shows typical examples of surface profiles from cuts made using cw and at each of ten laser pulsing frequencies. The 'mark-space' ratio of the pulses was kept constant at 1:1 (e.g. at 500Hz the pulse duration was 1ms). The profile assessment length was 7mm in each case (although only 4 mm is illustrated). It is quite apparent that there is a progressive improvement in edge quality as pulsing frequency approaches 500Hz. Figure 3 shows autocorrelation functions for a selection of the profiles of figure 2. From the information relating to cw cutting it can be established that the mean wavelength of the periodic cutting event was 0.121mm. At the cutting speed used (30 mm/s) this represents a temporal frequency of approximately 248Hz so that it appears that optimum results are obtained at a laser pulsing frequency of the order of twice this value.

At high pulse rates (i.e. 1000 to 800Hz) the material being cut appears to respond to the incident energy as if it were a continuous input. This may be attributable to the fact that the individual pulses are insufficiently powerful or too short in duration to effect material removal. The molten metal/oxide mixture therefore acts as a pulse integrator. Essentially, the 'natural' striation frequency overrides any superimposed pulsing effect. The autocorrelation function for the 900Hz pulsed profile (fig. 3b) shows a slight increase in striation wavelength over that for cw conditions (fig.3a).

At intermediate pulse rates (700 to 400Hz) a change in the laser material interaction is observed. In this region individual pulses have the capability of material removal. The striation frequency is determined by the laser pulsing frequency as can be seen from figures 1b, 2 and 3d. The striation wavelength for the profile produced with 500Hz pulsing can be estimated from fig. 3d to be 0.0594mm which represents a temporal frequency of 505Hz. In this range of frequencies the surface finish of the cut edges is much improved. Figure 4 summarises the measured Ra values and it is apparent that there is a reduction to approximately one quarter of the value obtained under cw cutting. This is observed for both the curves shown in fig. 4. The higher curve is for the unfiltered profiles (i.e. for sampling length or cut-off equal to the 7mm assessment length). The lower curve is for Ra values obtained after the profiles had been filtered with a phase corrected high pass filter (as described in ref. 3 and implemented digitally in the frequency domain). A non standard cut-off of 0.5mm was sued in an attempt to exclude a low frequency periodic roughness component which is most easily noticed in the smoother cut profiles but which can be observed to be generally present in the autocorrelation functions of fig. 3. Indeed, figure 3c (600Hz pulse frequency) is dominated by this component. Since the presence of long wavelength irregularities appears to limit the improvement in surface finish obtainable further investigation into this effect was undertaken.

Autocorrelation analysis of cuts carried out at 27, 30, 33 and 39 mm/sec established that the wavelength of the low frequency roughness component changed in proportion to the cutting speed. Division of the cutting speed (mm/sec) by the striation wave length (mm) in each case showed that the temporal frequency of this effect remained constant at 50Hz. This is attributable to an a.c.ripple associated with the laser h.t. power supply circuitry. Investigation of the laser output using a 'Laser Beam Analyser' (ref. 4) with fast response, coupled to an oscilloscope verified that the laser output did indeed exhibit output power fluctuations at this frequency. Communication with the laser manufacturer has indicated that current power supply designs have minimised such effects.

Figure 4. compares high magnification (X 1200) electron micrographs of the C.W. cut edge and the edge produced with a laser pulsing frequency of 500Hz. The improvement in the flatness of the pulsed cut is obvious. At this magnification it is evident that the surface of the cut edges are extensively microcracked. These microcracks are approximately parallel to one another and perpendicular to the general striation direction. The cutting process which generates the striations leaves a very shallow covering of oxide covered melt on the surface of the cut edge. Rapid solidification and quenching by conduction to the substrate builds up stress along the lines of the striations which results in eventual cracking of the oxidised surface.

At low pulse rates (300 to 100Hz) the surface finish of the cut edge becomes a great deal coarser due to the increasing size and separation of the unit interactions resulting in a repeated 'drilling' event. Figure 5A shows a general view of the cut edge quality at a laser pulsing frequency of 100 Hz. The appearance of this cut gave rise to speculation about the direct production of saw blades by laser cutting and as a pilot study the same cutting parameters were used to cut a sample of 1mm thick high carbon steel. This steel, of the following composition; .95%C, .25% Si, 1.25% Ma, .55% Cr, .5% W and .15% was cut in the fully annealed condition. Figure 5B shows the cut edge quality. The decreased thickness of the material compared to the mild steel has facilitated melt ejection and thus eliminated the adherent dross effects evident in fig 5A. Metallographic analysis revealed that the heat affected zone beneath the oxidised surface was fundamentally martensitic to a depth of 70µm at the top of the cut and 90µm at the bottom. These differences are due to the more prolonged heating cycle at the bottom of the cut due to the fact that all the molten material passes through this zone prior to ejection. This hard (650 VHN), abrasive edge has been found to be capable of cutting wood and mild steel in ad hoc tests although further work is needed to assess any commercial feasability.

The kerf widths of the laser cuts were measured in all cases and were found to be unaffected by any of the process parameters used. The results can be summarised as follows:

a) Kerf width at the top of the cut; 370µm ± 5%
b) Kerf width at the bottom of the cut; 210µm ± 5%

This uniformity of Kerf widths reflects the domination of the laser spot diameter (325um) at the top of the cut and the similarity of the fluid flow ejection at the bottom. The difference between the top and bottom Kerf widths indicates that the cut face is inclined at an average angle of $3°40'$ to the perpendicular of the sheet steel surface.

The effects of pulse width variation:

To investigate the effect of pulse width variation several cuts were made at constant frequencies of 400 and 500Hz over a range of mark-space ratios from 1:1 to 7:1 and also under cw conditions. As it was desired to maintain the constant average power output of 350W employed throughout these experiments, it was considered unwise to attempt to produce mark space ratios less than 1:1. The production of the extremely intense but short lived, pulses needed to keep the average power at the correct level was considered to cause an unacceptable level of load on the laser.

Figure 6 shows the variation of Ra roughness with pulse width for cutting with a 500Hz pulse frequency. Similar results were obtained at 400Hz. Two facts emerge from observation of fig. 6. Firstly, within the limits employed, the mark space ratio does not appear to have a particularly strong influence on the measured Ra values of the cut edges. Secondly, there seems to be a significant difference between cw and pulsed conditions even when the pulse mark space ratio is large (e.g. 7:1). This seems surprising, it appears that a relatively slight modulation of what would otherwise be cw output is sufficient to ensure the dependence of striation generation on pulsing frequency rather than material property dominated periodic ignition. This preliminary investigation identified the mark-space ratio of the pulsed laser output as being equal to that of the externally input laser arc trigger signal. This is perhaps misleading in that the lasing action takes time to establish a steady state and further time to decay away. The exact correlation between the arc trigger signal, laser pulse output and striation characteristics is undergoing further investigation.

It was thought that the differing loading conditions on the laser with varying pulse widths might produce different degrees of the 50Hz power ripple previously identified. The three curves shown in fig. 6 are for differing high pass filter cut-off values chosen to enable any such effects to be observed. The general similarity of the curves suggests that these effects were insignificant in practice.

It should be mentioned at this point that for the purposes of direct comparison the profiles analysed were all traced close to the top face of the cut edge. This location was chosen because periodic effects are most marked in this region. In some cases this single trace evaluation was not representative of the cut edge as a whole

Figure 7 shows two cuts taken at different mark space ratios at 400Hz laser pulse frequency. Figure 7a, for a mark-space ratio of 4:1 shows a marked superiority to fig. 7b which, like the majority of the cuts examined, had a ratio of 1:1. The laser pulsing domination of the topography of the top edge of the cut contrasts strongly with the fluid flow/material property dominated lower portion. Note that this effect is not present in the 500Hz, 1:1 ratio cut edge (fig. 1b). It can be postulated that the larger amount of material needing removal for the generation of the longer wavelength 400Hz striations needs a pulse of longer duration than is available with a 1:1 mark-space ratio.

CONCLUSIONS

It has been demonstrated that, when cutting ferrous materials modulation of the output from a high power laser can lead to a cut edge the topology of which is controlled by laser pulsing parameters rather than material constants. At pulsing rates above a limiting value for each laser-steel combination (in this case 700Hz) the material tends to integrate out the pulses and responds in much the same way as it would to a continuous wave input. Below this value it is possible to generate cuts of much higher quality than are available by C.W. means. In this case a pulse frequency of 500Hz reduced the Ra value of the roughness to approximately one quarter of its original value. At still lower pulsing frequencies a cut edge can be produced which has a deliberately roughened topography. In the case of the cutting of high carbon steels it is possible to produce a potentially useful hardened saw-blade type finish. Pulse widths can be manipulated to facilitate material ejection from the cut zone to ensure dross free edges.

ACKNOWLEDGEMENTS

This work was carried out as part of a larger SERC research programme (GR /B57057) The authors would like to express their appreciation for the support of this research and particularly to the late Professor John Butters of the Department of Mechanical Engineering.

REFERENCES

1. Arata, Y., Maruo, H., Miyamoto, I. and Takeuchi, S. "Dyanmic Behaviour in Laser Cutting of Mild Steel". Trans. JWR1 (1979) Part 2, pp 15-26.

2. Arata, Y. and Miyamoto, I. "Some Fundamental Properties of High Power Laser Beams as a Heat Source (Parts 1, 2 and 3)". Trans. Japan. Weld. Soc. Vol. 3 Pt 1 pp 143-180 (April 1972).

3. Whitehouse, D.J. "Improved Type of Wavefilter for use in Surface Finish Measurement" Proc. IMechE. 182, Pt. 3k, pp 306-318 (1967-68).

4. Lim, G.C. and Steen, W.M. "Measurement of the Temporal and Spatial Power Distribution of a High Power CO_2 Laser Beam". Opt & Laser Tech. 14, pp 149-153 (June 1982).

LIST OF FIGURES

Fig.1. a) Scanning electron micrograph of typical striated laser cut edge of a mild steel specimen.
   b) S.E.M. of improved quality cut edge as a result of pulsing at a frequency of 500Hz.
Fig. 2.   Surface profiles from cuts produced under continuous wave and pulsed conditions.
Fig. 3.   Autocorrelation functions of selected profiles from fig.2.
Fig. 4.   S.E.M. comparison of cw and 500Hz pulsed cuts showing evidence of surface micro cracking.
Fig. 5.   S.E.M.'s of cuts made in mild steel and high carbon steel at a pulse rate of 100Hz.

Fig. 6. The relationship between Ra roughness and laser pulse length (mark-space ratio).

Fig. 7. a) S.E.M. of cut produced using 400Hz pulsing with 4:1 mark-space ratio.
   b) S.E.M. of cut produced using 400Hz pulsing with 1:1 mark-space ratio showing inadequate material ejection.

TABLE 1:

Process parameters used during cutting.

Cutting speed (mm/s)	Laser pulse frequency (Hz)	Laser pulse duration (ms)	Mark-space ratio	Relevant figs.
Laser pulse frequency effects				
30	CW	—	—	1a,2,3,4,6
30	1000	.500	1:1	2
30	900	.555	1:1	2,3
30	800	.625	1:1	2
30	700	.715	1:1	2
30	600	.835	1:1	2,3
30	500	1.000	1:1	1b,2,3,4,6
30	400	1.250	1:1	2,7b
30	300	1.665	1:1	2,3
30	200	2.500	1:1	2
30	100	5.000	1:1	2,3,5
Power ripple identification				
27	400	1.250	1:1	
30	400	1.250	1:1	2,7b
33	400	1.250	1:1	
39	400	1.250	1:1	
Mark:space ratio effects				
30	500	1.000	1:1	1b,2,3,4,6
30	500	1.200	1.5:1	6
30	500	1.500	3:1	6
30	500	1.600	4:1	6
30	500	1.750	7:1	6
30	500	2.000	C.W.	1a,2,3,4,6
30	400	1.250	1:1	2,7b
30	400	2.000	4:1	7a

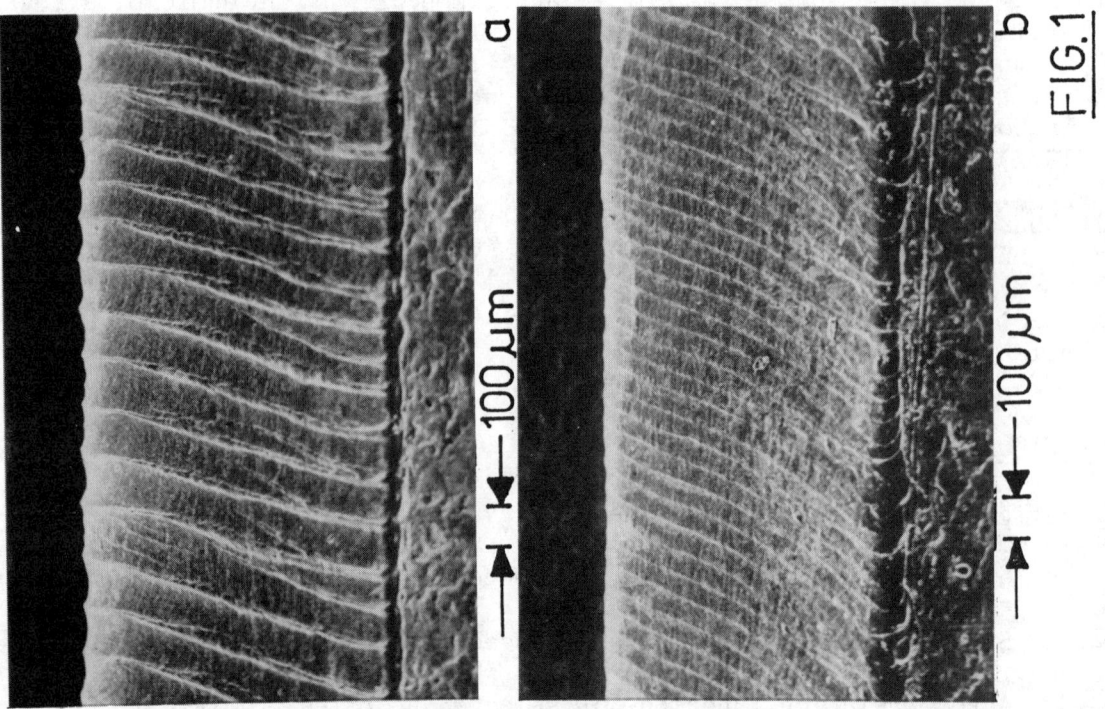

Fig.1. a) Scanning electron micrograph of typical striated laser cut edge of a mild steel specimen.
   b) S.E.M. of improved quality cut edge as a result of pulsing at a frequency of 500Hz.

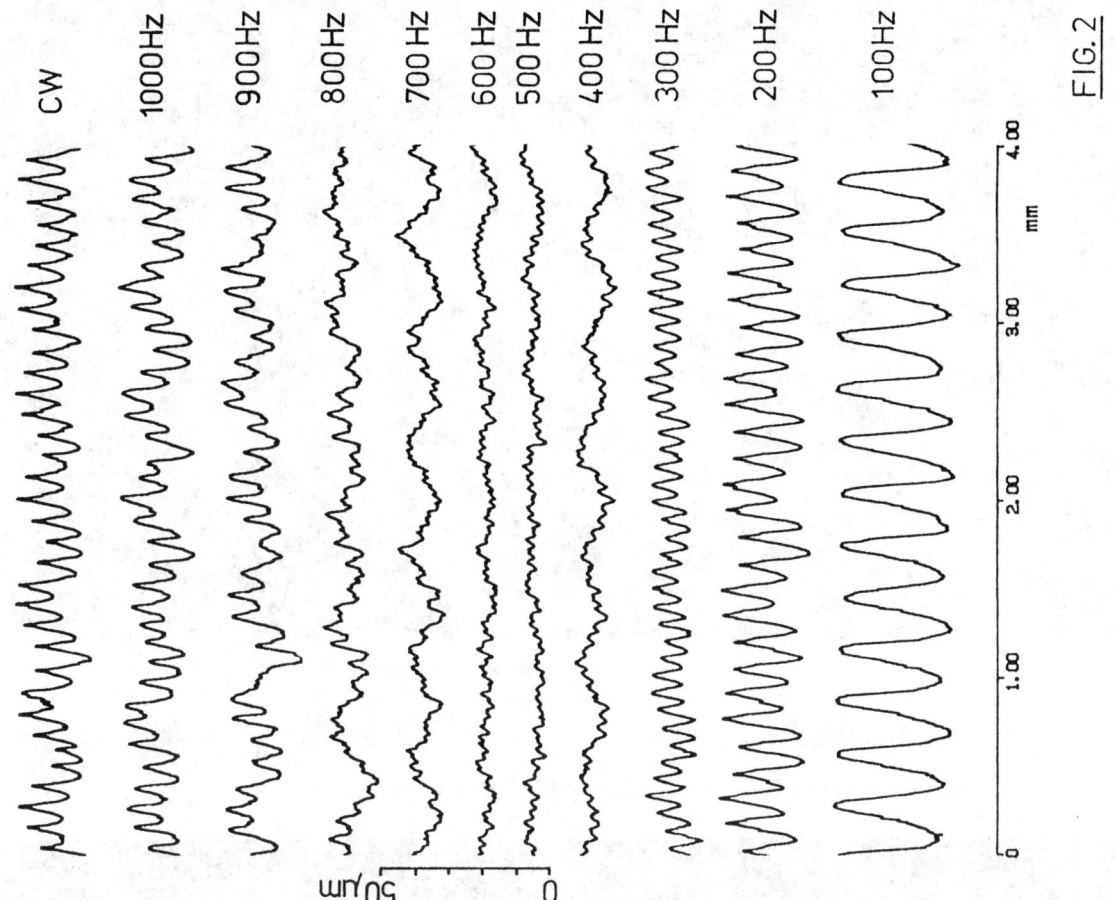

Fig. 2. Surface profiles from cuts produced under continuous wave and pulsed conditions.

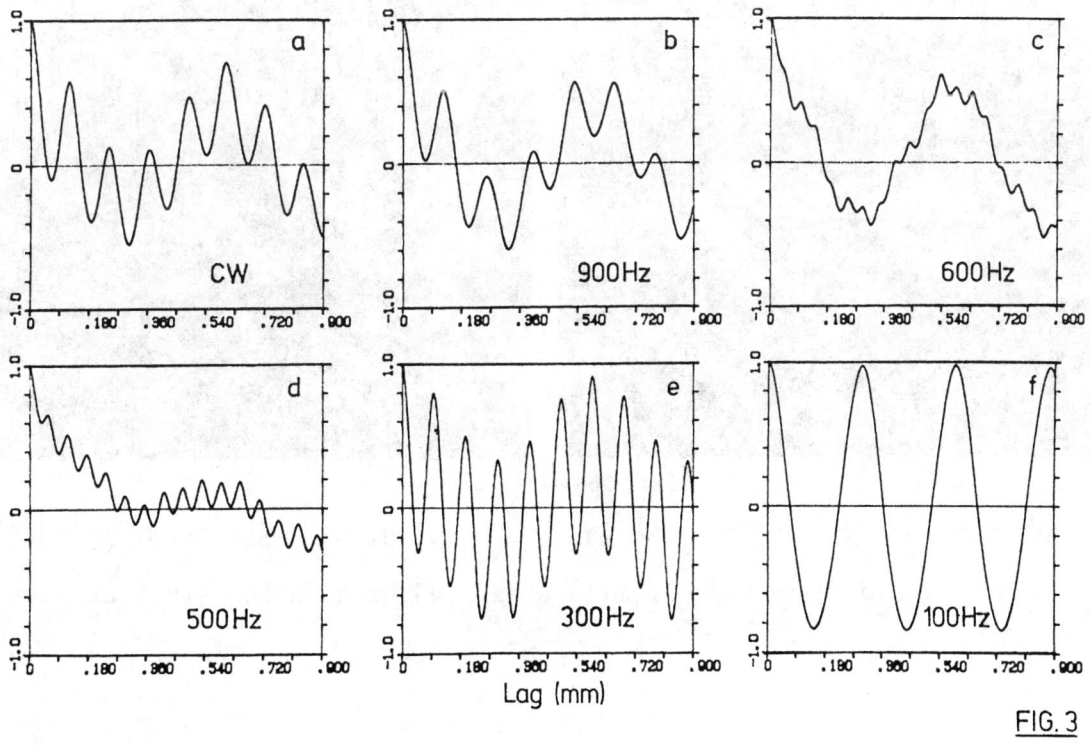

Fig. 3. Autocorrelation functions of selected profiles from fig.2.

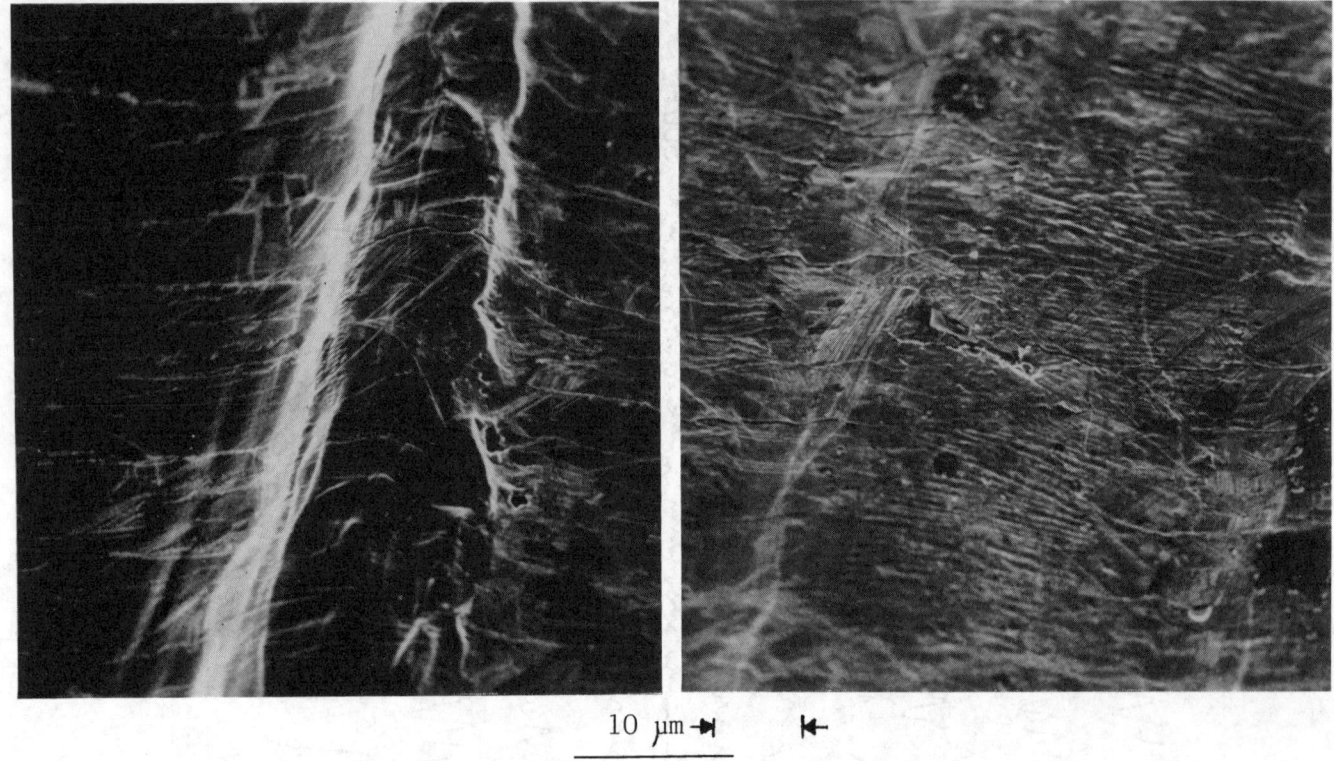

Continuous Wave   Pulsed at 500Hz

Fig. 4. S.E.M. comparison of cw and 500Hz pulsed cuts showing evidence of surface micro cracking.

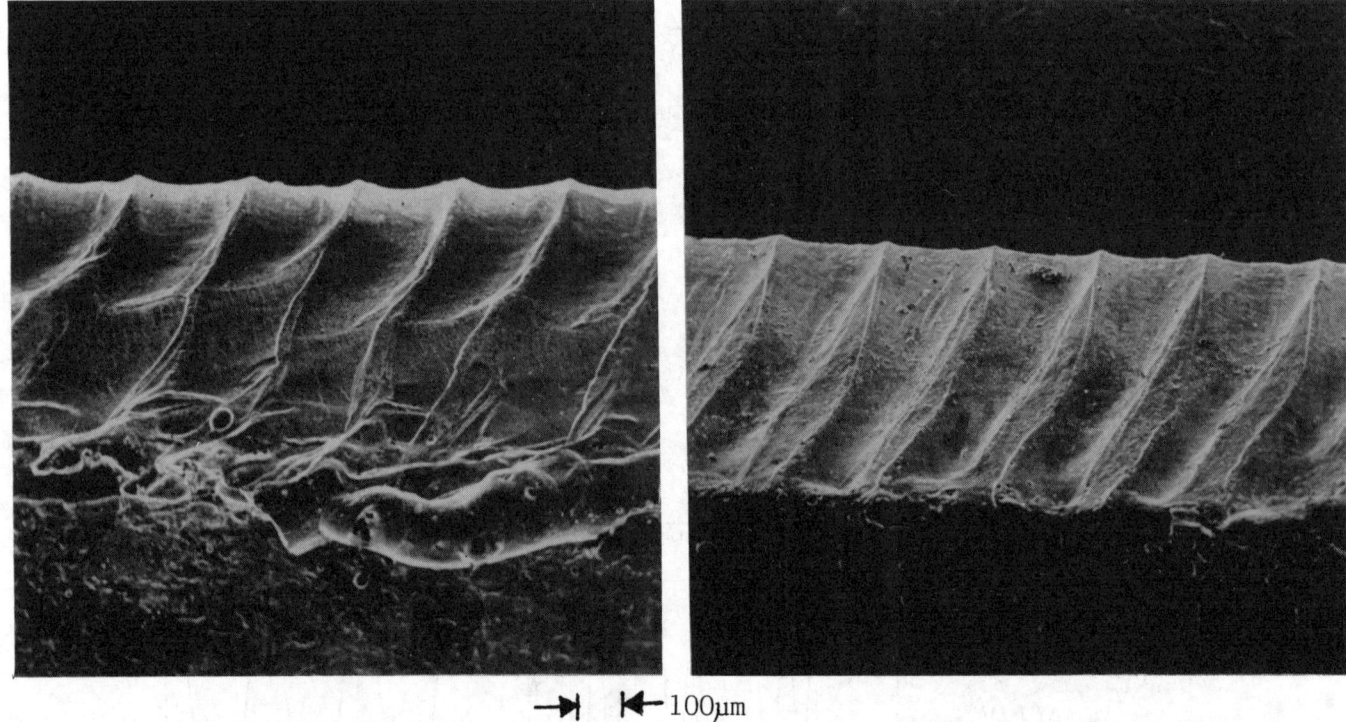

Mild Steel (1.25mm thick)   High carbon Steel (1.0mm)

Fig. 5. S.E.M.'s of cuts made in mild steel and high carbon steel at a pulse rate of 100Hz.

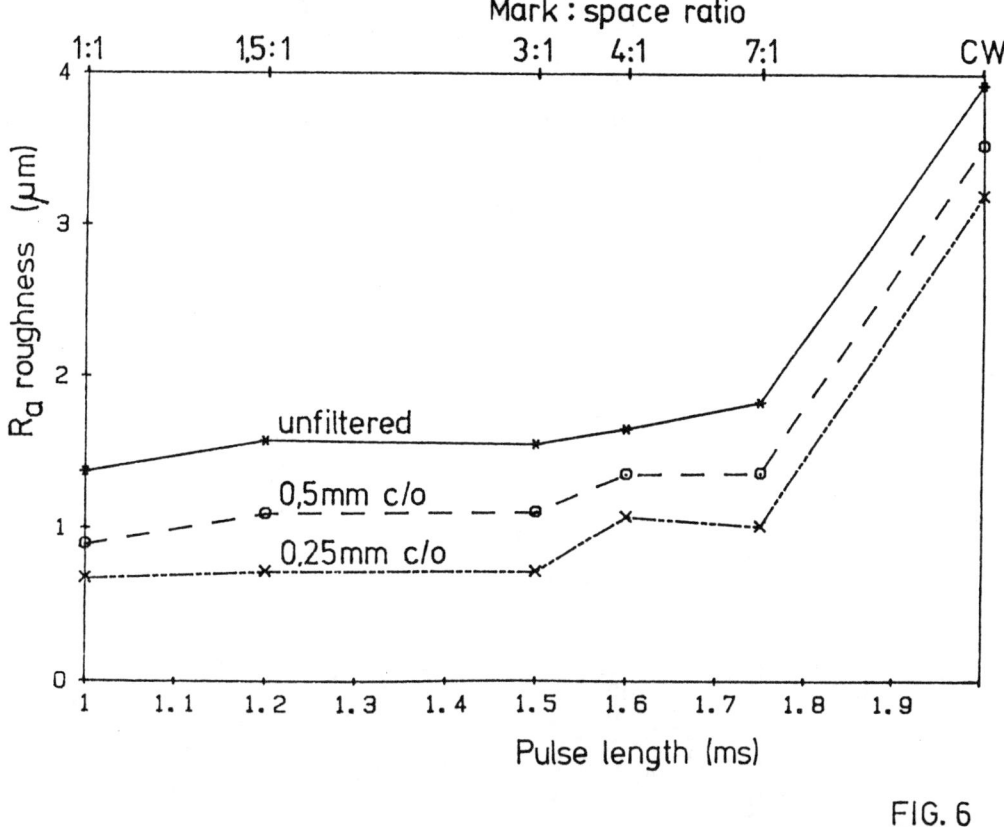

Fig. 6. The relationship between Ra roughness and laser pulse length (mark-space ratio).

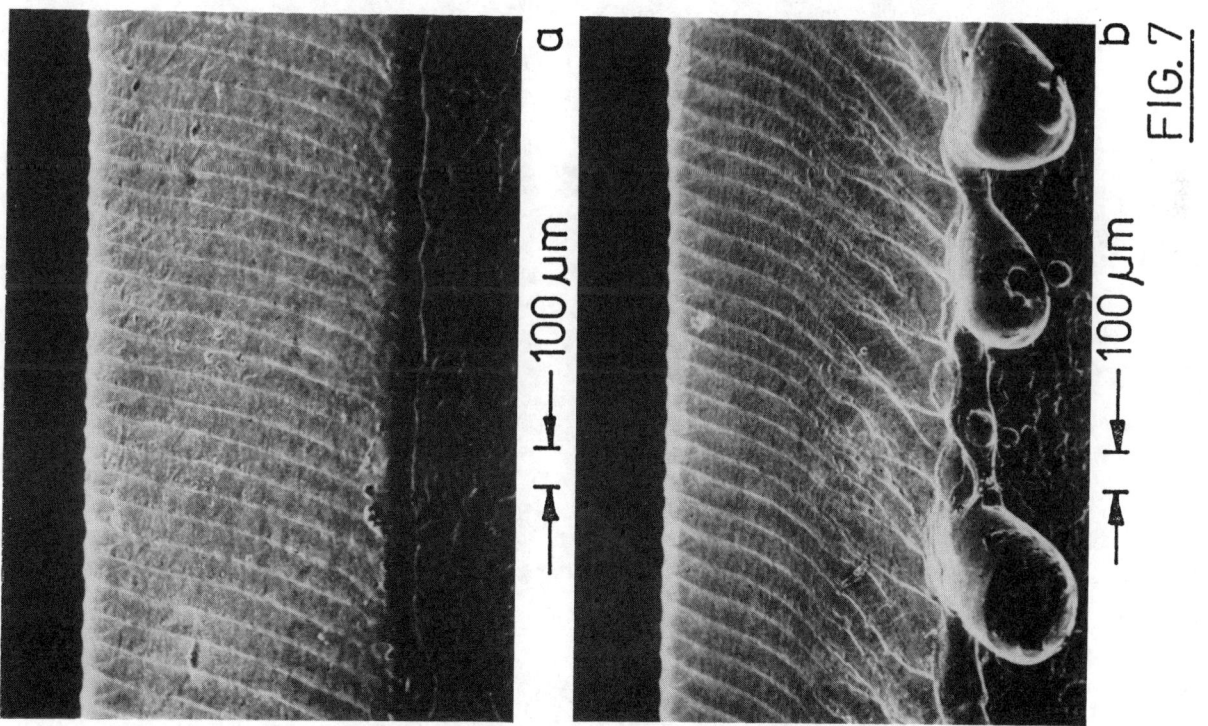

Fig. 7. a) S.E.M. of cut produced using 400Hz pulsing with 4:1 mark-space ratio.
b) S.E.M. of cut produced using 400Hz pulsing with 1:1 mark-space ratio showing inadequate material ejection.

Laser cutting of sheet metal in modern manufacturing

J. M. Weick

and

R. Wollermann-Windgasse

Trumpf GmbH & Co, West Germany

The state of the art of laser cutting is described. Cutting parameters
for 500 W, 750 W and above 1,000 W and typical applications
for the different power levels are demonstrated. Limits, caused by material
thickness, material composition and geometry of the workpiece are discussed.
The advantages of pulse mode for higher accuracy are shown. The benefits
of the combination punching/laser cutting for the manufacturing process
and the gain of flexibility are shown by discussing some typical workpieces.

General

Before the laser has moved from the laboratory to a viable machining method, punching and nibbling, flame cutting and plasma cutting were the processes used for small and medium production runs. All these operations have their particular application. If, however, the emphasis is on machining exact contours, laser cutting produces a better surface quality and a higher feedrate than nibbling, flame cutting or plasma cutting..

In 1979 resonators were developed which incorporate an adequate output power for laser cutting as well as the necessary reliability. They enabled the design of the first combination laser cutting/punching machine launched on the European market. This machine was featured by a 500 W laser. The combination of laser cutting and punching made the laser technique equally suited to workshop application. Lasers ranging from 500 to 1,500 W (slow flow, fast axial flow and transverse flow) are being used for cutting steel sheet from 0.5 to 10 mm thick.

There is no significant difference between the beam handling equipments of the cutting machine used for different output power values (figure 1): Beam bending mirrors directing the laser beam to the work area, a lens focussing the laser beam with diameters of 10 to 20 mm down to 0.1 to 0.2 mm, a nozzle through which the focussed beam and the oxygen are led to the workpiece. The suction system mounted beneath the workpiece removes both dust and slag.

To achieve an optimum cut quality, the components must be handled carefully: The beam bending mirrors must be protected against distortions caused by their holders, the coating of both mirrors and lenses must have a low absorption to prevent thermal distortion, the oxygen must not contain any contaminations such as water, nitrogen or solvents. The design of the nozzle has an impact on cut quality, too. The design of the beam handling equipment must be such that the optics are protected against contaminations which would diminish the laser output power.
In addition the alignment system must not be exposed to vibrations deriving from other machines on the shop floor to such a extent that the adjustment might be lost.

Cutting Specifics with Different Output Power Values

The first CO_2 lasers used for cutting steel sheet were slow flow lasers with a maximum output power of 500 W. Since the typical power extraction is 50 to 80 W per meter of active length with these systems, the total length of the resonator is approx. 8 meters. The typical mode quality is TEM00 which allows a perfect edge quality particularly in sheet thicknesses of 1 to 3 mm. Laser with other modes would supply rougher edges in that range.
The possible cutting speeds are shown in figure 2, referring to a 2.5" lens from 1 to 3 mm and a 5" lens from 4 to 6 mm. The 5" lens would produce the same cut quality, however, lessens the cutting speed by approx. 20 % as compared with the 2.5" lens.

The advantges of the 5" lens over the 2.5" lens are as follows: It may be used for the complete range of thicknesses. Furthermore, its service life is considerably higher since the distance between lens and workpiece is longer. The tolerance of the focus alignment is ± 0.5 mm compared to ± 0.2 mm with the 2.5" lens. The kerf width in a 2 mm thick sheet is 0.1 mm with the 2.5" lens and 0.15 mm with the 5" lens.

The 500 W laser can also be operated in 6 mm mild steel. This would, however, entail a considerable speed reduction (0.4 m/min - as shown in figure 2), and any deviation from optimum conditions would diminish the cut quality.

State-of-the-art slow flow lasers for metal cutting have a maximum output power up to 750 W and are equipped with a resonator of approx. 16 m in optical length. Using the 2.5" lens in 1 to 4 mm material provides an increase in cutting speed. Using the 2.5" lens in 2 mm material, the kerf width is 0.15 mm and 0.25 mm with the 5" lens.

These figures show that the 500 W laser provides a better focussing of the beam. Anyway, the edge quality is still excellent with the 750 W system.

Considering the fact that this laser type produces an excellent cut quality up to 6 mm and that the cutting speed is increased by 100 % in the range from 4 to 6 mm compared with the 500 W laser, this is up to now the ideal laser available for the material range from 1 to 6 mm.

Figure 3 shows the surface roughnesses of a typical laser cut.

To allow the application of CO_2 lasers in thicker material, laser output powers between 1,000 and 1,500 W must be available. The systems concerned are fast axial flow and transverse flow lasers. As to the mode quality, the fast axial flow lasers have certain advantages over transverse flow lasers.

Although the cutting speeds are higher for the complete range of material thicknesses as compared with the 750 W laser, the edge quality in terms of surface roughness below 2 mm sheet thickness is not as good as the quality obtained by 500 and 750 W lasers.

Considering the price per meter cut, the 1,250 W laser would prove more efficient thanks to its high feedrates. For high-precision parts and small parts with thicknesses up to 2.5 mm to be machined in the pulse mode at reduced cutting speeds, it is more expedient to use the 750 W laser. Some customers who require an optimum surface finish, particularly in the smaller thickness range, opted for the 750 W laser.

Laser Limits in Sheet Metal Production

Using the laser output power of 1,250 W in mild steel at a suitable feedrate (higher than 0.5 m/min), the maximum thickness is 10 mm. 12.5 mm could be cut with higher output power values in the laboratory. Carbon steel and certain tool steels with reduced thermal conductivity would allow better cut edges in thicknesses up to 12.5 mm. No burr will occur in the mentioned steel type through the complete range of thicknesses.

When cutting stainless steel, a burr-free cut is only ensured in 1 or 2 mm thick material. When cutting thicker material, burr would occur at the cut edge due to chromic and nickelous oxides. The thicker the sheet, the higher the burr. CW lasers provide for a good cut quality up to 5 mm (with 1,250 W). In 6 mm material no parallel cut can be obtained since it spreads conically towards the sheet bottom. The pulse mode permits parallel cuts in 6 mm material, however, the cutting speed must be reduced to a third of the CW value in conformity with the reduced average power.

Aluminium is confined to 3 mm with the 1,250 W laser. The burr which consists of aluminium oxide is very hard and causes scratches in the material if particles of the burr break off and wander between sheet and workpiece table.

The oxygen assisted laser cutting operation produces a thin oxide layer on the cutting surface. Sometimes it is necessary to remove the oxide layer. The application of inert gas would not eliminate the formation of oxide layers since a considerable speed reduction would be involved and extremely hard burr would form.

Application of the Pulse Mode

There are two limits due to the geometry of the part:

First, high-precision parts and complicated forms in thin sheets which do not allow the application of the optimum cutting speed in CW. Figure 4 shows two examples of parts with small radii and tolerances below those applicable to optimum cutting speeds in CW (+/- 0.1 mm). To this effect, the cutting speed must be reduced to 2 m/min and cutting is done in the pulse mode. The full CW output power would cause the burning away of the cut edge, and the reduction of the CW power would have an impact on the cut quality. Only the pulse mode permits an optimum cut quality at reduced speed since merely the average output power is reduced and the output power per pulse is either maintained or yet enhanced.

Figure 5 illustrates the machining of a complete contour, not including the six small holders, at full speed in CW. As it is possible to change over from CW to pulse mode during the cutting process, the piece rate can be minimized.

Second, in the case of small shapes with a tendency to burn out in areas that have been overheated. Figure 6 shows different laser cut holes of 1 mm diameter. Punching would not be feasible in the present case.

Figure 7 illustrates the production of small cutouts. 7a shows the difference between starting with CW and starting with the pulse mode to provide for a slow traverse through the material with the machine standing still. Cutting takes place in CW. The first method causes a crater in the material, the second method produces a thin start hole with a diameter equal to the cut width.

Figure 7b shows small circular cut-outs (top and bottom view) produced in the same way: Piercing takes place in the pulse mode, cutting in CW. The cut-outs are identical.

As to figure 6, the piercing process is necessary to obtain a small start hole. However, the circular cut-out is too small and cannot be realized in CW. So cutting must be done in the pulse mode. Otherwise the contour would burn out and an irregular shape would be generated.

The capability of the laser to produce small start holes brings further benefits: If prepunching is not possible, as in the said case, or if a laser-only machine is used, laser cutting may be started without the risk of damage to the focussing lens. There is a great risk of damage to the lens due to spattered particles if the material exceeds 5 mm (5" lens) and if cutting must be started inside the material.

Combination Laser Cutting and Punching

Laser cutting is the ideal method to produce contour parts. Fine shapes such as thin slots with 0.2 mm in width, tapered cut-outs, extremely narrow webs and small radii can be laser cut but not punched or nibbled. Although punching and nibbling may be superseded by lasing in some cases, the majority of the parts must also be punched for economic reasons.

This statement is justified by the following calculation:

In the case of e.g. hole diameters of 50 mm and hole distances of 100 mm, up to 160 holes per minute can be produced. Apart from the piercing process, laser cutting must be done at 25 m/min to cope with the punching process.

Using realistic data for laser cutting (sheet thickness 2 mm, piercing process 1 s, feedrate 6 m/min), 23 holes with 50 mm diameter can be produced per minute.

The combination machine incorporates the benefits of both procedures: Round and rectangular holes may be punched with standard tools (see figures 5 and 6). The production method of non-standard shapes is determined by the number of holes. If only a few parts are to be machined and if the production of these special tools is not justified, laser cutting will be applied. To eliminate the time for piercing, start holes are always prepunched when laser cutting. That's why the combination machine is a very flexible and efficient solution for sheet metal production.

If only a few holes are to be made in the overall range of parts, it is expedient to use a laser-only machine. As compared with the combination machine, it is a moderate-prized system at less flexibility.

Laser Outlook

What will be the laser cutting systems of the near future be like?

(1) The system design will be improved: Laser resonators will be more compact and the beam output will be increased to higher power levels, e.g. 2 kW, which allow higher cutting speeds and the production of thicker material. Since all these lasers will be fast axial or cross flow systems, the mode quality will be improved to attain the edge quality of slow flow lasers in thin material.

(2) Furthermore, cutting 3-dimensional parts on 5-axes machines will be feasible.

Figure 1

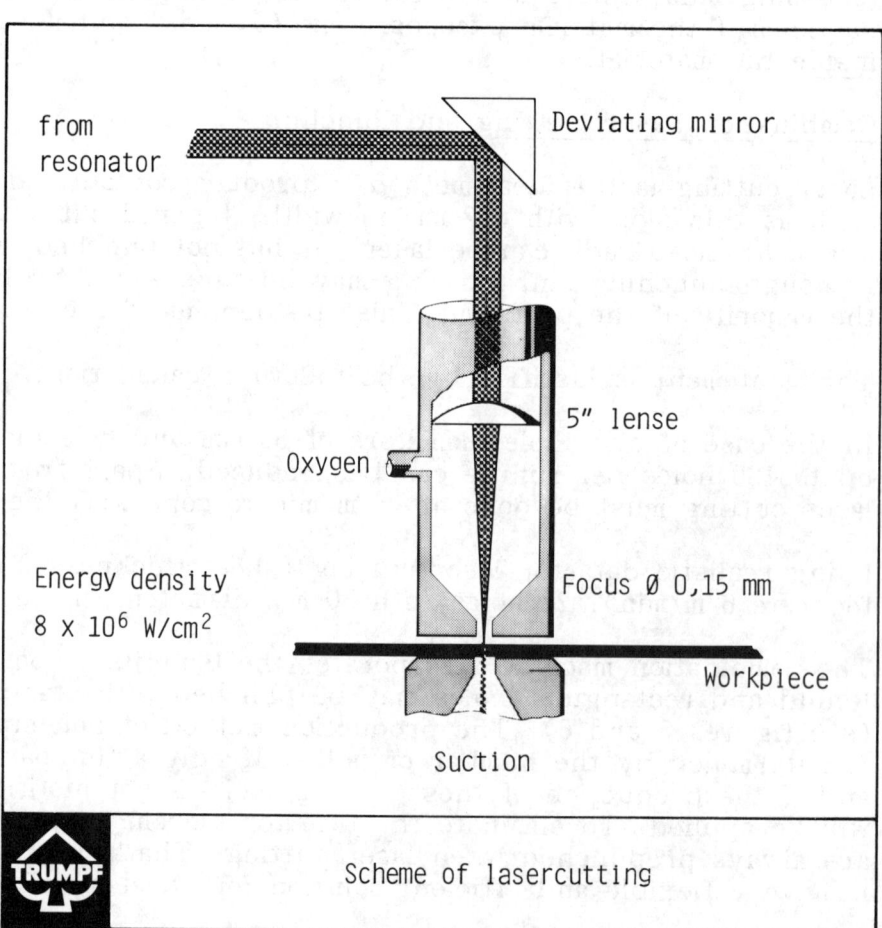

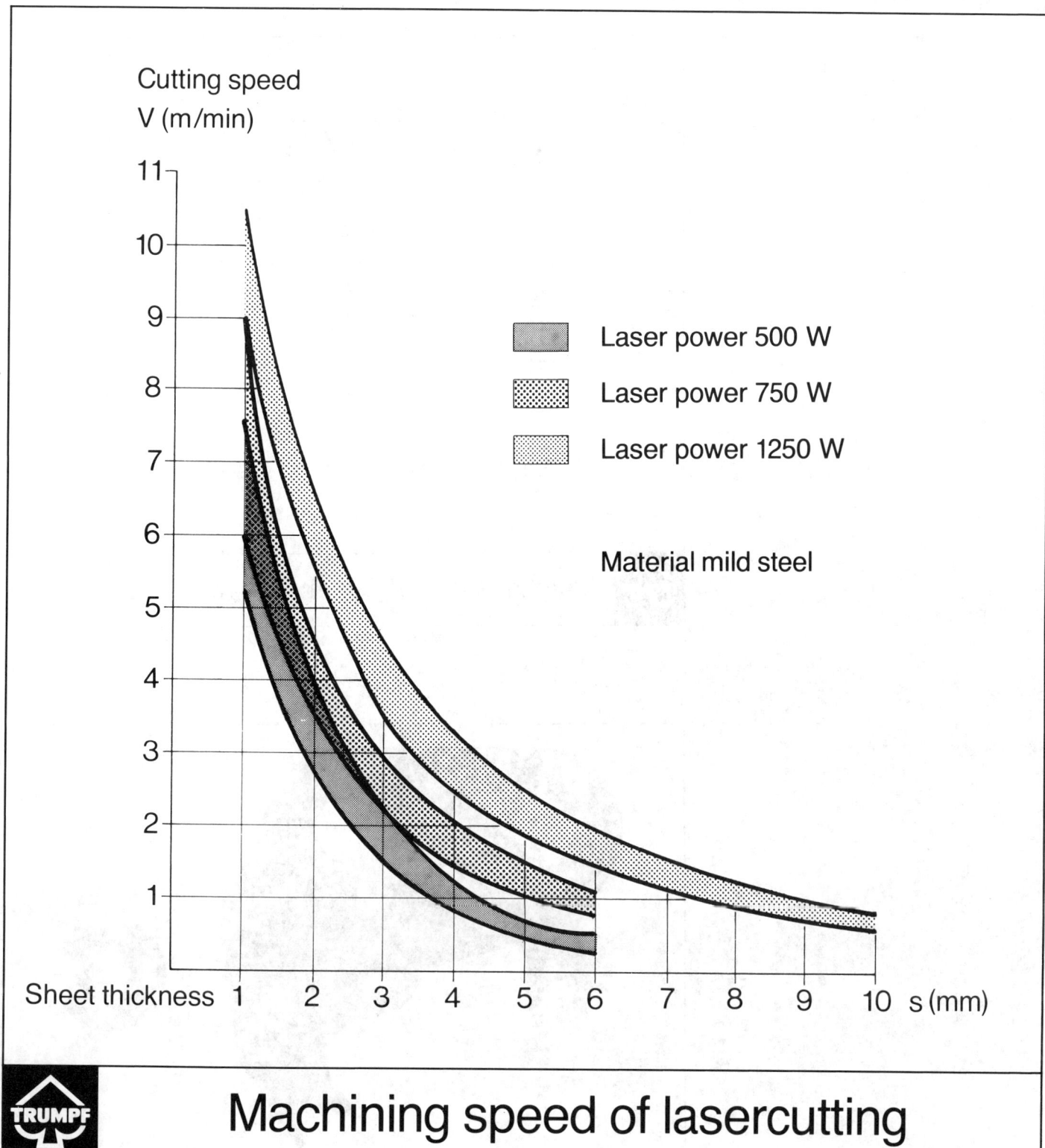

Figure 2

Figure 3

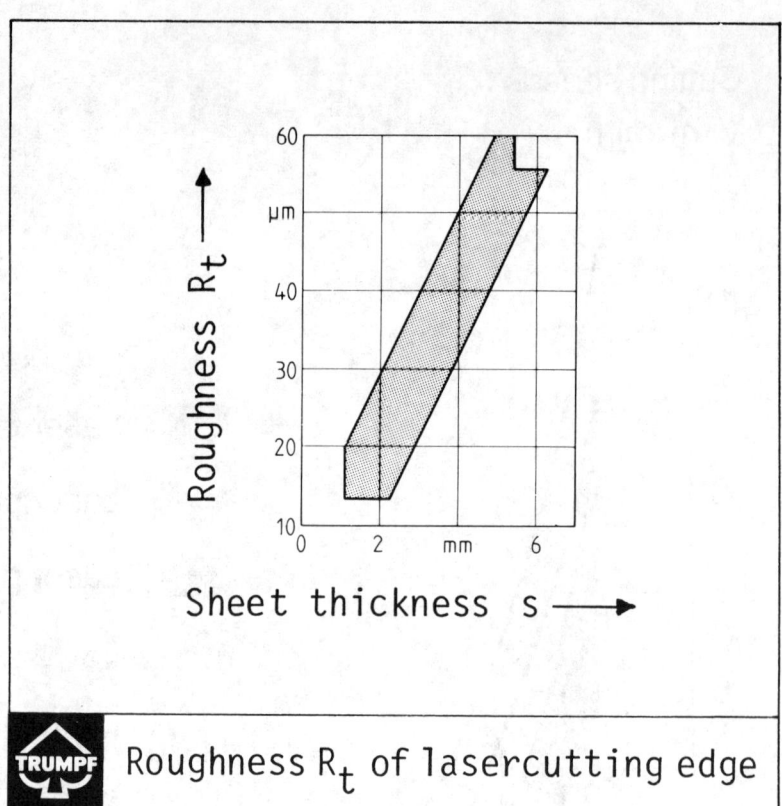

Figure 4

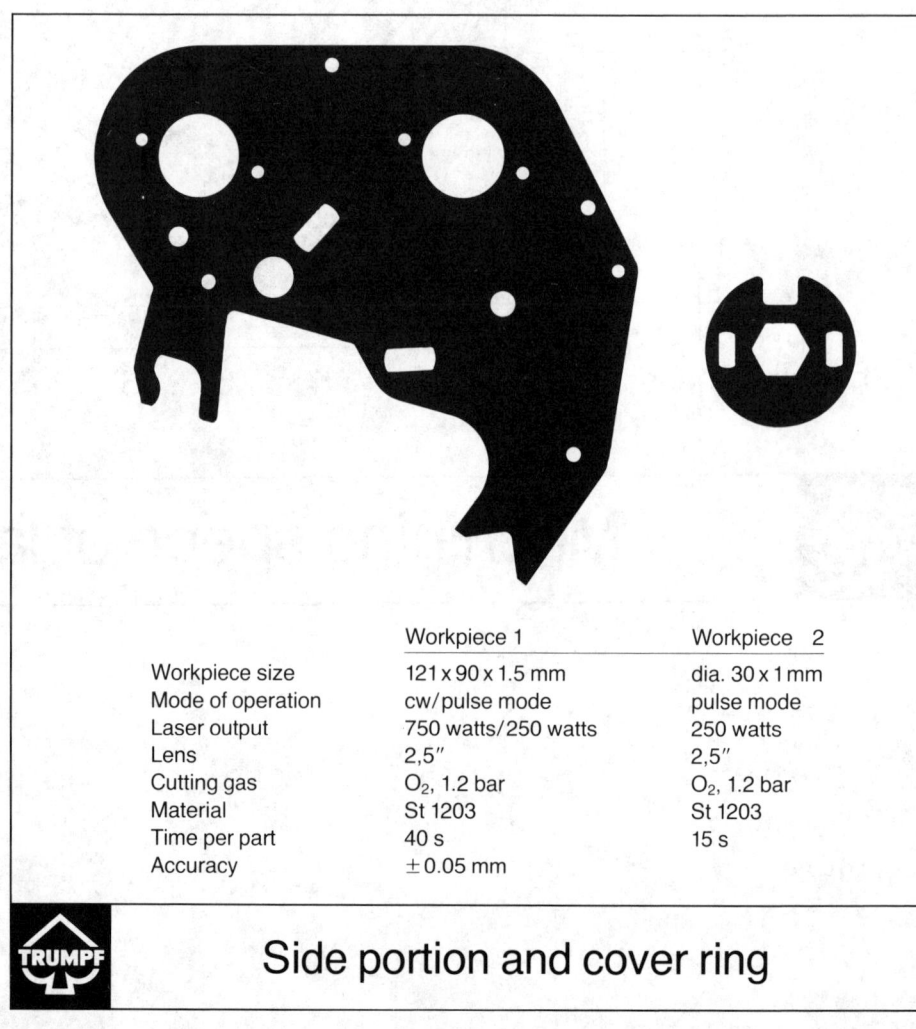

Figure 5

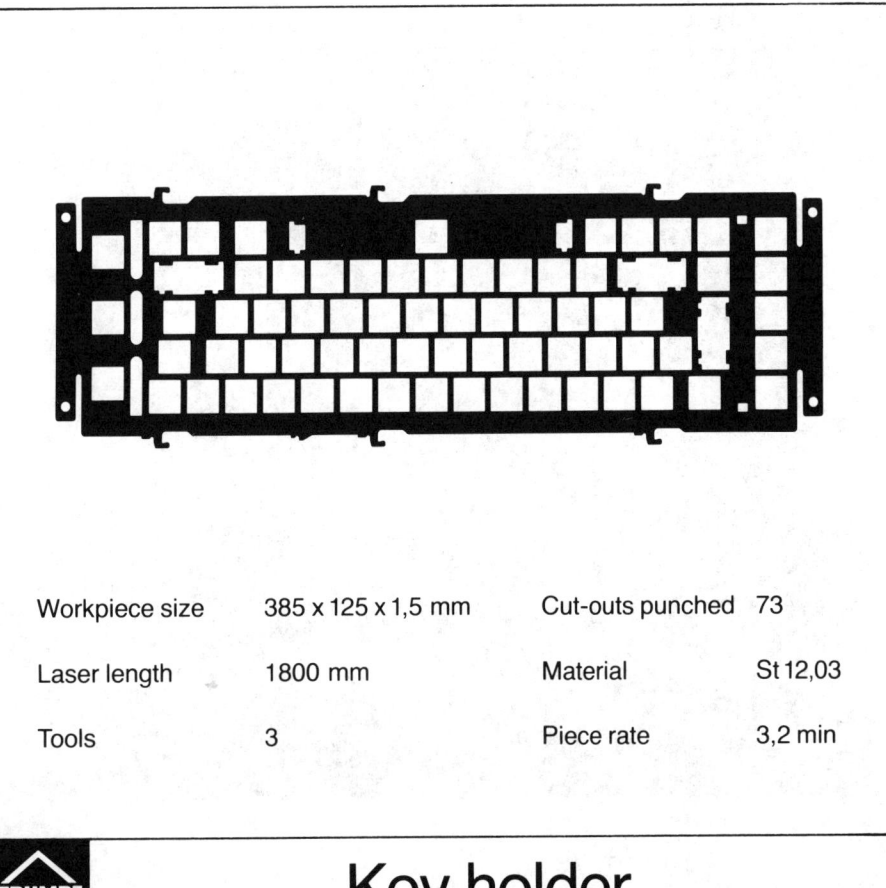

Workpiece size	385 x 125 x 1,5 mm	Cut-outs punched	73
Laser length	1800 mm	Material	St 12,03
Tools	3	Piece rate	3,2 min

Key holder

Figure 6

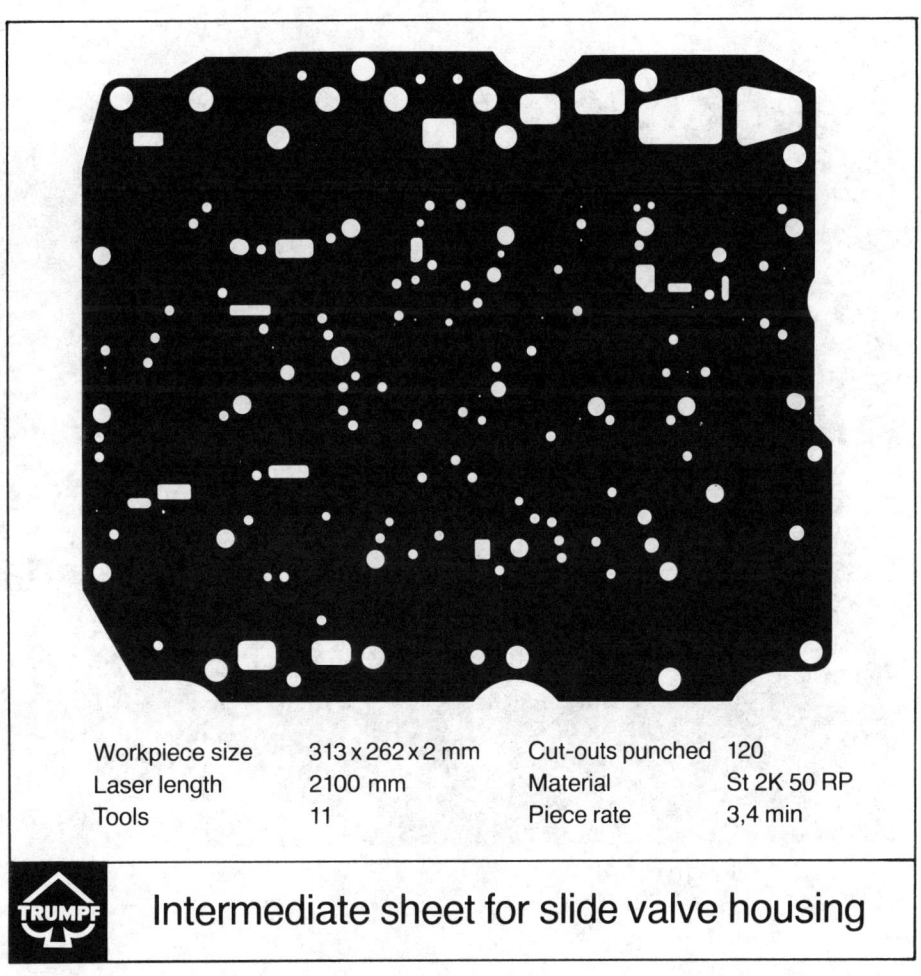

Workpiece size	313 x 262 x 2 mm	Cut-outs punched	120
Laser length	2100 mm	Material	St 2K 50 RP
Tools	11	Piece rate	3,4 min

Intermediate sheet for slide valve housing

Figure 7aFigure 7b

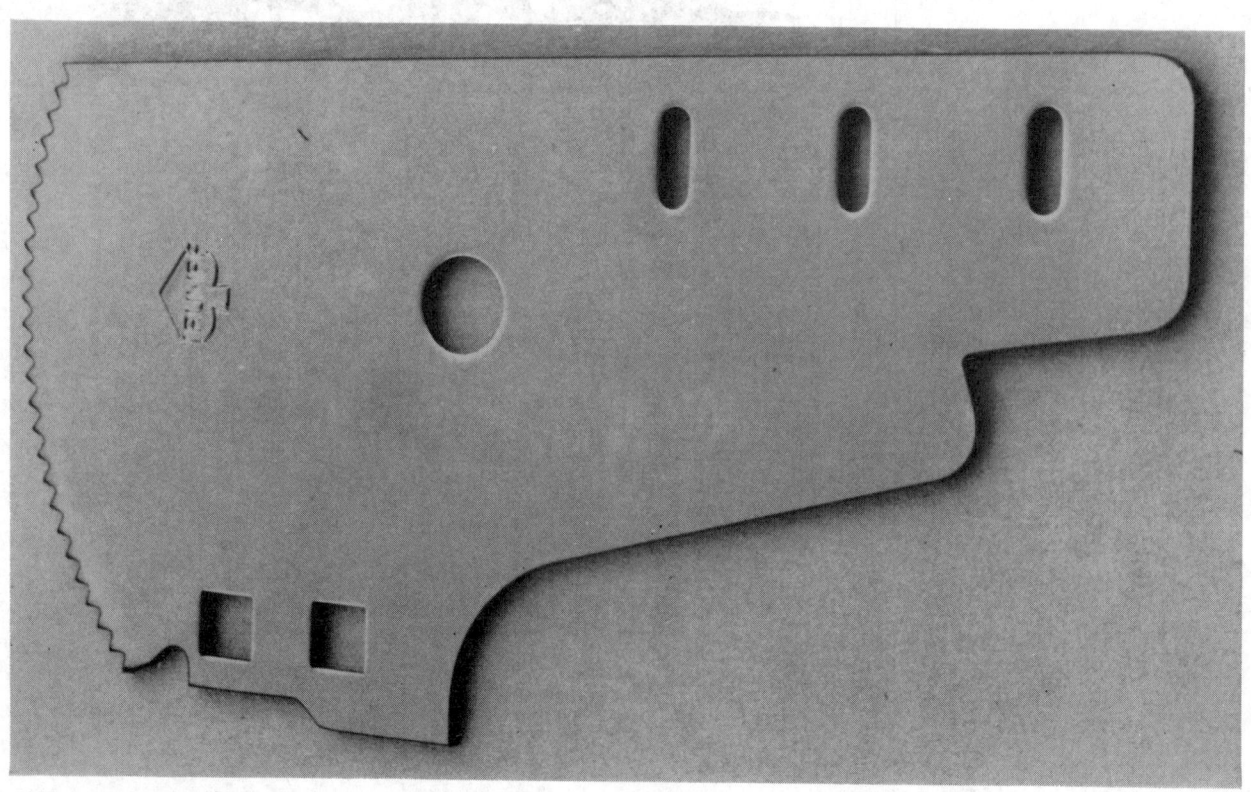

Figure 8

Laser cutting of Al 7075 sheets

A. Di Ilio
Universita dell'Aquila
G. Dionoro
Universita di Cagliari
and
F. Memola Capece Minutolo
and
V. Tagliaferri
Universita di Napoli, Italy

This work is concerned with surface morphology and changes in mechanical and structural properties in laser cutting of aluminium alloy sheets. 1.2, 2 mm thick Al sheets for aircraft construction were cut with a 2.0 kW cw CO_2 laser using two different shielding gases: O_2 and He. Surface roughness was measured adopting mechanical holographic techniques on cut surfaces. These tests allow to determine the most important morphological parameters: roughness index R_a bearing area Sp and number N of profile peaks intersected by the mean line. Microhardness tests were conducted in order to evaluate mechanical properties and metallographic tests to identify structural modifications. The morphological characterization of the cut surface was completed by evaluating cut path width. The results obtained show that it makes no significant morphologic difference between the two shielding gases used, whereas optimum cutting speed is higher when O_2 is employed.

1. INTRODUCTION

The use of lasers in metal cutting has been extensively studied in latter years, as the numerous works published testify. However, little attention has been focused up to now on laser cutting of aluminium alloys, currently used in a variety of applications. This is probably due to their high reflectivity and thermal conductivity that determine a more complex mechanism of interaction between laser beam and metal than in other materials [1].

The object of the present work is to identify optimum process parameters of an ALCLAD 7075-0 type aluminium alloy and evaluate the effect of such parameters on the morphology of the cut surfaces and on the size of the heat affected zone using two different shielding gases.

2. MATERIALS AND TEST PROCEDURES

Through-the-width and interrupted cuts were performed on 50 x 100 mm^2 strips of aluminium alloy type ALCLAD 7075-0 of two different thicknesses (1.2 and 2 mm). A 2 kW cw BOC CO_2 laser was employed. Table I shows chemical composition of the material.

The laser beam was focused onto the surface through a lens with focal length of 100 mm; spot diameter was 0.3 mm.

Two shielding gases, oxygen and helium, were used. These were conveyed to the cutting zone through a nozzle of diameter 1.5 mm placed at 1 mm from the surface of the material and coaxial with the laser beam. Oxygen flow rate was $0.03 m^3$/min at a pressure of 0.22 MPa, that of helium $0.09 m^3$/min at the same pressure. Table II gives the process parameters (beam power P, cutting speed V) covering a sufficiently wide range in order to identify optimum cutting conditions.

All cuts were then subjected to the following:
- visual evaluation of cut quality;
- measurement of cut path width;
- morphologic characterization of cut surfaces by means of roughness measurements;
- microhardness measurements.

3. EXPERIMENTAL RESULTS AND DISCUSSION

3.1 Visual Examination of Cut Quality

The cuts obtained were divided into three classes:
a) through cuts with good surface finish, of high quality;
b) through cuts with poor finish;
c) not through cuts.

Figures 1 and 2 show cut quality as a function of beam power and cutting speed. As can be seen, the range of acceptable cuts increases with beam power. For the 1.2mm thick sheet at the same power cuts of acceptable quality are obtained over a much wider range of cutting rate.

Moreover, with oxygen as shielding gas at the same power the minimum cutting speed required for obtaining through cuts is slightly enhanced. This is evidently due to the energy released during the oxidation reaction and seems to be more pronounced in the thinner aluminium sheet[2,3].

3.2 Width of Cut Path

The width w of the cut path was measured for all interrupted cuts on the top surface of the material at at least 10 mm from the beginning and end of the cut. This was done in order to eliminate errors due to strain in the attack zone and to the delay of the laser beam in the region of interruption.

The behaviour of w versus V and P is plotted in Fig. 3 which refers to the two shielding gases used. As can be observed, cut path widths are small and vary between 0.27 and 0.35 mm for oxygen shielded cutting and 0.25 and 0.31 mm for helium; in both cases w fluctuates around laser beam diameter (0.3 mm).

This result seems to confirm that oxygen does not have a determining effect on cut path width. In fact, the oxidation reaction which takes place in the presence of oxygen is not self-sustaining owing to the formation of refractory oxides which melt at high temperatures[2]. Hence the energy supplied by the laser beam must necessarily melt these oxides in order that the oxidation reaction might always continue on new material. This mechanism probably only involves the zone directly affected by laser radiation and cannot propagate outside the focused spot. From the previous statements it emerges that the energy supplied by the oxidation reaction enables a higher cutting rate at the same power but no noticeable increase in cut path width.

3.3 Morphological Characterization of Cut Surfaces

The morphologic characteristics of the cut surfaces were identified by measuring roughness index Ra and the number of profile peaks N intersected by the mean line.

In Figs. 4 roughness measured at the beam inlet is plotted against cutting rate and beam power for the two gases employed. As can be seen roughness tends to increase with speed, except in a zone comprised between 40 and 80 mm/s where it diminishes. The Ra values obtained in helium and oxygen-shielded cutting do not differ appreciably but in the former case (Fig.4b) spread is greater and no well-defined pattern emerges.

In Figs. 5 N is plotted versus V. Examination of these figures reveals that in oxy-laser cutting N tends to decrease with cutting speed: surface wave frequency is increasingly lower and amplitude almost unchanged. In helium assisted cutting N varies between 15 and 30 but here too no clear cut dependence on V can be detected.

It would appear that in helium ambient morphological parameters are independent of process parameters. This is corroborated by Fig.6 which shows the behaviour of N as a function of power. In laser cutting using helium N is independent of beam power as opposed to oxy-laser cutting where N decreases with increasing P. This could be explained by the fact that at higher powers viscosity of the molten matter is reduced thus promoting the fusion of one or more rivulets of this material resulting in the formation of wider cords, further apart.[4,5]

3.4 Microhardness Measurements

Figures 7 a,b,c,d show, as an illustration, Vickers microhardness (245.2 mN) at two different power and speed combinations for the two shielding gases employed. The hardness of the base material is in the region of 600 MPa. The maximum value attained in the vicinity of the cut path (50 μm) is about 1000 MPa. At higher cutting rates, at the same power, a reduction in the size of the heat affected zone, comprised between 100 and 300 μm is observed for both gases.

The heat-affected zone appears to be consistently larger in oxygen-shielded cutting; this can be attributed to the energy contribution furnished by the oxidation reaction.

4. CONCLUSIONS

The following conclusions can be drawn from the results obtained:
- cut path width and size of the heat-affected zone confirm that the quality of laser cuts obtained on 7075-0 aluminium alloy sheets is acceptable, within the limits of this technology, both in oxygen- and helium-shielded cutting;
- morphology of the cut surfaces is characterized, for both gases employed, by an Ra value of 8-10 μm. The number of peaks tends, in the case of oxygen, to diminish with increasing cutting speed and beam power;
- the use of oxygen determines, at equal power, an increase in the critical cutting rate but does not produce a rise in cut path width.

The results confirm that energy supplied by the laser source is sufficient to effect the cut. Under the present experimental conditions the oxidation reaction does not determine appreciable differences in cut quality but only contributes to enhance critical cutting speed.

REFERENCES

1) Mondolfo L.F. Aluminum Alloys: Structure and properties. Butter Worths, London 1976 pp. 56-63.
2) Ref. 1, pp. 343-346.
3) Dell'Erba, M., Daurelio, G. " Processi di taglio con laser CO_2 di metalli altamente riflettenti ", Inter. Rep., Centro Laser Bari - Italy.
4) Goldsmith, A. et al. Handbook of thermophysical Properties of Solid Materials, Arman Research Foundation, Macmillan Company, N.Y. 1961.
5) Gallo, A. et al. " Problemi di taglio dei metalli con raggio laser ", La Meccanica Italiana n.148, Feb. 1981, pp.27-40.

Al	Cu	Mg	Cr	Zn	Others
90	1.6	2.5	0.23	5.6	0.7

Tab. I - Chemical composition of Al 7075

	He			O_2		
P (kW)	1.9	1.5	1.0	1.9	1.5	1.0
V (mm/s)	5-120	5-80	5-60	20-150	10-120	5-40

Tab. II - Cutting parameters.

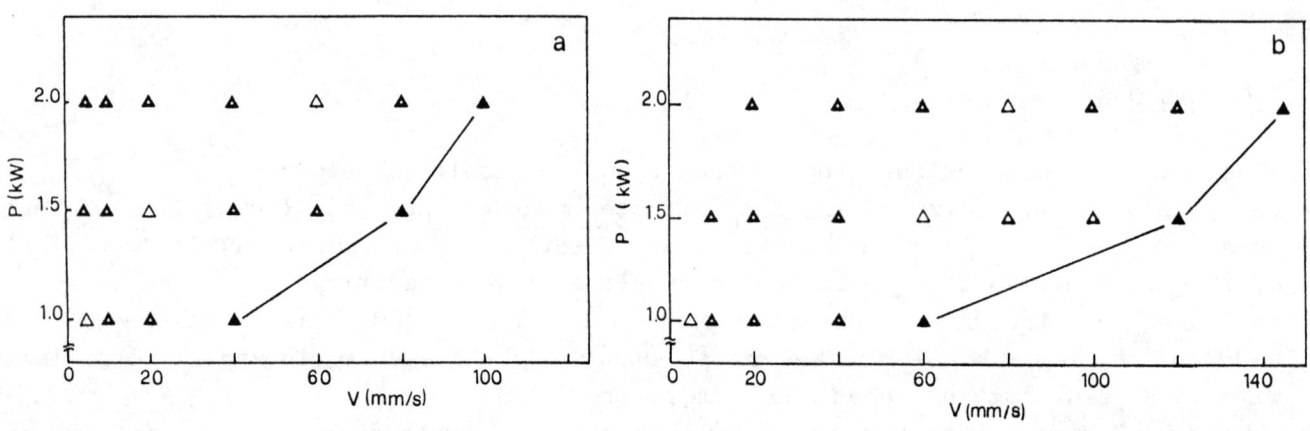

Fig.1 - Effect of laser cutting parameters on cuts quality. Shielding gas:Oxygen. On the left of the continous line through-the-width cuts are obtained. ▲:cuts with occlusions at the bottom; ▲: cuts without occlusions; △: samples that are dross-free and evidence a very regular kerf width. a) sheets 2.0 mm thick; b) sheets 1.2 mm thick.

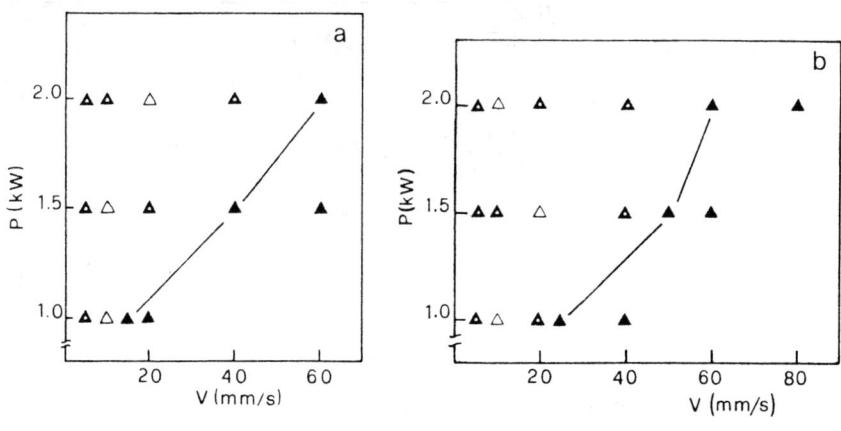

Fig.2 - Effect of laser cutting parameters on cuts quality. Shielding gas: Helium.
a) sheets 2.0 mm thick
b) sheets 1.2 mm thick
For symbols see Fig.1

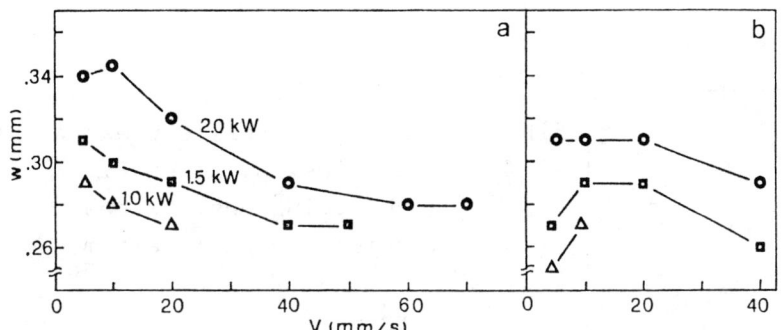

Fig.3 - Cut width w vs. cutting speed V for various powers P. a) shielding gas: Oxygen; b) shielding gas: Helium. Sheets 2.0 mm thick.

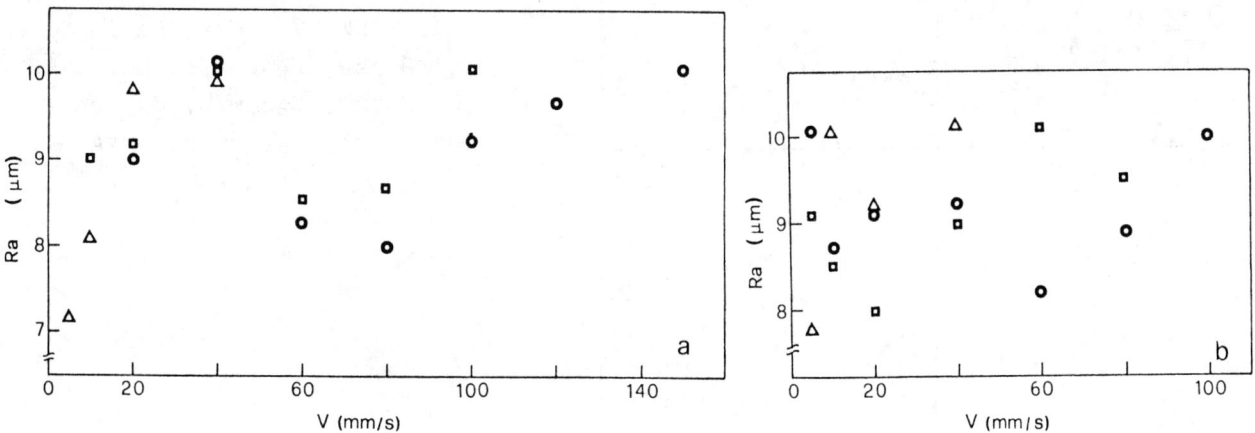

Fig.4 - Cut surface roughness R_a vs. speed V. a) shielding gas: Oxygen; b) shielding gas: Helium. Sheets 2.0 mm thick.

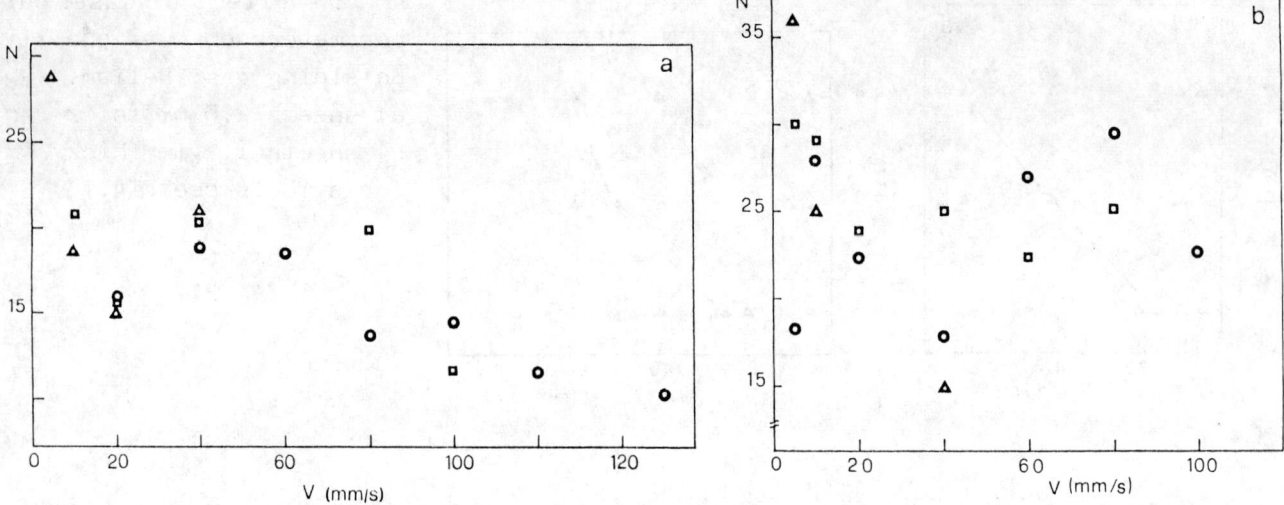

Fig.5 - Number of profile peaks N intersected by the mean line vs. speed V. a) shielding gas: Oxygen; b) shielding gas: Helium.

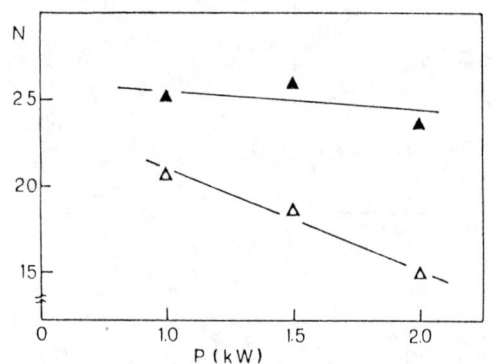

Fig.6 - Number of profile peaks N intersected by the mean line vs. power P.
△ shielding gas: Oxygen;
▲ shielding gas: Helium.

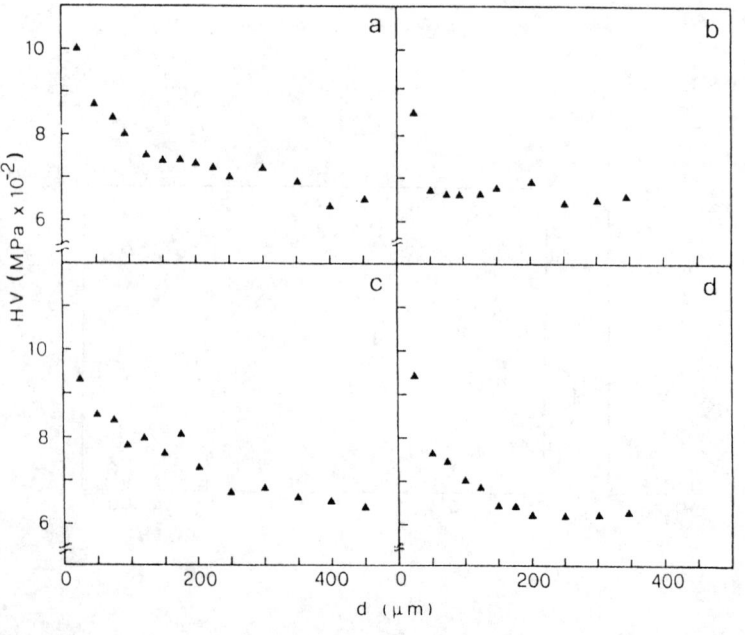

Fig.7 - Microhardness values HV vs. distance from the cut path edge.
a) P=1kW, V=5mm/s, Helium
b) P=1kW, V=40mm/s, Helium
c) P=1kW, V=5mm/s, Oxygen
d) P=1kW, V=40mm/s, Oxygen

Prevention of dross attachment during laser cutting

F. N. Birkett, D. P. Herbert
and
J. Powell
Loughborough University of Technology, UK

ABSTRACT

When certain materials are cut using medium power ($\cong$ 500 W) carbon dioxide lasers, dross is formed which adheres strongly to the underside of the cut material and which requires an additional removal process. The paper describes a process which gives dross-free cut components when used with stainless steel and similar materials.

INTRODUCTION

Adherent dross occurs when cutting stainless steel or similar materials with medium power carbon dioxide lasers. The dross is attached to either side of the bottom edge of the kerf and is generally difficult to remove from the cut component. Arata et al[1] recognised the need for dross removal or prevention and proposed the techniques of pile-cutting and tandem nozzle cutting to deal with the problem.

In laser cutting of metals a cohesive force is generated between the molten layers on the surface of the cutting front and non-reacted sub-layers. This cohesive force must be overcome, usually by the momentum from the cutting gas, to allow the dross to separate, but with stainless steels the low fluidity of the molten oxide suppresses the smooth separation of the molten dross from the bottom edge of the kerf.

Arata's pile-cutting technique involves covering the stainless steel material to be cut with a thin mild steel sheet, lightly constrained, before cutting in the normal manner. This technique improves the separation of molten dross from the bottom kerf edge, but is wasteful and inconvenient since it uses additional material and introduces the requirement to lay on and remove the mild steel sheet.

For tandem nozzle cutting, Arata introduces a second off-axis gas jet behind the normal coaxial cutting jet, with the two jets used together. While this technique reduces the amount of adhesive dross, the problem is not completely alleviated and results are not as good as those obtained using the pile-cutting method.

From our examination of the process, we concluded that adherent dross is a problem because it remains firmly attached to the underside of the kerf edge on cut component and waste material alike, and that a technique which caused the dross to adhere to the waste material only was a viable solution.

THE "DROSSJET"

In the normal laser cutting process, the momentum imparted to the reacted layer from the cutting gas jet is usually sufficient to overcome the cohesive force described earlier. While Arata's tandem nozzle introduces additional momentum by using a second nozzle on the top surface of the cut, the best position to add extra momentum is on the lower surface where the dross actually gathers.

An experimental arrangement was constructed, as shown in Figure 1, which consisted of a single nozzle placed beneath the workpiece with the gas jet directed onto the lower surface of the workpiece at the point where the kerf would be formed. The jet was arranged to flow transversely to the direction of the cut, as Figure 1 shows. The experimental equipment was designed to allow variation in the inclination of the nozzle (α) and in the separation of the nozzle tip from the workpiece lower surface (δ).

Tests were made with this equipment to cut 2 mm thick stainless steel using a Coherent 525 laser with a 2.5 inch focal length lens, and early results showed that the system forced dross to attach to the opposite side of the kerf from the nozzle. Further tests established that consistent results were obtained using a nozzle inclination $\alpha = 25°$ and a nozzle tip separation $\delta = 3$ mm, using nitrogen. Figure 2 shows clearly that the dross has been "blown" onto one side of the kerf edge, leaving the edge nearest to the nozzle completely free of dross.

MULTIPLE "DROSSJET"

Having established the basic parameters of nozzle inclination and separation, a second experimental arrangement was constructed as shown in Figure 3. Eight identical nozzles are arranged in a ring, with each nozzle connected to a solenoid valve. The solenoid valves are opened and closed in a sequence determined by the profile shape to be cut. It was considered that eight nozzles were the minimum number necessary to deal with all forms of straight line and circular arc cutting.

When used to cut complex profiles the multiple "drossjet" gave quite remarkable results (Figure 4) with only small adjustments needed from the initial settings. The correct sequencing of the solenoid valves is produced by using the M-function commands from the Computer Numerical Control device used to move the material, and requires some care in the preparation of the control program, but once the program is finalised, repetitive results are easily obtained.

The multiple "drossjet" technique as developed is best suited to laser cutting systems which move the material on an x-y table for profiling purposes. Further developments to make the technique suitable for use with moving optics systems are being investigated together with a simple method of solenoid valve sequencing by sensing cutting direction.

ACKNOWLEDGEMENTS

The authors wish to thank Mr. C. Eley and Mr. M. Bramley of the Department of Mechanical Engineering at Loughborough University of Technology for their work in the construction of the experimental equipment and associated controls. We are grateful to S.E.R.C. for financial support.

REFERENCES

1. Arata, Y., Maruo, H., Miyamoto, I. and Takeuchi, S. "Quality in Laser Gas Cutting of Stainless Steel and Its Improvement". Trans. J.W.R.I., Vol. 10, Part 2, pp. 1-11 (1981).

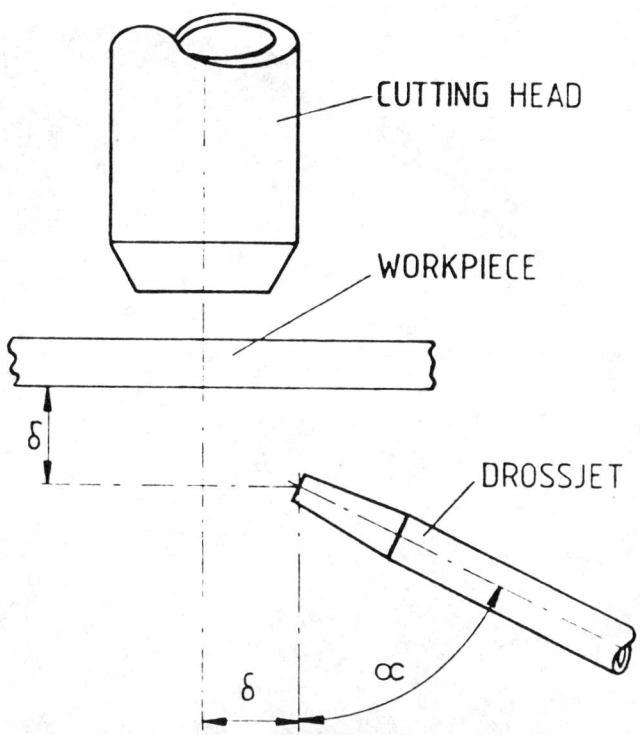

Figure 1. Arrangement of single experimental "drossjet" showing nozzle inclination α and offset from workpiece lower surface δ: cutting direction is normal to the plane of the diagram.

Figure 2. View of lower surface of 2 mm thick stainless steel cut using "drossjet", showing dross clearly attached to one side only of kerf edge. Approximately 3 x full size.

Figure 3. Multiple drossjet arrangement.

Figure 4a. Lower surface of circular arc cut in 2 mm thick stainless steel, as cut. The dark line indicates limit of dross edge, with dross removed by drossjet action. Approximately 12 x full size.

Figure 4b. External corner of 2 mm thick stainless steel sample, showing freedom from dross. Approximately 12 x full size.

Excimer lasers in photolithography

M. C. Gower
SERC Rutherford Appleton Laboratory, UK

ABSTRACT

Because of their high pulsed powers in the ultraviolet spectral region, excimer lasers are of interest as lamp sources in photolithographic mask aligners. Their short wavelengths enable higher resolution and packing densities to be achieved on the silicon chip, while their high power should allow large wafer areas to be processed with a rapid throughput of devices.

1. INTRODUCTION

The minimum feature size which can be replicated by optical lithography in a photoresist material from a master mask in close proximity or contact is approximately $\sqrt{\frac{\lambda Z}{2}}$ where λ is the wavelength of the illumination source and Z is the separation between mask and photoresist. On the other hand, projection of an image of the mask by lenses or mirrors can produce features sizes as small as $\sim \frac{\lambda}{2NA}$ where NA is the numerical aperature of the optical system (NA = $\sin \theta$ where θ is the light acceptance half angle of the optical imaging system). The ever increasing demand to achieve larger packing densities of circuit elements on silicon wafers demands ever higher resolution from optical lithography. For both proximity and projection printing of mask patterns this high resolution is most readily achieved by decreasing the wavelength of the lamp source. The use of excimer lasers as alternative sources to the Hg (-Xe) arc lamps currently used in photolithography is being vigorously investigated in a number of laboratories around the world. Excimer lasers produce intense bursts of ultraviolet light in the wavelength range 150 – 350 nm with repetition rates of up to 500 pulses/sec (see Table 1). In this paper we will discuss some of the advantages and disadvantages of using excimer lasers for photolithography.

	WAVELENGH (nm)	ENERGY/PULSE (mJ)
F_2	157	40
ArF	193	500
KrF	249	1000
XeF	351,353	500
KrCl	222	100
XeCl	308	500

Table 1 Excimer laser wavelengths and typical single pulse energies

EXCIMER LASERS

The most common type of excimer laser uses rare gas halide molecules such as ArF, KrF, XeF or XeCl as the working media. These molecules are produced in the excited upper laser level by high voltage discharges in gas mixtures of rare gases and halogen bearing molecules. A complicated sequence of electron and ion reactions in the discharge is responsible for forming the appropriate excited rare gas halide molecule from the gas constituents originally added to the laser vessel. The wavelength of the laser emission depends upon the type of rare gas halide molecule created. It can be selected by simply changing the original gas mixture added to the laser (see Table 1).

Because rare gas atoms are inert in their ground state, the lower laser levels of rare gas halide molecules are unstable. This instability is responsible for producing lasing over a large wavelength range (up to 20Å). Thus the temporal coherence of most excimer lasers is more akin to that obtained from lamp sources (see Table 2).

		EXCIMER LASER	Ar^+ LASER	Hg-Xe LAMP
TEMPORAL COHERENCE	$\Delta\lambda$(nm) L_c(cm)	1 $\sim 10^{-2}$	≤ 0.01 ≥ 10	10 $\sim 10^{-3}$
SPATIAL COHERENCE	Mode-n_o (N)	3×10^5	1	5×10^7
SPECKLE CONTRAST	$1/\sqrt{N}$	1.8×10^{-3}	1	1.4×10^{-4}

Table 2 Coherence and speckle properties of excimer lasers compared to lamps and common lasers

Furthermore, the extremely high gain obtainable over a large area from excimer molecules allows many spatial modes to oscillate within the laser cavity. Hence the laser output has the beam divergence properties more like that of an aperatured lamp than a single transverse mode laser. This lack of spatial and temporal coherence makes the excimer laser extremely attractive as a lamp source for photolithography. Highly uniform beams can be produced from the laser and interference effects such as laser speckle, which may arise from the presence of dust or other optical imperfections, are not observed. Thus excimer lasers retain the beam uniformity of the lamp but produce exceedingly large powers at short wavelengths.

3. OPTICAL LITHOGRAPHY

The dominant method for replication of mask patterns on silicon wafers as used in the production of electronic chips is photolithography. Placing the mask in close proximity with the photosensitive resist on the wafer can yield resolutions down to $\sim$ 4 μm over a large area of the wafer (see Fig 1). This proximity printing can produce

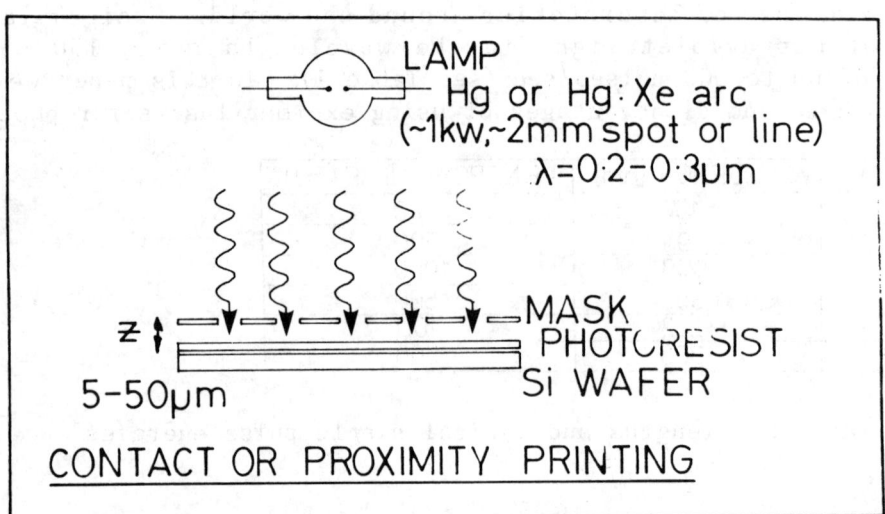

Fig.1. Contact and/or proximity printing

a relatively high throughput of devices (greater than 100 silicon wafers/hour) with moderate resolution. On the other hand if the mask is placed in vacuum contact with the wafer, extremely high resolution can be obtained – down to 0.2 μm. However since the contact and the possibility of trapping particulate matter between mask and wafer often leads to damage of the mask only a few wafers/day can be processed.

Resolutions of ~ 1 μm without damage to the mask can be obtained by using lenses or mirrors to project an image of the mask onto the wafer. Obtaining this high resolution over as large an area as possible is at the forefront of optical lens and mirror design techniques and is presently limited to an area (field of view) of about 1 cm^2. Thus to expose the full 4-6" diameter wafer many identical masks on a single 'reticule' are often scanned by the (slit) light source (Fig 2(a)) or alternatively the image stepped and repeatedly exposed across the wafer (see Fig 2(b)). The imaging

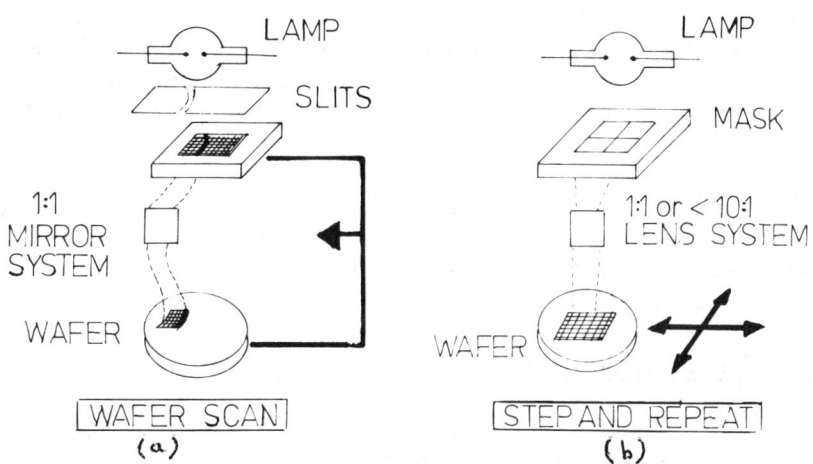

Fig.2. Projection printing (a) Wafer scanning (b) Step and repeat systems

system in projection printing may be such as to give unit magnification or up to 10:1 image reduction. Throughputs of up to 60 wafers/hour, can be achieved.

The most common types of lamp used in optical lithography are Hg or Hg-Xe ~ 2mm wide arc sources which have electrical inputs of 0.5 - 2 kW. The majority of the emitted light is at wavelengths longer than 300 nm (low-powered deuterium or Cd-Xe arc lamps are sometimes used to produce shorter wavelengths in contact printing). In projection printing, for a given resolution ($\sim \lambda/2NA$) it is preferable to use as short a wavelength as possible in conjunction with a small numerical aperature so that image focussing is more readily achievable (depth of focus ~ λ/NA^2). Even with the inherent uniformity of incoherent lamp sources, beam homogenisers such as light guides or fly's eye lenses are often used to maintain a uniformity of illumination of < $\pm$ 1% intensity fluctuations across the exposed area.

4. EXCIMER LASER PRINTING

(a) Contact and Proximity

Linewidths down to 0.2 μm have been produced in a variety of resist materials and excimer laser combinations. Jain et al[1,2] have reported linewidths down to 0.5μm using contact printing with XeCl and KrF lasers at 308 and 248 nm respectively in Shipley AZ2400 and IBM experimental diazonaphthoquione-Novalak (DNN) resists. Steep wall angles were obtained and because of the lack of coherence from the laser, speckle and standing wave effects on the walls were absent in the photoresist. This IBM group has also found[3,4] that there is little dependence of the resist sensitivity on the 308 nm laser peak power (ie no reciprocity failure) for Shipley AZ1450J, 2400 and several IBM experimental resists. This behaviour is quite remarkable considering that compared to lamp sources the instantaneous laser image was about eight orders of magnitude more bright. Kawamura et al have investigated the exposure of polymethyl methacrylate (PMMA) resist to KrF laser radiation at 249 nm[5]. A group at Bell Laboratories[6] have used an F_2 laser to perform similar studies at 157 nm in PMMA, AZ2400, HPR204 and copolymers of MMA and MAA, and were able to write lines by contact printing as narrow as 0.15 μm. In most cases integrated doses in excess of ~ 10 mJ/cm^2 are required to expose the resists. Karl Suss KG-GmbH & Co of W Germany have recently begun to manufacture a contact printer which has an excimer laser illuminator and can produce linewidths as narrow as 0.2 μm.

(b) Projection

By modifying a Perkin-Elmer model III 1:1 wafer scanning mirror projection printer (Fig 1(a)) to accept a XeCl 308 nm laser source, Jain and Kerth[7] have recently shown that 1 μm features could be printed in DNN resist with a full 8.2 cm diameter wafer scan time of only 15 seconds. This experiment demonstrates that the higher power available from the laser in the spectral region around 308 nm allowed the wafer scan time to be reduced by a factor of ~ 5 over that achieved using a Hg lamp. Thus the throughput of the machine could possibly be increased accordingly. Clearly it may also be possible to adapt step and repeat machines for use with excimer laser sources provided that the optical imaging system can be made to be sufficiently achromatic at the laser wavelength.

Dubroeucq and Zahorsky at Thomson-CSF[8] have used a fused silica microscope objective with an NA of 0.2 to project images into AZ2400 resist using a 249 nm KrF laser. They found that even with this relatively crude optical projection system they could write lines as small as 0.7 μm - albeit only over a small area.

5. DIRECT ETCHING WITH EXCIMER LASERS

As well as offering the potential of higher resolutions and a greater throughput than can be achieved using lamps, excimer lasers can also be used to directly etch a photoresist polymer without the need for wet chemical development. The short burst of UV laser photons can produce a rapid scission of the polymeric bonds in the photoresist and cause vaporised material to be ejected from the irradiated site with very little heating of the surrounding resist[9]. Thus direct etching by the photons can occur with extremely high definition. Using F_2, ArF and KrCl lasers at 157, 193 and 222 nm respectively we have used contact printing to write ~ 0.5 μm lines by direct etching in PMMA resist[10]. Similar results have been obtained using a simple projection lens[11].

6. CONCLUSION

It seems clear that the ever increasing trend of the electronics industry to produce small devices will push the technology of optical lithography to shorter wavelength sources. X-ray, electron or ion-beam lithography while posessing the possibility of higher resolution are as yet nowhere near as developed (and as cheap) as optical techniques. Excimer laser sources appear to offer in the near future a means of achieving in production a relatively high throughput of submicron devices fabricated using optical lithographic techniques. Some developments of laser technology will be necessary before this will happen on a wide scale. The best beam quality available from a commercial excimer laser has a uniformity of ~ ± 5% fluctuations over 50% of its pulse energy. This is far short of the ~ ± 1% source fluctuations which can be tolerated in resist exposure. Various optical devices[12] may have to be employed to further uniformise the laser beam. Although excimer lasers can currently produce average powers up to 100W and can operate for up to 10^8 pulses under certain conditions without having to replenish the spent laser working gas mixture, this type of performance should be routinely produced with the laser power remaining at a <u>constant</u> value. Above all else in a production environment the laser must be reliable. Excimer laser manufacturers are currently making great strides in these directions.

REFERENCES

1. Jain,K., Wilson, C.G and Lin, B.J. IBM J Res Dev, 26, 151, (1982)
2. Jain, K., Wilson, C.G. and Lin, B.J. IEEE Electron Dev Letts, EDL-3, 53, (1982)
3. Rice,S and Jain, K. IEEE Trans Electron Dev, ED-31, 1 (1984)
4. Jain,K., Rice,S and Lin,B.J. Polymer Eng and Sci, 23, 1019, (1983)
5. Kawamura,Y., Toyoda,K and Namba,S. Appl Phys Letts, 40, 374 (1982)
6. Craighead,H.G., White,J.C., Howard,R.E., Jackel,L.D., Behringer,R.E., Sweeney,J.E and Epworth,R.W. J Vac Sci Technol B1, 1186 (1983), Appl Phys Letts, 44, 22 (1984)
7. Jain,K and Kerth,R.T. Appl Optics, 23, 648, (1984)
8. Dubroeucq,G.M and Zahorsky,D. Proc of Microcircuit Engineering '82, p73 (1982)
9. Srinivasan,R and Mayne-Banton,V. Appl Phys Letts, 41, 576 (1982)
10. Davis,G.M and Gower,M.C. SERC Rutherford Appleton Laboratory Report RAL-84-049 B33 (1984)
11. Latta,M., Moore,R., Rice,S and Jain,K. J Appl Phys, 56, 586, (1984)
12. Latta,M.R and Jain,K. Optics Comm, 49, 435 (1984)

LASER PROCESSES – SURFACE TREATMENT

Laser surface alloyed Fe-Cr-C
S. Das, I. Dumler
and
J. Mazumder
University of Illinois, USA

ABSTRACT

Surface related failures such as corrosion and wear can be remedied by modifying the sufrace chemistry by laser surface alloying. Using this process a wide range of alloys can be generated with novel microstructure. Microstructure and composition of laser surface alloyed Fe-Cr-C alloy is characterized by TEM, scanning Auger microprobe, SIMS, and Microprobe. Highly refined microstructure with metastable crystalline, and amorphous phases are observed. Both Cr and C are found to be distributed uniformly. Impurities such as P, S, Si, etc., seems to be responsible for amorphous phases. Due to rapid solidification even high chromium content alloys often retained F.C.C. structure. Chromium carbides are observed inside the grain. Their microstructure and their implications are discussed.

INTRODUCTION

Laser surface alloying is a recent process of surface modification. In this process the surface characteristic of a bulk material is changed by alloying with suitable metal powders. Laser beam acts as a source of heat. The inherent rapid solidification in this process produces very refined microstructure with extended solid solution and sometimes amosphorus phases when composition permits. This paper presents the study of surface alloying of Cr onto AISI 1016 steel substrate. In our previous works the effect of variation of process parameters on the shape and size of molten metal pool and the solute element distribution had been studied. Moreover corrosion and wear resistance properties of laser alloyed surfaces had been studied. This paper mainly deals with microstructural characterization of laser surface alloyed AISI 1016 steel with chromium in order to understand the effect on microstructure on wear properties.

---

   *S. Das,   Graduate Research Assistant, Department of Metallurgy
  **I. Dumler, Resident Microscopist, Materials Research Laboratory
***J. Mazumder, Associate Professor, Department of Mechanical and
               Industrial Engineering

The use of powders in laser surface modification has been reported by several authors. Gnanamuthu[1] applied Ni and Cr powders onto steel substrates as a slurry or by spraying. To minimize the porosity and to achieve uniform mixing this process needs very high power (12.5 kW) or oscillating beam at lower powers (<6KW). Ayres[2] used a special type of nozzle to inject Si powder into the molten pool while alloying 5052 Al. Mechanically vibrated gravity flow system and wire feeding were adopted by Breinan et al.[3] Powell and Steen[4] used a modified spray gun for laser cladding. Here powder feedrate was changed by varying the carrier gas flow rate.

Some surface property study has been reported. Moore et al.[5] suggests that under high power density (10^7 W/cm^2) and short interaction time (0.1 - 1.2 ms) surface roughness can be reduced by decreasing laser power and speed. Esquivel et al.[6] carried out surface roughness measurements for laser surface alloys produced with low powers and high speeds (up to 111 cm/s) condition. They found surface rippling at all speeds. The corrosion resistance of laser surface alloyed Fe + Cr alloys has been measured by Moore and McCafferty[7] in Na_2SO_4. Lumsden et al.[8] used 1 N H_2SO_4 solution for corrosion testing Fe + Cr + Ni alloys. Both of them found corrosion resistance of surface alloys to be far superior to that of the substrate.

Mazumder et al.[9] reported some amorphous phase and BCC crystalline phase in the alloyed layer under low power density (0.8 x 10^6 W/cm^2) and a traverse speed of 50 mm/s.

The objective of this present work is to explain the improved wear and corrosion resistance properties Fe-Cr and similar alloys from microstructural point of view.

EXPERIMENTAL PROCEDURE AND METHODS

A 10 kW CW CO_2 laser was used for the runs. Beam diameter was set at 2 mm. An overlap of approximately 50 percent was used for all samples. An inert gas shield was used to minimize oxidation of the sample and combustion of powder particles under the laser beam. A flow of helium gas at 1.13 - 1.7 m^3/hr (40 to 60 ft^3/hr) was piped into the shield. Argon gas-assist for powder feed was set at 0.085 - 0.14 m^3/hr (3 to 5 ft^3/hr). Commercially available Cr powders of 2 μm diameters were fed to the molten metal pool created by the interaction of laser beam with the substrate by a screw feeder. Figure 1 represents schematic diagram of powder delivery system. The delivery systems used 10 mm bore copper tubing, with 10 mm bore flexible polymer tubing used to make connections requiring small bends. The copper delivery chute was positioned about 8 mm ahead and 15 mm above the point of laser-material interaction. This is done to avoid overheating of copper chute.

Bars of AISI 1016 steels of 150 mm length, 50 mm wide and 6 mm thick were used as substrates. Up to 20 overlapped alloying traces were made to coat an area of approximately 30 mm wide and 50 mm long.

For microstructural characterization transmission electron microscope (TEM), microprobe, energy dispersive x-ray analysis (EDAX), scanning auger microprobe and secondary ion mass spectroscopy (SIMS) were used.

Samples for TEM, SIMS, and scanning auger microprobe were cut parallel to the surface of alloyed layer. But for microprobe analysis samples were cut perpendicular to the surface of alloyed layer.

For making TEM samples special precautions were taken. At first the alloyed surface was ground flat. Then a section of about 3/4 mm thick was cut parallel to the surface. That section was then ground to 4 mils from base material-alloyed zone interface towards the top surface of the alloyed material. By this technique any chance of retained base material was avoided. Both jet polishing and ion milling techniques were employed. Electrochemical solutions attack the crystalline phases more than amorphous phases. In case of ion milling situation is just reverse. We were also interested to see the solute distribution at a different part of molten metal pool. For this we punched discs of 2 mm diameter near the edge of the molten metal pool and center of the pool respectively. Generally the holes created by the jet polishing and ion milling occur at the center of the disc thus we can see different regions of molten metal pool depending on from which part we punched out the sample. Ion milling was done in a cold chamber to avoid phase transformation at higher temperatures.

RESULTS

Table 1 presents the Cr content of alloyed layers at different operating conditions. The results are in agreement with our previous findings. It implies that concentration of Cr increases with increase in speed and with decrease in power of laser beam.

Table 1

VARIATION OF COMPOSITION WITH PROCESS PARAMETERS

Sample #	Power	Traverse Speed	Feed Rate g/s	Wt%Cr
CR-1	5 kW	15 mm/sec	0.22	25
CR-2	6 kW	15 mm/sec	0.22	12
CR-3	5 kW	25 mm/sec	0.22	35
CR-4	6 kW	25 mm/sec	0.22	14
CR-5	5 kW	35 mm/sec	0.22	40
CR-6	6 kW	35 mm/sec	0.22	22

The microstructures of laser surface alloyed region can mainly be classified into two groups--crystalline phases and amorphous phase. Again the crystalline phases can be classified into three groups: (i) low Cr content phases (%Cr $\leq$ 17), (ii) medium Cr content phases (% Cr = 25 to 50), and (iii) high Cr content phase (% Cr $\geq$ 50). Figure 2 represents the above observations.

Figure 3A shows the amorphous phase. The EDAX spectra Fig. 3B shows that amorphous phase consists mainly of impurities, like S, Si, Al, etc.

Table 1 represents that we have produced mainly two types of alloys--low Cr content ($\leq$ 17%) and medium Cr content ($\leq$ 50%). The main feature of low Cr content crystalline phase is presence of martensite with high dislocation density. Some twins are also present in those martensite platelets. Figures 4 and 5 show the bright field and dark field images of those martensite laths. One interesting feature of this low Cr content crystalline phase is very thin grain boundary and negligible amount of Cr segregation at grain boundary. By EDAX technique grain boundary segregation was studied. In this low Cr alloy the Cr-carbide ppt.has some amount of Mn in it.

The most important and interesting crystalline phase is the medium Cr content. The Cr content generally varies from 25 to 40% depending upon the processing parameters. The general structure of this phase is high density of dislocations in a BCC matrix. In some cases we have seen some FCC phases present up to 25% Cr content level. But mostly the crystalline matrix is BCC. Figure 6A shows the presence of dislocations in the BCC matrix. The diffraction pattern of the matrix is shown in Fig. 6B. The most significant finding is the Cr-carbide precipitation on the grain body in general. Figure 7 shows the Cr carbide precipitates on the grain body. The grain boundary segregation was studied using EDAX. It has been found that there are little segregation in the grain boundary. The grain boundary thickness is slightly higher than that of low Cr content samples.

Due to mechanical trouble in the powder feeder we had irregular powder flow in the pool for some samples. In those samples fluctuation of Cr concentration was observed. In some cases due to high Cr powder entrapment we got high Cr content phases. Those phases are full of stacking faults (Fig. 8).

Scanning auger electron microprobe study was done to see Cr and C distribution in the alloyed zone in general. By Cr line scan we have found more or less uniform Cr distribution in the matrix (Fig. 9). Dot maps of C and Cr distribution show more or less uniform distribution (Figs. 10 and 11).

SIMS study also shows uniform distribution of Cr, C, and Cr carbide. It also shows high local concentration of impurities in the matrix. These patches of high concentrated impurities are distributed throughout the matrix (Fig. 12).

DISCUSSION

It is hard to explain the formation of amorphous phases under the present processing condition, where cooling rate is not more than 10^6 to 10^7 °K/sec. But careful study of the material with EDAX and SIMS reveals that these amorphous phases are sinks of impurities. Most probably these impurities (Si, S, P, Ca, etc.) produce local Constitutional Supercooling and thus form some glassy phases. Under TEM it has been observed that these amorphous phases are uniformly distributed throughout the matrix. The SIMS result shows that highly concentrated localized impurity regions are uniformly distributed in the matrix (Fig. 12).

The amount of chromium in the alloyed region decreases with decrease in speed and with increase in beam power. From Table 1, we can comment that with increase in power from 5kw to 6 kw the % Cr in the alloy decreases by about 50 percent. We have found martensitic structure in the material processed at higher power (6kw) and at lower speed (15 to 25 mm/sec). This is due to the fact that at these process conditions molten pool is bigger and for a fixed powder delivery produces lower chromium content. Formation of martensite in low Cr content material is in agreement with the result obtained under equilibrium condition.

We have found higher solute content at the edge of the pool than at the center. This is due to the surface tension driven flow and its effects on mass transfer as explained in a previous publication[11].

In medium Cr content crystalline phase the Cr-carbide precipitation is on the grain body in general. The crystallographic orientation of the matrix does not change on both sides of the precipitates while we traverse the sample from one side to the other under the electron beam in diffraction mode. This proves that precipitation is on the grain body in general. This uniform Cr carbide precipitation throughout the matrix is responsible for much improved wear properties of laser surface alloyed material (reported earlier)[10]. High dislocation density also gives rise to increased hardness of this material.

Again equilibrium phase diagram shows that we can get austenite at higher temperatures for Cr content less than ≈ 16%. But in our case extension of γ loop has been observed.

The scanning auger microprobe and SIMS study both represent uniform distribution of alloying elements and carbon. That is very essential for improved corrosion resistance property.

ACKNOWLEDGMENT

This work was made possible by a grant from UIUC Materials Processing Consortium. The electron optics facility of UIUC MRL has also contributed in this work.

CONCLUSION

The following are conclusions from these preliminary studies.

1. Unique microstructures can be formed by choosing appropriate processing condition.
2. A uniform distribution of chromium carbide can be generated within the grain body which is beneficial for wear properties.
3. Fairly high chromium content (up to 40 percent) alloys can be easily generated. Chromium distribution is also uniform. These properties contribute to the better corrosion resistance.
4. More or less uniform distribution of amorphous phases among the crystalline areas were observed. This does not affect the mechanical properties of the alloyed components.

REFERENCES

1. Gnanamuther, D. S., Optical Engineering, 19 (5), 1980, pp. 783-799.
2. Ayers, J. D., Thin Solid Films, 84, 1981, pp. 323-331.
3. Breinan, E. M., D. B. Snow, C. O. Brown, and B. H. Kear, Rapid Solidification Processing Principals and Technologies II, R. Mehrabian, B. H. Kear, and M. Cohen, eds.; pp. 440-452, Claitors Publishing Division, Baton-Rouge, LA, 1980.
4. Powell, J., and W. M. Steen, Lasers in Metallurgy, K. Mukherejee and J. Mazumder, eds.; pp. 93-104, AIME, Warrendale, PA, 1981.
5. Moore, P., C. Kim, and L. S. Weinman, Applications of Lasers in Materials Processing, E. A. Metzbower, ed.; pp. 221-224, ASM, Metals Park, OH, 1979.
6. Esquivel, O., J. Mazumder, S. M. Copley, and M. Bas, International Conference of Rapid Solidification, Claiton, VA, 1980.
7. Moore, P. G. and E. McCafferty, J. Electrochem. Soc., 128, 1981, pp. 1391-1393.
8. Lumsden, J. B., D. S. Gnanamulthu, and R. J. Moore, Corrosion of Metals Processed by Directed Energy Beams, C. R. Clayton and C. M. Preece, eds.; pp. 129-134, AIME, Warrendale, PA, 1982.
9. Chande, T., A. Ghose, and J. Mazumder, Laser Processing of Materials, K. Mukherjee and J. Mazumder, eds.; Published by AIME, 1984.
10. Mazumder, J., C. Cusano, A. Ghose, and C. Eiholzer, Laser Processing of Materials, K. Mukherjee and J. Mazumder, eds.; Published by AIME, 1984.
11. Chande, T. E., and J. Mazumder, J. Appl.-Physics, (1985).

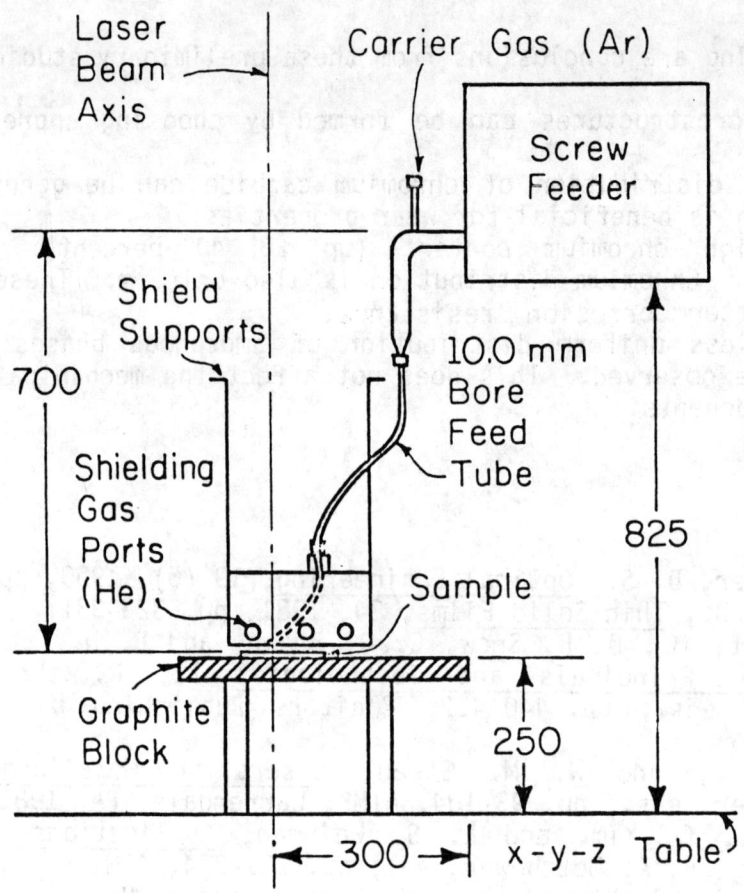

Figure 1 Schematic diagram of screw-fed, gravity-flow, carrier gas-aided powder delivery system. All dimensions are in mm.

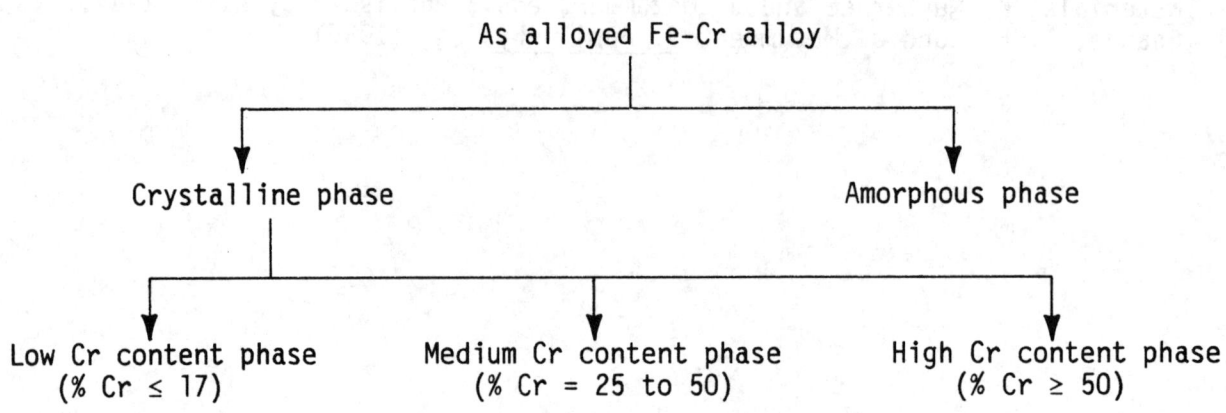

Figure 2

Classification Of Microstructures

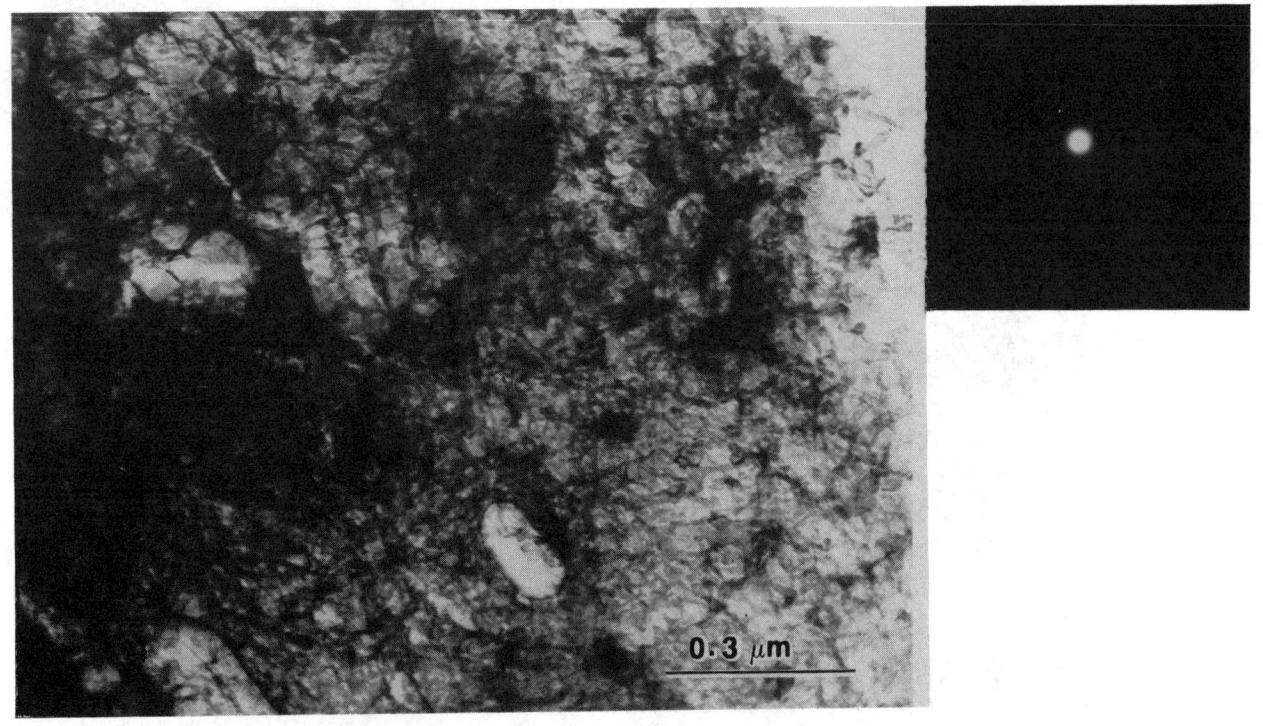

Figure 3A An Amorphous Region in the Alloy (5 kW, 25 mm/s)

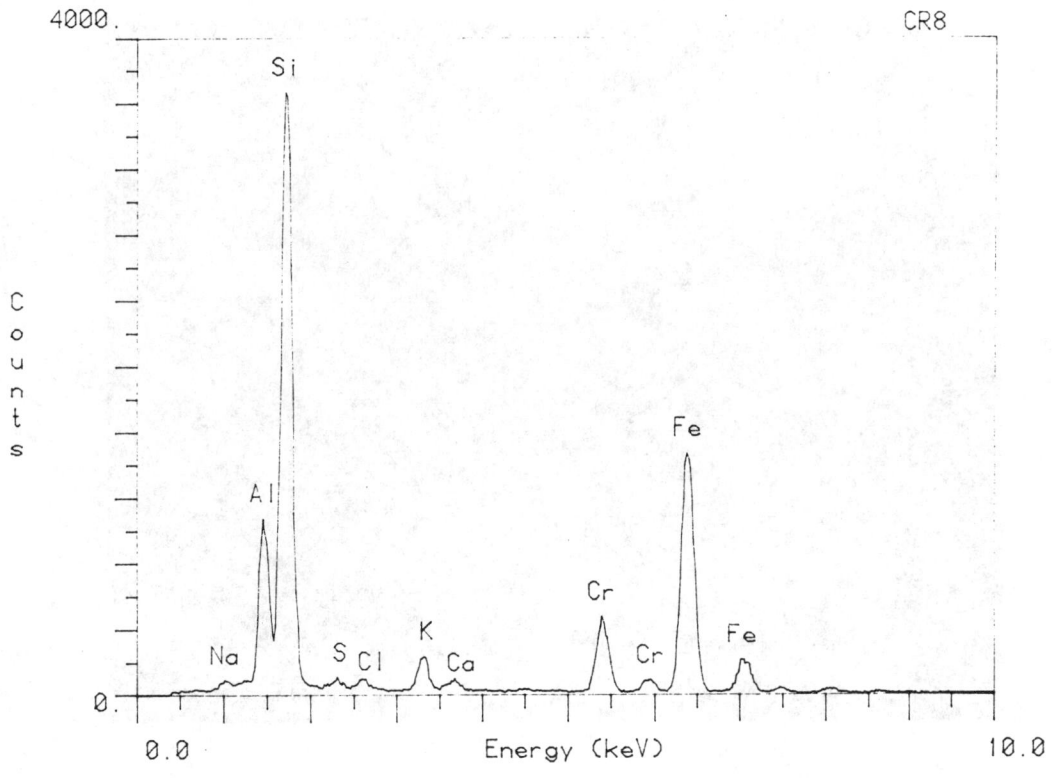

Figure 3B EDAX Spectra from the Amorphous Region

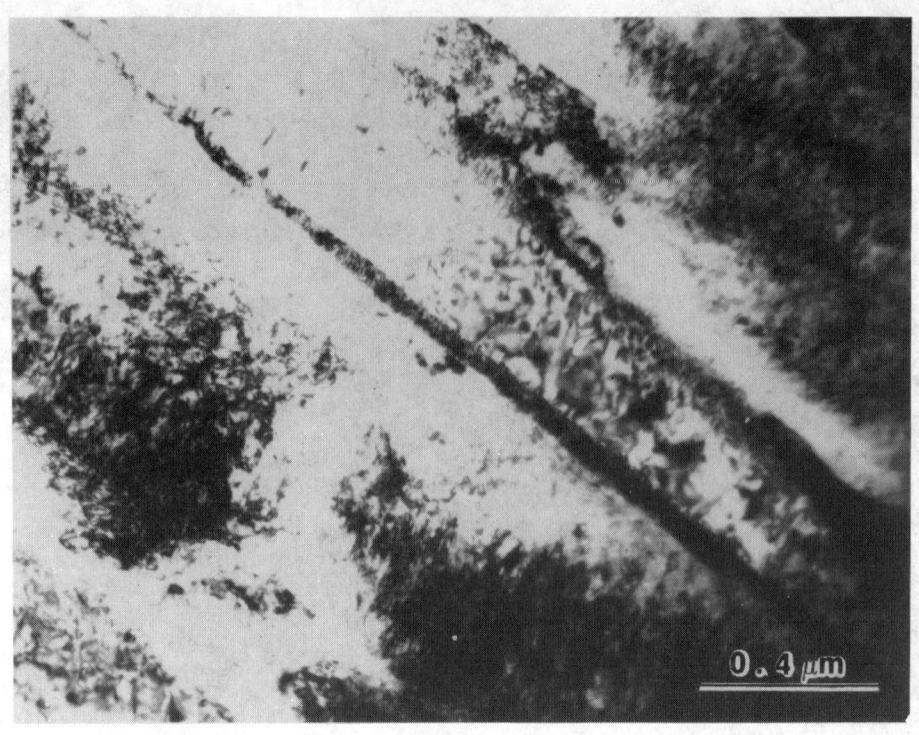

Figure 4   Bright Field Image of Martensite in low Cr Alloy (6 kW, 25 mm/s)

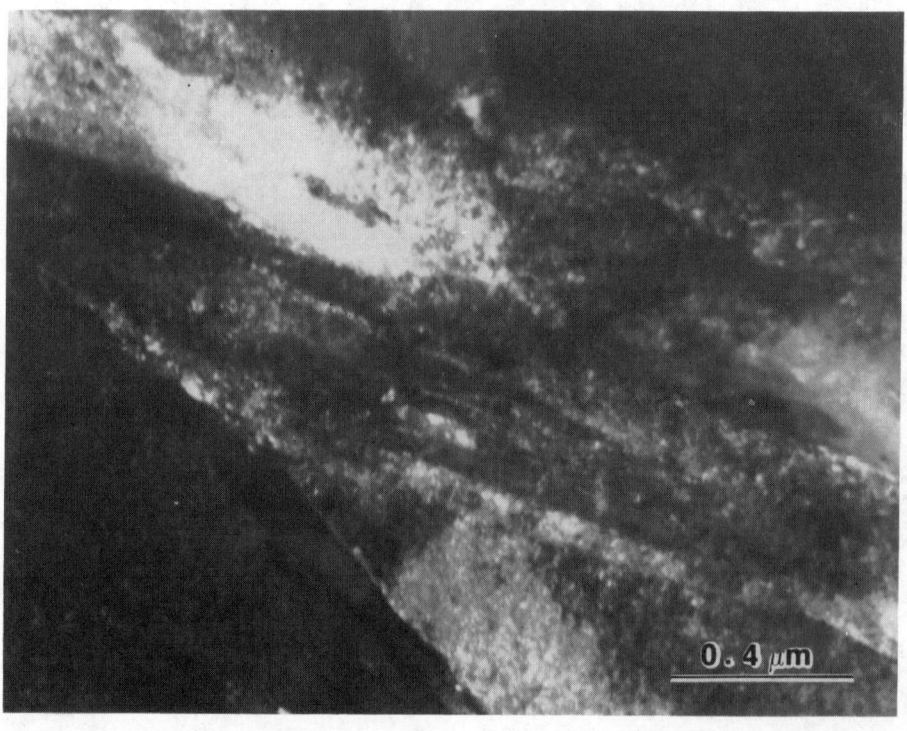

Figure 5   Dark Field Image of Martensite in low Chromium Alloy

Figure 6A   Dislocations in the Crystalline Phase (5 kW, 25 mm/s)

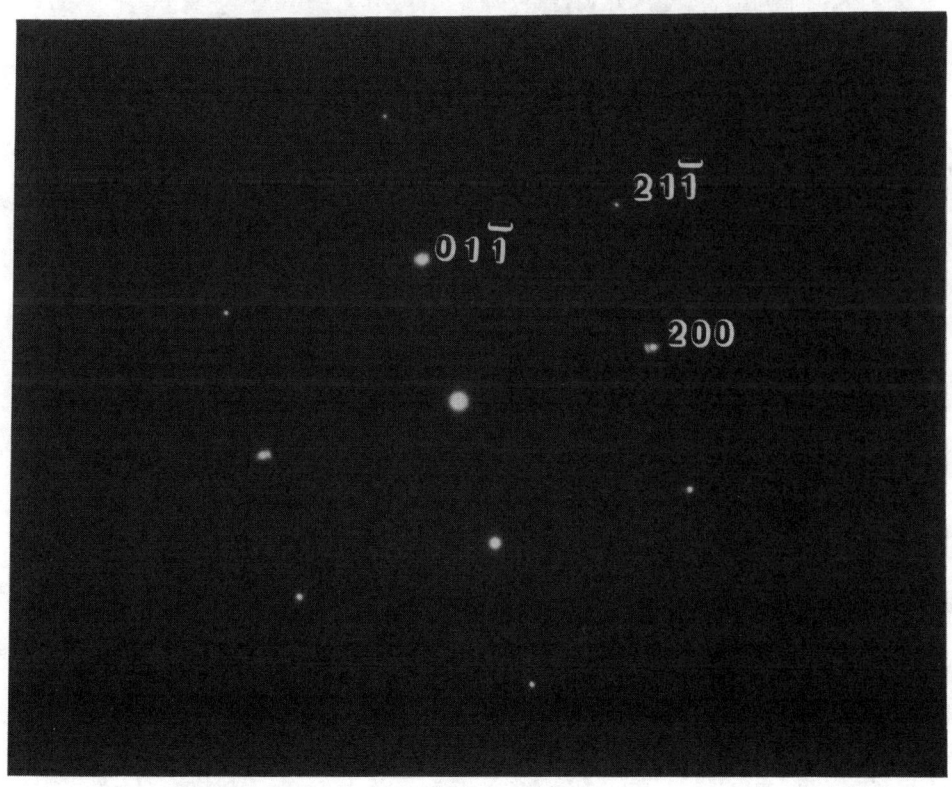

Figure 6B   Diffraction Pattern for Region in Figure 6A.

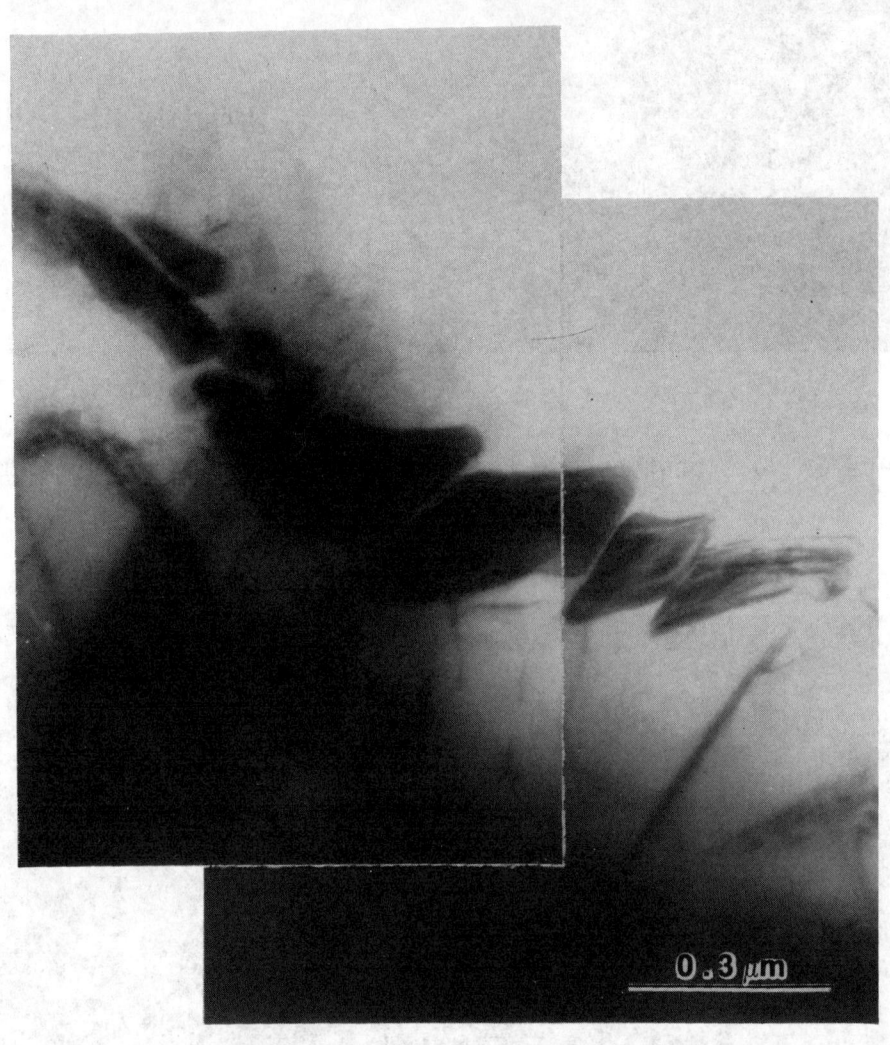

Figure 7   Chromium Carbide Precipitates in the Grain Body (5 kW, 25 mm/s)

Figure 8   Stacking Faults in the High Chromium Alloy

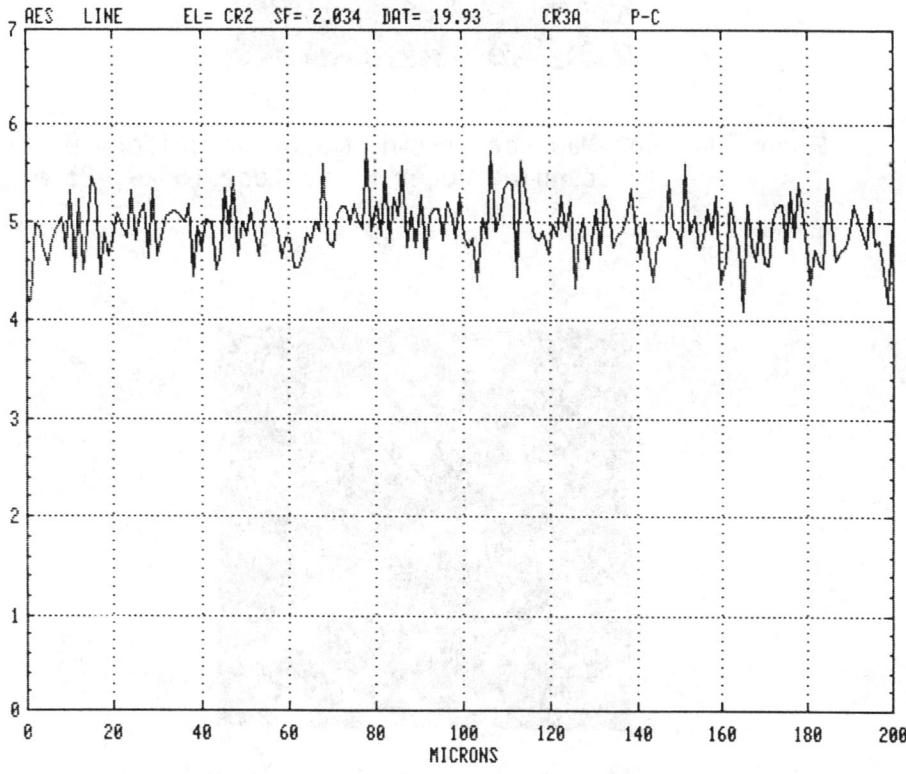

Figure 9   Line Scan for Chromium in Scanning Auger Microprobe (5 kW, 25mm/s)

Figure 10 Dot Map for Carbon Showing Uniform Distribution, in Scanning Auger Microprobe (5 kW, 25 mm/s)

Figure 11 Dot Map for Chromium Showing Uniform Distribution, in Scanning Auger Microprobe (5 kW, 25 mm/s)

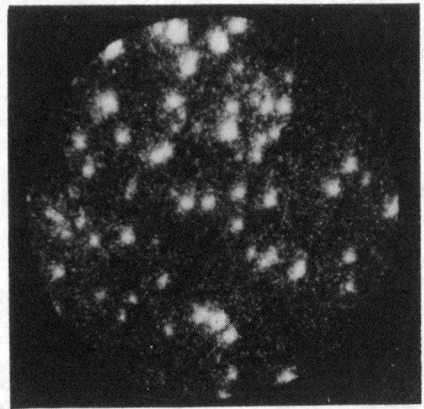

Figure 12 Distribution of Sulphur in the Matrix Obtained Using SIMS (5 kW, 25 mm/s)

In situ clad alloy formation by laser cladding
T. Takeda, W. M. Steen
and
D. R. F. West
Imperial College of Science and Technology, UK

Abstract.

Laser cladding of a metallic surface can be achieved by either melting a preplaced powder bed or by blowing powder into the laser generated melt pool (1). In this paper a process variation is studied in which the blown powder consists of a mixed powder feed, produced in three ways. In the first method a single hopper containing a mixed powder, feeds a single argon blown delivery pipe. In the second method a multiple hopper system feeds the individual alloying elements into a single argon blown delivery pipe. In the third method a multiple hopper system feeds a multiple delivery pipe arrangement.

The preliminary results using Fe/Cr/Ni show that insitu alloy formation of the clad layer is possible below a certain speed. At higher speeds some inhomogeneity is found due to the lack of mixing.

The extent of the mixing is considered to be dependent on the time a given location is molten and the convective stirring action. Some quantitative understanding of this complex situation was found by using copper markers embedded in the substrate.

Experiments, using the third method of a multiple feed system, were found to be able to produce layered cladding; the depth of the layered interface was found to be dependent upon the impact location of the powder streams. There was considerable mixing between layers.

Introduction.

The laser clad process using blown powders has been shown (1,3) to give a controlled level of dilution and in particular controlled heat penetration into the substrate. This latter point means that it is one of the few processes capable of cladding small areas on fins or thin sections. Thus Rolls Royce has been using this process in production on turbine blades since 1982 (2). It is a process which they have found costs around 10% of the cost of the previous process.

This cladding process has been developed further at Imperial College by the work of Weerasinghe and others (1,3). In this process powder is blown into the laser generated melt pool. It has been shown by calculation and experiment that the powder arriving at the pool is not molten. Thus unless there is a melt pool on the surface there will be no sticking of the powder and therefore no clad. On the other hand if there is a melt pool then there will be a fusion bond. It has been shown (3) that the fusion depth into the substrate can be accurately controlled to produce good fusion bonded clad layers with very low levels of dilution. Adding this advantage to the well known fact that laser heating offers low thermal penetration and therefore low distortion, the attraction of laser cladding becomes apparent.

There have been various developments of this process. Optical feedback devices have increased covering rates by approximately 40% making the process economically competitive with plasma cladding (3). Specialist hopper systems for the steady flow of low powder feed rates with built in powder feed rate monitoring have been developed (4) and are now marketed by Quantum Laser Corporation USA. Optimal covering patterns have been developed by Rolls Royce Ltd., while Quantum Laser Corporation has been successful in designing processes for generating crack free deposits with a wide range of hardfacing alloys.

The present paper is concerned with three ways in which in situ alloying can be achieved; it re-presents and extends results recently reported (5). The advantage of in situ alloying by avoiding pre-alloying is to possibly reduce the cost of materials; to allow patterned and variable composition clads to be laid down in a single pass with the composition variation being through the thickness of the clad or along the length of the clad; and to allow the producton of clad layers in which one component does not have to melt e.g. WC, TiC, SiC.

Experimental.

Cladding with mixed powder feed onto an En3 steel substrate (0.2 %C) was performed in three different ways:

(a) Using premixed powder in a single hopper with a single feed pipe (fig 1a).

(b) Multiple hoppers feeding into a single delivery pipe (fig. 1b); only three hoppers were used each separately controlled by its own stepping motor and gas delivery system.

(c) Multiple hoppers feeding separate delivery pipes (fig 1c); only two hoppers and two feed pipes were used.

In each experiment, a single track was laid down, using a Control Laser 2kW laser, under carefully monitored conditions of laser power, beam diameter, mode structure, powder feed rate, traverse speed and powder injection position and angle.

The tracks produced were generally examined for size and shape, for microstructure, as seen by optical microscopy, for chemical composition, as measured by electron microprobe analysis (EPMA) using a spectrometer on a JEOL JSM 35 instrument and for microhardness, as measured on a Leitz Miniload 2. In the EPMA analysis the centre lines of processed tracks and edge zones were scanned at $10\mu m$ intervals and a continuous plot of the composition obtained (ZAF corrections were not applied). Microsegregation of fine scale cellular/dendritic structures is not necessarily detected by the procedure used. The levels of dilution were calculated as the percentage of the substrate melted into the whole clad region. Nearly all the experiments reported in this preliminary study were performed in a regime of very low dilution.

Further tracks were produced in which the clad was laid down over a copper marker wire which was embedded in the substrate surface. These tracks were only examined by EDAX with a view to establishing data on the flow pattern within the melt pool while cladding.

The alloy system studied is that of Fe/Cr/Ni, for which interest in relation to stainless steels and metallurgical data are available with which to compare the results (6,7). The powders were fed as separate elemental powders.

In the twin feed system, method 'c' only two powders were used. They were nickel and Colmonoy 5 (a nickel based hardfacing alloy).

Results.

Premixed powder feed with a single feed pipe:

Using a mixture (wt.%) of 18:9::Cr:Ni with the balance Fe a series of tracks were made at different traverse speeds. The size and shape of these tracks are shown in fig 2. For a laser power of 1.7kW and beam diameter 7.9mm with a Gaussian mode (approximately TEM00) and a powder feed rate of 0.293 g/s, it was found that the composition on random transverse sections was approximately uniform on the scale of the microprobe analysis up to a traverse speed of 7mm/s. The hardness values and composition through the centreline of such a track are shown in fig 3 together with a macrograph of a transverse section, fig 4. Some small composition variation of around $\pm$ 2% Cr was noted near the surface and lower interface of the track. The hardness showed a slight fall near the interface where some slight dilution would be expected; though here, on a random section, the chromium content was found to rise. The hardness values observed were significantly higher than that for the fully annealed steel of similar composition (see table 1). At double this speed i.e 13.6 mm/s and a slightly lower power of 1.5kW there is a marked lack of homogeneity shown by the variations in hardness and chromium content ($\pm$ 9% Cr), fig 5,6.

Three compositions of the powder feed were used 13:6, 18:9 and 25:20. The average hardness values are noted in table 1.

Table 1.

Comparison of the literature values (/) and observed hardnesses for the three compositions examined.

Cr/Ni Ratio	Structure Indicated by Schaeffler diag.(6)	Hv expected (7)	Hv observed.
13:6	martensitic	(type 414)* 450 (Hardened) (C 0.15 max)	330-380
18:9	austenite/martensite /ferrite	(type 301)* 400 (work hardened) (C 0.15 max)	322-386
25:20	austenite/ferrite	(type 310)* 180 (annealed) (0.25% C max)	158-183

Stainless Steel Type	Composition (wt %)				
	C	Cr	Ni	Si	Mn
414	0.15	11.4-13.5	1.25-2.5	1.0	1.0
301	0.15	16-18	6-8	1.0	2.0
310	0.25	24-26	19-22	1.5	2.0

All of the clad regions exhibited a cellular/dendritic solidification structure and reasonable compositional uniformity at speeds up to 7mm/s. The structures are shown in figs 7,8. Complex structural variations are suggested by the swirl effects in the microstructures of the 25:20 sample for various traverse speeds fig 9, 10, 11. Detailed structural studies have yet to be made.

The observed height and width of tracks showed the expected hyperbolic relationship with speed.

Triple Hopper with single feed:

As expected, for a given mass flow rate of powder the size of the clad region and the level of dilution were similar to those found with a single hopper single feed system. The composition also behaved similarly; that is above 7mm/s inhomogeneities became apparent. Hardness values for the 27:20 track, expected to show an austenite/ferrite structure, were 155 ± 15 Hv. The cellular/dendritic microstructure was similar to fig 8.

Twin Hopper and twin feed system:

Separately fed powders of Colmonoy 5 and pure nickel were used. The composition (wt%) of the Colmonoy 5 powder was 11.5 Cr, 2.5 B, 3.75 Si 0.65 C, 4.25 Fe, balance Ni.

A series of tracks were made with varied powder injection location as illustrated in fig 12 under processing conditions which achieved virtually nil dilution. The random transverse section of a deposit made at 6.7mm/s with 0.142 g/s Colmonoy 5 and 0.097 g/s Ni is shown in fig 13. There is a structural variation band running horizontally through the deposit. Fine dendrites (interpreted as nickel based solid solution) were observed; the upper region contained a smaller dendrite size. For comparison fig 15 shows a deposit solely of Colmonoy 5 (hardness 700-765 Hv). The hardness traverses across transverse sections made with various values of the powder impingement point defined by Δx showed a change in hardness at different depths depending on the impingement point of the powder stream. The height from the substrate at which the hardness rises above 500 Hv was found to vary with Δx (fig 16). The 500 Hv contour was chosen as a level of hardness substantially greater than Ni or the En3 substrate steel to indicate the location at which the Colmonoy 5 addition is exerting a considerable influence.

Even though an observable layer was noted both by hardness and composition, there was considerable mixing between the layers. EDAX scans from top to bottom of these layers showed a typical variation from >~12% Cr at the top to ~8 % Cr at the bottom. The Cr composition at the top showed large variations suggesting that these late arriving particles had insufficient time to become fully mixed (fig 17).

The EDAX scan shows a distinct band in the Cr concentration, fig 17, but there are indications of a number of poorly mixed particles. The results indicate that it may be possible to control the thermal experience of the particle entering the clad layer thereby allowing cladding with thermally sensitive materials such as SiC and SiN.

Flow studies in the melt pool:

In order to explore the flow mechanism in the melt pool clad tracks were made at various speeds over transversely embedded copper wires. The wires were 0.7mm in diameter and were let into tight fitting machined grooves in the specimen surface and were shot blasted to reduce their reflectivity. Copper was chosen for the ease of distinguishing it from Ni on an EDAX scan and for its solubility in nickel.
It did, however have a lower melting point than the base metal and different surface tension properties in the melt from the Ni base alloys. Thus these marker experiments may not totally reproduce the beam penetration into the substrate nor the exact flow pattern; however, a very good approximation should be shown since very little Cu is involved. The results of an EDAX scan of one of these runs is shown in figure 19. In this example the clad was stopped near the time that the copper first melted, so that the approximate shape of the melt pool is shown.

Discussion.

The process of mixed powder feed can produce either relatively homogeneous layers or layers of variable composition; the preferred type will not be known for any particular alloy system until various property analyses (eg wear) have been made. However, in both cases it is interesting to understand the mechanism involved in clad formation.

From the mixed powder deposits of Cr/Ni/Fe it has so far been found that a reasonably homogeneous deposit can be formed at speeds less than a certain value (7mm/s; 1.7kW; 8mm beam diameter). At greater speeds than this there was poor mixing particularly at the edges and substrate interface regions; it was also noted at the surface of samples where the powder had arrived late in the melting process, as in the twin feed with negative values of Δx.

The EDAX scans showed that the concentration peaks detected typically spanned around 100 μm. This is of the same order as the size of the particles in the powder stream (Cr 40-80 μm; Ni 80-160 μm; Fe 100-180 μm). This suggests that concentration peaks are due to particles which have not fully dissolved.
The mechanisms leading to homogeneity in these melt pools will be convection and diffusion, both occurring only while the pool is molten. Thus the time for which the material is molten and the steepness of the surface thermal gradients driving the convective flow are probably the most important parameters determining homogeneity.

The molten time can be assessed from the dimensions of the melt pool and the melt pool can be approximately mapped using copper marker wires embedded in the substrate (fig 19). This experimental result taken at 4.4mm/s shows a melt pool at least 5mm long from which it is estimated that a minimum fluid velocity of 48mm/s with a fast sideways mixing must have occurred. Similar experiments, made using different speeds have provided data on the melt pool size behind the beam.

The preliminary results are shown in fig 18. Detailed discussion of these flow patterns will be left to another paper. Here, though, we note that the pool is shorter for faster processing speeds and the time the material is molten is greatly reduced with increased speed. (molten time = pool length/speed). Approximate molten times found from these experiments are given in table 2.

Table 2.

Pool sizes and molten times for various traverse speeds for a laser power of 1.86 kW, 4 mm beam diam. and 0.315 g/s powder feed.

Speed mm/s	Pool Length mm	Molten Time s	Molten Time s
4.4	5	1.13	
7	3.5	0.5	0.3 fig (14
12	1.8	0.15	
19.5	1.7	0.09	

These centre line molten times compare reasonably well with the mathematical model of Weerasinghe (8) which is based on a stationary pool and heat flow only by conduction. However because the pool shape predicted is quite different from that found in the tracer experiments variation of molten times across the clad layer as found from experiment and theory are quite different, as shown in fig 14. From this data it appears that around 0.5s is required for a 100 μm particle to mix in the melt pool. If the diffusivity of Cr in Ni in the liquid state is taken as an average value of 2×10^{-5} cm/s then the diffusion distance would be of the order of 50 μm; that is to say the expected diffusion distance is approximately the value of the particle radius at the processing speed of 7mm/s with a 1.7kW beam. Hence it is not surprising to find that there is reasonable homogeneity in the central region of deposits made at speeds less than that value. However, if reference is made to fig 14 it will be seen that the molten time at the edges is only half that of the centre and hence the diffusion process is probably insufficient to explain the absence of concentration packets in these regions. Enhanced diffusion is required and this could be obtained from convective flow within the pool where the turbulent eddies are smaller than the particle size. Eddy diffusivities in liquids are known to be up to 100,000 times greater than the static diffusivity (10). For a velocity of only 48mm/s which is required for the copper to be transported a distance of 5mm from the wire in the time the beam was on the wire there is an expected Reynolds number, Re of 5×10^5, this is ample to explain a substantial increase in the diffusivity, even at the edges.

Concerning hardness values (table 1) comparison is hindered by the higher carbon levels of the commercial alloys compared with the very low carbon values expected with alloying using the pure elements. In the 18:9 in situ alloy the hardness was much higher than the annealed 301 steel but similar to the work hardened material; this contrasts with the 25:20 steel whose hardness corresponded to the annealed state and not the work hardened state. Laser surface melted tracks in 17:11 stainless steel (316) were found by Lamb (9) to have hardness values of only 220 Hv i.e. close to the annealed value. The reason for the higher values observed in the in situ 18:9 alloy is the subject of further study as is the analysis of the structure.

The banding observed in the twin feed method suggests that there is freezing from the base and so late arriving particles cannot mix throughout the depth. The EDAX scan in fig 17 shows that there is considerable mixing between the layers probably as a consequence of the mixing of the powder in the impinging powder streams. However the change in scale of the dendrites without any large compositional change being observed in the EDAX scans for either Ni or Cr is also suggestive of a variation in cooling rate. The flow structure in forming these deposits is the subject of future work.

Summary and Conclsions.

1. In situ alloys can be clad by blowing mixed elemental powders.

2. Homogeneity depends on having a certain molten time and so varies with traverse speed, location in deposit and impingement point of the powder.

3. There is a lack of homogeneity at clad speeds greater than 7mm/s for a laser power of 1.7kW, beam diameter 8mm and powder feed rate of 0.293g/s.

4. Layered structures can be prepared from a layered powder feed if the impingement point of the powder is correctly directed.

5. A range of stainless steels has been prepared by blowing a premixed powder feed.

Acknowledgements.

The authors wish to express their thanks to Komatsu Ltd, Tokyo, Japan for supporting Mr. T. Takeda's studies at Imperial College and to Quantum Laser Corp. USA for financial support for this work.
They would also like to acknowledge the technical help of Mr G. Briers with the EPMA measurements, Mr. R. Stracey with the operation of the laser, and Mr. A. Pace for the experimental work on copper markers.

REFERENCES:

1. V.M. Weerasinghe, W.M. Steen 'Laser Cladding with Pneumatic Powder Delivery' proc conf. 'Lasers in Material Processing' Los Angeles 1983 ed. E. Metybower publ. ASM. Metals Park Ohio 44073.

2. R.M. McIntyre 'Laser hard surfacing of turbine blade shroud interlocks'. Paper 8301-022 pp230-233 1983.

3. V.M. Weerasinghe, W.M. Steen 'The lasers other role' Weld & Met Fabr. Nov 1983.

4. V.M. Weerasinghe, W.M. Steen pat appl. No. 8425716 October 1984.

5. T. Takeda, W.M. Steen, D.R.F. West 'Laser cladding with mixed powder feed' ICALEO '84, Boston U.S. Nov 1984.

6. A. Schaeffler 'Constitution diagram for stainless steel weld metal' Metal Progress 56 (5) 680 and 680B 1949.

7. Metals Handbook vol. 1. 8th Edition p 414. 1961.

8. V.M. Weerasinghe, W.M. Steen, 'Computer simulation model for laser cladding' proc conf. 'Transport phenomena in materials processing' PED vol 10/HTD vol 29, ed: Chen Mazumder & Tucker Book No. H00283 ASME pub. 345 East 47 St. NY 10017. p15-23 1983.

9. Lamb. M., Ph.D. thesis, London University 1961.

10. Kalinske and Pien Ind. Eng. Chem. 36, 220, 1944.

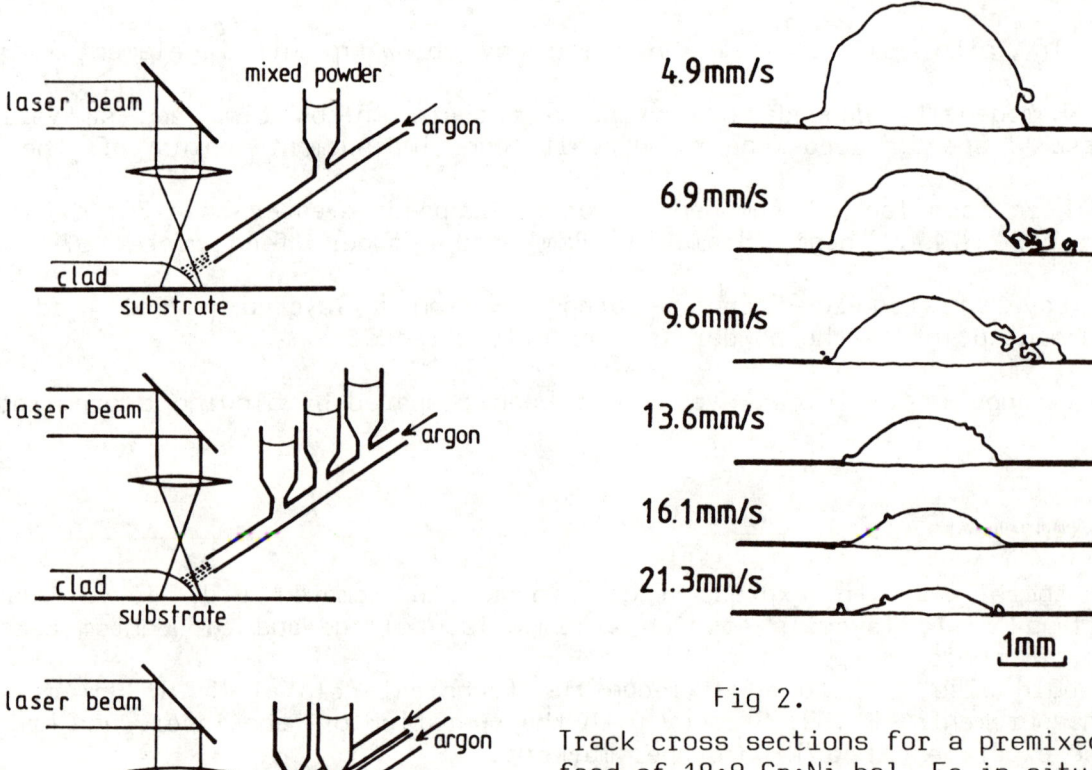

Fig 2.

Track cross sections for a premixed powder feed of 18:9 Cr:Ni bal. Fe in situ alloy. Processing conditions were: laser power 1.7 kW, beam diameter 7.9mm (TEM00), powder feed rate 0.293 g/s

Fig. 1. Experimental Arrangement
a) Arrangement for the single hopper single feed system.
b) Arrangement for the triple hopper single feed system
c) Arrangement for the twin hopper twin feed system.

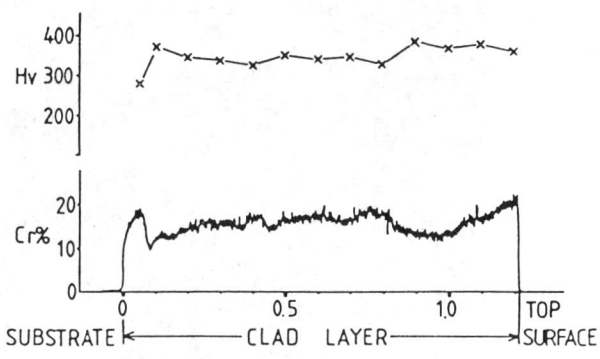

Fig 3.

Variations in hardness and chromium content, measured by EPMA through the centreline of a clad sample from premixed 18:9 feed. Processing conditions as in fig 2 except traverse speed which was 6.9 mm/s.

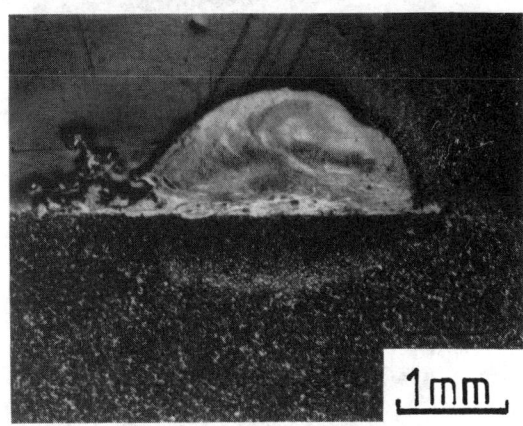

Fig 4.

Macrograph of the sample analysed in fig 3.

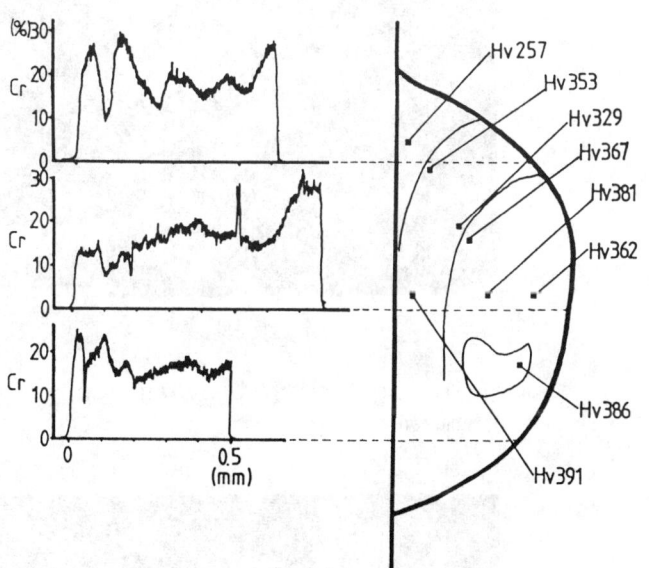

Fig 5

Chromium % from EPMA scan across a sample of 18:9:Cr:Ni premixed feed deposited at 13.6 mm/s. Laser power 1.5kW, beam diameter 7.9mm, powder feed rate 0.293g/s.

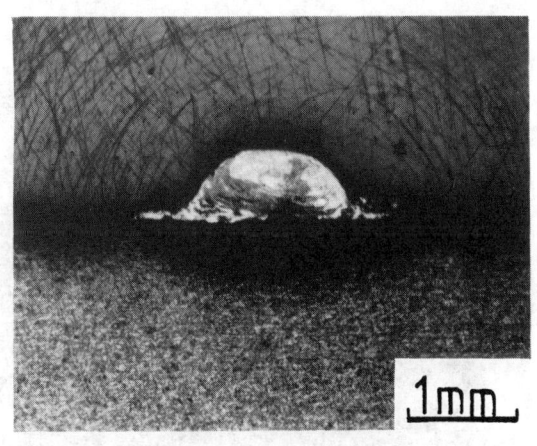

Fig 6.

Macrograph of sample analysed in fig 5.

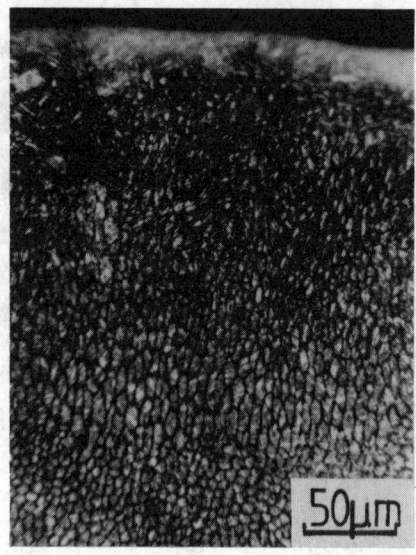

Fig 7.

Micrograph taken near the surface of a clad track made with premixed powder feed of 13:6:Cr:Ni. Process conditions Laser power 1.59kW, beam diameter 5mm, powder feed rate 0.262 g/s, traverse speed 5mm/s.

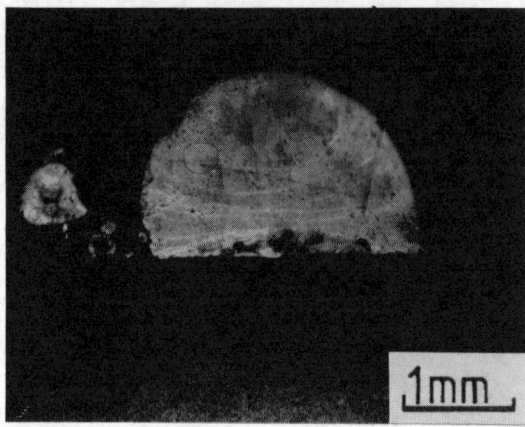

Fig 9.

Macrograph of the track cross section for the 25:20 specimen deposited at a traverse speed of 3.6mm/s, laser power 1.54kW, beam diameter 5mm, powder feed rate 0.244g/s

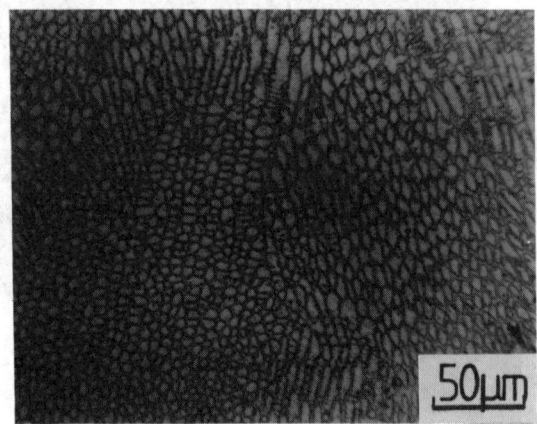

Fig 8.

As for fig 7 except the powder feed was of 25:20:Cr:Ni. Process conditions laser power 1.54kW, beam diameter 5mm, traverse speed 7.2mm/s powder feed rate 0.244g/s

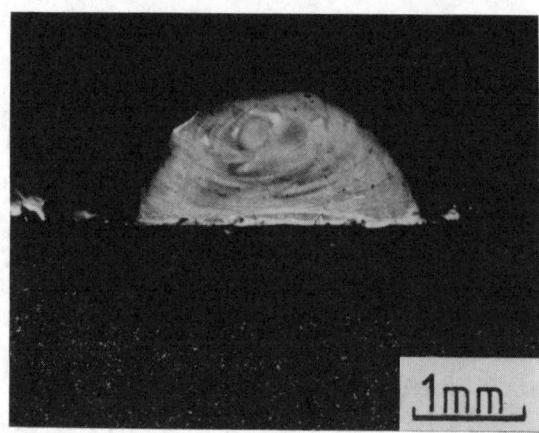

Fig 10

As in fig 9 with traverse speed at 4.8mm/s

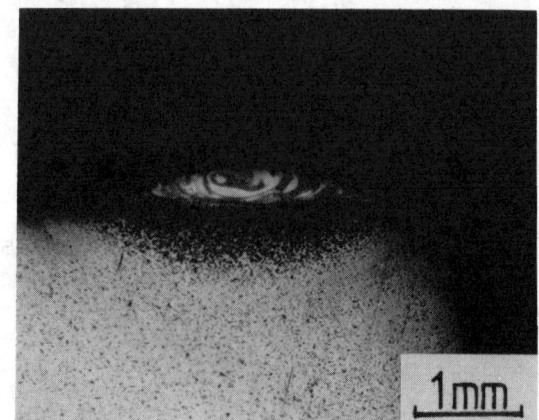

Fig 11.

As in fig 10 with the traverse speed at 13.0mm/s

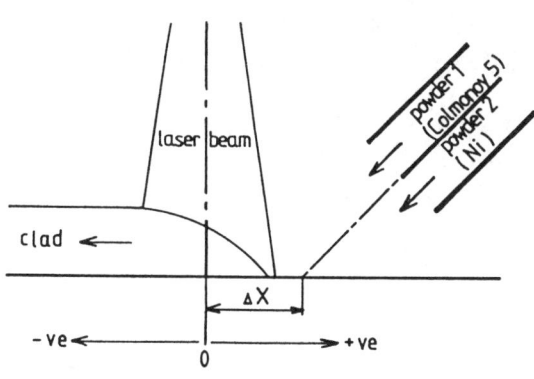

Fig 12

Diagram illustrating the definition of the powder injection parameter Δx.

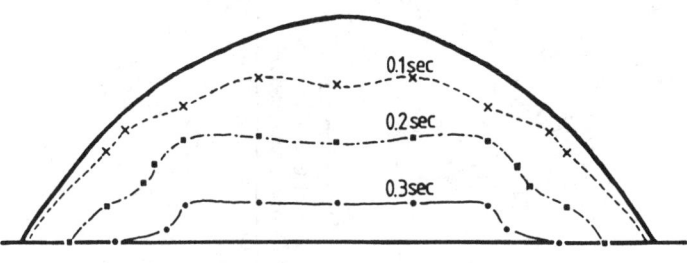

Fig 14.

Theoretical **molten** time contours for a laser power of 1.83kW, beam diam. 5.0mm, Traverse speed 6.67mm/s, powder feed rate 0.2g/s.

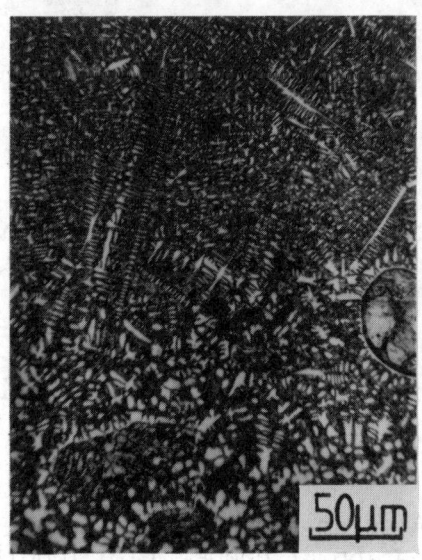

Fig 13.
Transverse section through a twin hopper twin feed sample. Process conditions: Laser power 1.72kW, beam diameter 5mm, traverse speed 6.7mm/s doughnut mode structure, powder feed rate upper powder: Colmonoy 5   0.142 g/s
lower powder: Nickel   0.097 g/s

Fig 15.
Micrograph of a single powder feed of Colmonoy 5 laser clad. Laser power 1.72kW, beam diameter 5mm, powder feed rate 0.233g/s, traverse speed 6.7mm/s

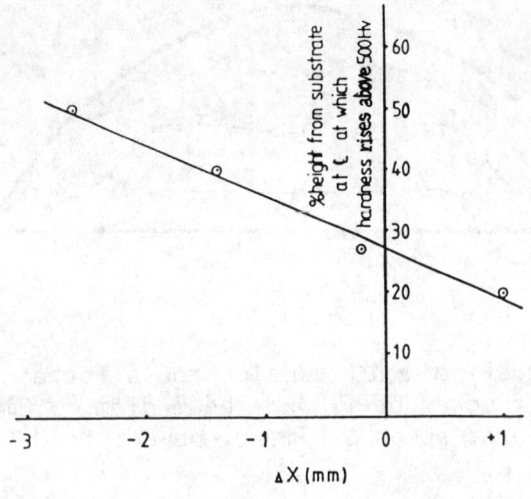

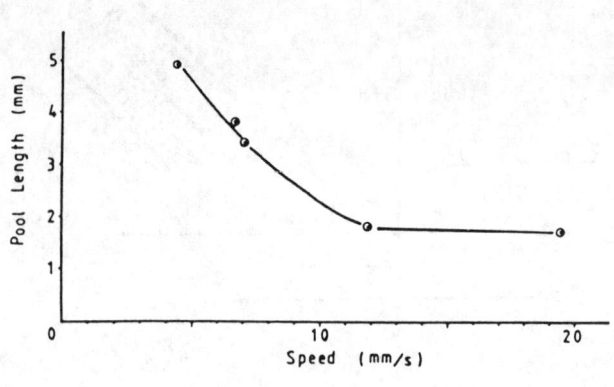

Fig 16
Variation of the % height above the interface at the centreline above which the hardness first rises above Hv 500, indicative of the location of the 'band'

Fig 18
Variation in melt pool length with speed.

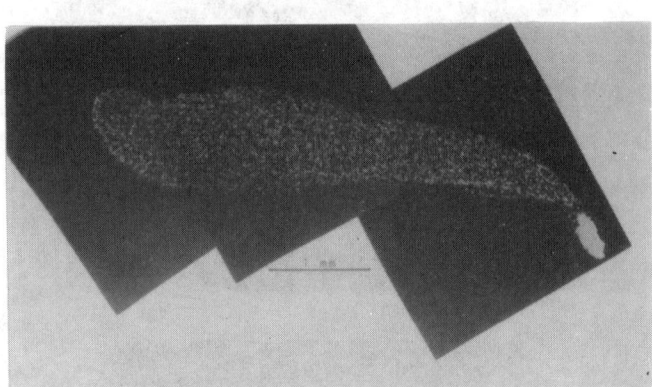

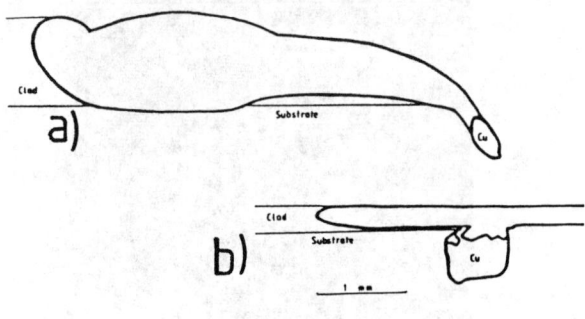

Fig 19.
EDAX scan and diagram of the location of copper in a longitudinal section of a clad. The diagram shows the location of copper in tracks taken at two speeds:
a) 4mm/s   b) 12mm/s

Fig 17

EDAX scan of a twin feed clad layer made with Colmonoy 5 above and nickel feed beneath. The track was made at 6.7mm/s; Laser power 1.72kW and beam diameter 5mm. (Photograph shows part of transverse vertical section.)

Laser surface alloying

E. F. Semiletova
and
T. H. Dumbadze
Georgian Polytechnic Institute, USSR

The present paper outlines the results of investigation of surface alloying of carbon materials with WC, WC + Co type powders by the effect of impulse and continuous laser emission.

It has been shown that the process of alloying and the quality of irradiated surface change depending on the method of powder supply to the heating zone and the type of emission used.

Preliminary application of powder on the surface being irradiated (independently of the type of irradiation) causes in general small depth of alloying (of the order of 0.010-0.06mm) when impulse emission is used. The depth of alloying significantly increases when continuous laser emission is used with simultaneous supply of alloying element to the heating zone. Thus, using emission power P= 1-3kw, it is possible to obtain on carbon material alloyed surfaces from 1 to 5mm deep with hardness Hrc =30-70 and manipulating heating conditions it is possible to change phase composition of alloying zone, obtaining mixture of different combination W, α - phase + intermetallic + carbide of η type $(W + \alpha + Fe_7 W_6 + \eta)$.

In the recent years scientists of many countries carry out investigations in the field of surface alloying of parts, tools with various elements to improve their operation characteristics.

Different types of alloying elements and methods of their application on surface for laser alloying are known. Many of them involve additional technological operations, such as, for instance, foil or specimen manufacture, fixing it to a surface, etc. All this increases the manufacturing cost of parts. In our opinion, the most convenient for use and economically advantageous is alloying element in the form of powder. Powder can be preliminarily applied on the irradiated surface /1-4/ or introduced into the heating zone simultaneously with the effect of laser emission /5/.

For wide introduction of the mentioned method of alloying into practice it is necessary to know the effect of change of properties of the object being strengthened.

This paper outlines the results of investigations of alloying zones obtained by the effect of emission on surface, covered beforehand with powder of the alloying element, as well as of alloying zones obtained by the effect of laser emission and simultaneous supply of powder to the surface.

Investigation of laser surface alloying of carbon iron alloys with two and three component mixtures, one of which is carbon, is of great scientific and practical interest.

Methods of investigation:

Carbon iron alloys with different carbon content (α-Fe, st.3, st.45, st.Y8A, Y12) and powder mixtures of industrial production of WC and BK8, BK15 (WC + Co) type were used for investigations. Specimens prepared in advance were irradiated on impulse laser unit in the energy range E= 5-30J., as well as on the continuous laser emission unit with emission power P= 0.1kw, P= 0.5kw, P= 1-3kw and rate of specimen shifting V= 2,2- 17mm/sec.

Technological indices of property changes of the irradiated specimens were studied: heating parameters, structural and phase transformations, chemical composition, microhardness, residual stresses, heat resistance. These indices were investigated using the method of metallographic, X-ray, electron probe X-ray analyses and gradual reheating.

Results and discussions:

Macroanalysis of alloying zones has shown that impulse heating always produces a heating spot of limited volume. Using continuous laser emission it is possible to obtain tracks of unlimitted length. Tracks' parameters, as well as these of heating spots, change significantly depending on the conditions of emission effect (Fig.1a). In the cross-section, a zone of specific parameters and shape is formed on the irradiated surface, independent of the method of powder application, type of emission, conditions of its interaction. This zone, depending on the matrix original structure, conditions of emission interaction may have several regions of different character and etching. Such a situation indicates that these regions had been under different temperature conditions.

Metallographic investigations of the zones obtained using different emissions show that these zones differ significantly (Fig.2). Martensitic structure can be seen on the surface of control specimens, irradiated by the impulse emission. Structure of specimens alloyed under similar conditions differs sharply by carbon content and indicates that a new composition alloy has been obtained on the surface of specimens, irradiated by continuous laser emmission with simultaneous supply of powder. White dendrites linked with each other by the base can be clearly seen in the structure.

It is established that hardness changes with depth in a stricktly regular pattern, which is an indication of different phase transformations occurring in different regions of alloying zone (Fig.3.). Moreover, it significantly varies depending on carbon content in the material, conditions of obtaining alloying zone and type of emission. The depth of alloying significantly increases for specimens alloyed by continuous laser emission and regions of the zone which underwent phase transformations in solid state increase.

Study of a large group of specimens irradiated (in preliminarily heat treated state) with impulse emission showed that kinetics of alloying process is significantly effected primarly by the height of alloying powder applied to the surface and with optimum height of powder layer on the surface - by the energy value in the impulse.

It is established that with low energy impulses of the order $E \leq 10J$

only a trace of heating zone remains on the irradiated surface. Apparently, the heat released by emission in the heating zone is absorbed by a layer of alloying element and is insufficient not only for matrix melt, but also for powder melt. On the surfaces irradiated with impulses with energy $E \geqslant 10J$ formation of melt of powder with the material is observed and simultaneously with this traces of molten metal splashing from heating zone occur. Results of chemical analysis and plots of hardness change indicate that new composition alloys are obtained in these zones (Fig.5.).

Determination of tungsten percentage on irradiated surfaces showed its relatively non-uniform distribution in the obtained zones. Non-uniformity is manifested in the formation of individual units of alloying element (their size being 300-600mu) on irradiated surfaces, the number of units and percentage of tungsten in them significantly change depending on energy value in the emission impulse (Fig.6.). Thus from one to two units (with content ~6.7%) are formed on heating surface, obtained by impulse $E \geqslant 12J$. The number of units increases with energy increase in the impulse. This occurs up to $E=16J$, and then a decrease of the number of units, as well as the reduction of tungsten content is observed. In other equal conditions, it is established that the greatest number of units with minimum percentage of tungsten is formed microvolumes obtained with impulses having the energy $E=15-16J$.

Non-uniform distribution of alloying element on the surface is, obviously, due to a non-uniformity of powder layer applied to the surface, imperfection of powder application method and different degree of heating of microsurfaces causing different mechanisms of alloying. Investigations of the character of tungsten distribution over the depth of heating showed that its dissolution proceeded significantly uniformly and it should be said that insignificant change of heating conditions decreases tungsten solubility and consequently causes decrease in hardness (Table 1,2).

Microhardness of alloyed microvolumes.

Table I.

Specimen	Microhardness kgf/mm^2			
	After irradiation		after annealing	
	Focus	Defocus $F=\pm\triangle 2mm$	Focus F	Defocus $F=\pm\triangle 2mm$
St.45 +WC	1000	800	300	260
St.Y8A Contr.sp.	840	-	190	
St.Y8A +WC	1180	1000	350	300

Table 2.

Specimen	Specimen surface position	Tungsten content, %	Microhardness kgf/mm^2
St.Y8A + BK8	Focus	3	1180
	$+\triangle 2mm$	2.8	1100
	$+\triangle 4mm$	2.3	900

Comparison of data on tungsten solubility depending on the condition of emission effect with the obtained hardness parameters shows that during alloying, as well as surface impregnation, it is necessary to differentiate material heating conditions, rational conditions and overheating conditions.

To check this, further experiments were carried out with Y7A + BK8 pair. The surface was preliminarily coated with powder. Emission of impulse, as well as of continuously operating lasers were used for irradiation.

Results of the investigations presented in Fig.4. and Table 3. confirm the above stated. These results show that increase in emission power (at other equal conditions) causes formation of overheating structures, reduction of hardness, tungsten solubility and, hence, a decrease.

Results of chemical analysis of alloying zones.

Table No 3.

Specimen	Content, % W	Treatment conditions	Distribution of alloying element
St.Y7A+BK8	1.5	$E= 15J$ $=6 \cdot 10^{-3}$	non-uniform
	54	$P= 0.1 kw$	uniform
	45	$P= 0.5 kw$	uniform
	38	$P= 1 kw$	non-uniform

It is established that in the zones, obtained by irradiation of laser continuous emission, solubility of allowing element increases, as compared to impulse laser. Under certain operation conditions a film of sintered carbides is formed.

Such a situation shows the advantages of continuous emission lasers for alloying of surfaces of any size in rational conditions, arranged in advance for each pair- material+alloying element.

Setting rational conditions of emission effect on surface is important not only from the viewpoint of chemical composition of material, powder, but also size, shape of part, tool, surface being alloyed.

The results of X-ray analysis given in Table 4 show that structure of microvolumes received with preliminary application of powder is represented by martensite; residual austenite and M6C type carbides have not been detected.

Results of X-ray analysis

Table 4.

Specimen	Treatment	Phase composition	Lattice parameters A^o
Y8A	Origin.	Fe, Fe_3C	2.87
Y8A	Therm.Treat.	$Fe, -Fe, Fe_3C$	2.88
Y8A + WC	Therm.Treat. irradiated	-"-	2.89
Y8A + BK8	-"-	-"-	2.89

Subsequent gradual heating of alloyed microvolumes to different temperatures showed that they undergo certain structural and phase transformations leading to hardness change. Hardness sharply decreases with the increase of heating temperature. However, even after heating up to $800^o C$ hardness in alloyed microvolumes is higher than in the original material (Fig.8), which may serve as a confirmation of formation of new composition alloy with high heat resistance in microvolume.

Decrease in microhardness at subsequent gradual heating is connected with relief of residual stresses, occurring in microvolumes due to high temperature, high heating and cooling rates, as well as due to formation of martensite of different saturation.

It is established that under certain conditions porosity is a characteristic feature for surfaces alloyed by laser emission, with simultaneous supply of powder. Occurrence of pores is obviously due to the fact that the surface being irradiated is for a long period of time exposed to high temperature and heat removal proceeds slowly, which produces formation of gas pockets and their emergence to the surface. In addition, appearance of pores on the surface may be caused in the process of solid alloy sintering if the binder is burnt out. In the cross-section, i.e. over the heating depth the number of gas pockets decreases, and in some cases these are completely missing. This situation indicates that porosity significantly varies, depending on the conditions of specimen irradiation, rate of specimen shifting and powder feed conditions to the emission zone.

Microhardness of these alloying zones is unstable over the depth and has a wide spread, which is clearly seen from Fig.6. Wide spread of microhardness may be caused by the formation of inhomogeneous structure, either by phase, or by chemical composition.

Hardness of alloying tracks changes depending on the conditions of their formation and is within HRc= 30÷70. Such hardness spread depending on the conditions of obtaining zones of alloying may be an indication either of different saturation of the structure with alloying element, or their exposure to different heat conditions, or combination of these factors.

Metallographic analysis of alloying zones, both over the depth and and on the surface, showed that formation of structure in the tracks, obtained under certain conditions, mainly occurs from liquid state and it is inhomogeneous by phase. Structure obtained in heating zone consists of three different by etching phases, which are also different with regard to hardness (Fig.2.) and that can be seen from chemical composition. The structure, independently of the conditions of its formation, is generally stable and presents a mixture of grains with torn off edges, different etching properties and graininess.

Light, poorly etched grains of relatively high hardness are fringed with blocks of another colour. There are dark strongly etched sections between these grains, which are characterized by low hardness and high etching capacity. Such inhomogeneity in the structure causes spread of microhardness over the depth and surface of zones.

Analysis showed that blocks consisting of two structures of different etching capacity represent small grains of tungsten fringed with intermettalic material. These blocks are bound with one another with α-iron solid solution. It should be pointed out that white grains, as a rule, also migrate into transition layer and in some cases they do it in large quantities. This causes formation of complex structural transformations in the transition layer (Fig.9.).

Diffraction pictures, given in Fig.10 and results of Table 5., confirm that a three phase mixture W, α-phase and type Fe_7W_6 intermetallic material, is actually formed. The large grains and blocks observed in the structure represent smaller grains of tungsten and tungstenite briquetted into larger blocks. α-phase represents tungsten solid solution in α-iron. Tungsten solubility in these conditions exceeds its maximum solubility during gradual heating.

Results of investigation of alloying zones.
Table 5.

Material brand	Zone phase composition	α-iron lattice parameter A°	Lattice parameter A°
YI2 origin.		2.866	
YI2+BKI5	α-phase, tungsten, intermetallic.	2.879	3.168
	-	2.866	-
Y8 origin.	α-phase, tungsten intermetallic		
Y8+BKI5		2.876	3.165

These conclusions are well supported by the data of chemical analysis, which showed that only tungsten and iron are present in the fused zone. Tungsten practically uniformly is distributed in the zone heated to liquid state and only in the transition layer its non-uniformity becomes evident, which agrees well with the results of metallographic analysis.

Sections, located above the transition zone were formed from the liquid phase, while those below are formed from the solid phase. Tungsten particles are detected below the transition zone, i.e. tungsten diffusion into the solid phase is also obsered (Fig.9b.).

Tungsten quantity in the transition zone is insignificant. Here the structure has partially dendritic arrangement. It seems that temperature in this zone exceeded iron melting point, but was below that of tungsten. Therefore, structure characteristic of pure iron hardened from melt has been formed.

Its distribution pattern and percentage is determined by the conditions of zone heating, material chemical composition, powder supply rate and in some cases there is about 90% of tungsten.

These results of the investigations allow us to assume that high temperature in the heating zone causes decomposition of tungsten carbide of BKI5 metalloceramic mixture, exposed to emissin zone, simultaneously causing burning out of cobalt and carbon not only from the mixture, but even from the main material. Diffractogram of individual specimens shows lines of carbide of $Fe_3W_3C-Fe_4W_2C$ composition, which is usually formed when tungsten is introduced into carbon steel. When carbon is deficient its structure is unstable and it also decomposes, forming tungstenide Fe_7W_6.

Analysing parameters of tungsten and α-phase (Table.5) lattice, it is possible to observe that tungsten lattice size is similar to that of pure element, while α-phase lattice parameter is somewhat larger than that of pure α-iron. This evidently can be explained by the fact that a higher-melting tungsten crystallizes before iron and its structure does not contain atoms of low-melting elements. Tungsten dendrites have time to coagulate with each other in the process of melt crystallization, forming particles with dimensions of the order of 10mu and above. This is due to insignificant rate of cooling of alloyed zone.

The increase of -phase lattice parameter may be explained by the fact that certain quantity of tungsten disolved in this phase. Judging by the lattice parameter, this quantity varies from 5 to 10%. A somewhat high solubility of tungsten may be explained by specificity of laser heating, during which intensive mixing of metals takes place in the liquid state.

Analysis of stresses and microdistortions in disperse mixture shows the following picture (Table 6). Tungsten lines on diffractograms are significantly less broadened, than α-phase lines. Obviously, tungsten grains are relatively unstrained in the mixture and have greater size as compared to iron grains.

This can also be explained by the fact that tungsten crystallization proceeds in the surroundings of still liquid matrix and for that reason it is practically free from any pressure.

Xray analysis of alloyed zones.

Table 6.

Material	Microstress G kg/mm^2	Line width (2II) of α-phase 10^{-3} rad	Line width (2II) 10^{-3} rad
YI2	-300	20	6
48	-250	22	7

Significantly smeared diffraction lines of -phase point to substantial microdistortions occuring in this phase. Moreover, sections containing -phase undergo compression stress from tungstenides and tungsten particles surrounding them.

Presence of tungstenides also explains high microhardness of zone sections in comparison with microhardness of pure tungsten and -phase.

CONCLUSIONS:

-Using laser emission (independently of its operation principle) it is possible to carry out surface alloying of parts made of iron carbon alloys with high-melting elements;

-Solubility of alloying element significantly varies depending on matrix chemical composition and conditions of emission effect. Alloying process can be controlled for a concrete pair (part+powder);

-tungsten carbide and alloys of BK8 type (applied to the surface in the form of powder) significantly increase not only hardness, but also heat stability of alloyed surfaces.

Forced introduction of powder of metalloceramic mixture of BK15 type into the zone of laser emission causes, under the conditions used, the formation of alloyed layer of a three-phase disperse mixture: tungsten, α-solid solution and tungstenide of Fe_7W_6 type.

-To obtain the required combination of technological indices of improving surface and chemical composition it is necessary to arrange rational conditions of emission effect for every pair of materials, depending on the type of laser used.

REFERENCES

1. УI-я Тбилисская научно-производственная конференция по новой технике и технология в машиностроении, приборостроении и электротехнике (1968 г.): тез.докл./Семилетова Е.Ф., Сирадзе А.М. "О локальном легировании лучом лазера" - -Тбилиси, 1968, с 36-37/.

2. Миркин Л.И. "О возможности насыщения железа углеродом под воздействием светового импульса лазера" - Докл.АН СССР, 1969, т.186, №2, с 305-307.

3. Семилетова Е.Ф. "Способ локального легирования металлов, В кн.Труды ХУ научно-технической конференции ГПИ, Тбилиси, 1970, вып.II.

4. Бетанели А.И., Даниленко Л.И., Лоладзе Т.Н., Семилетова Е.Ф. и др. "Исследование возможности дополнительного легирования поверхности стали Р18 с помощью луча лазера"-ФХОМ, 1972, №6, с 22-26.

5. Григорянц А.Г., Сафонов А.И., Шибаев В.В. "Влияние режимов порошковой лазерной наплавки на условия формирования и размеры наплавленных валиков. "Сварочное производство", 1983, №6 с II-I2.

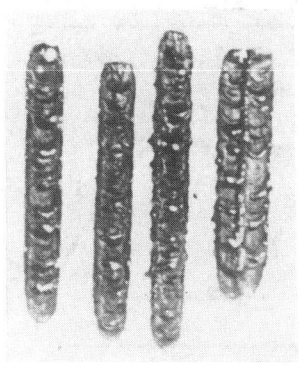 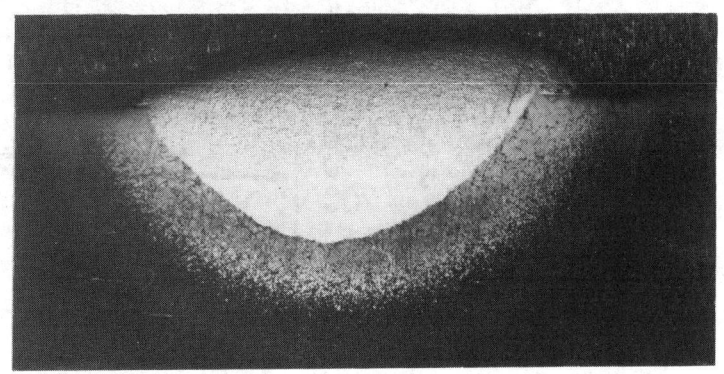

a b

Fig.1. Alloying zones obtained with CO_2-laser emission

a- view from above   b- cross-section

a b c

Fig.2. Fragments of structures

a-irradiated specimen with impulse emission
b-alloyed specimen with impulse emission
c-alloyed specimen with CO_2-laser

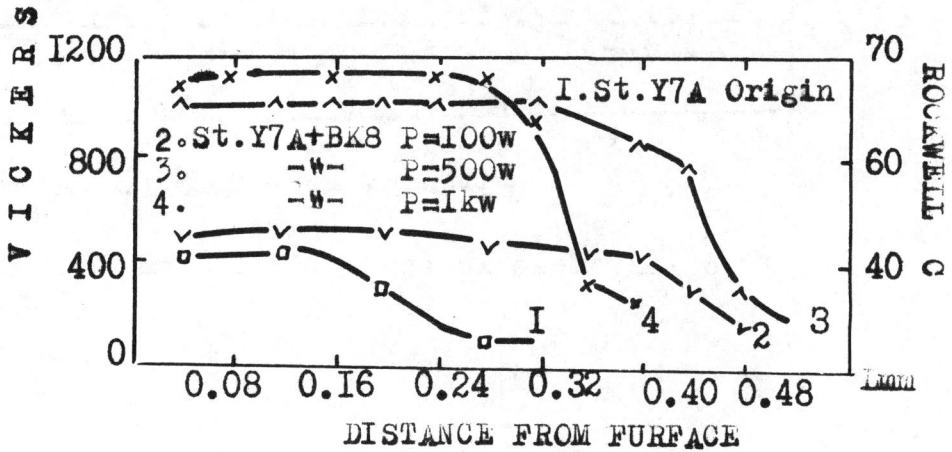

Fig.3. Hardness over depth of alloying zones using impulse emission

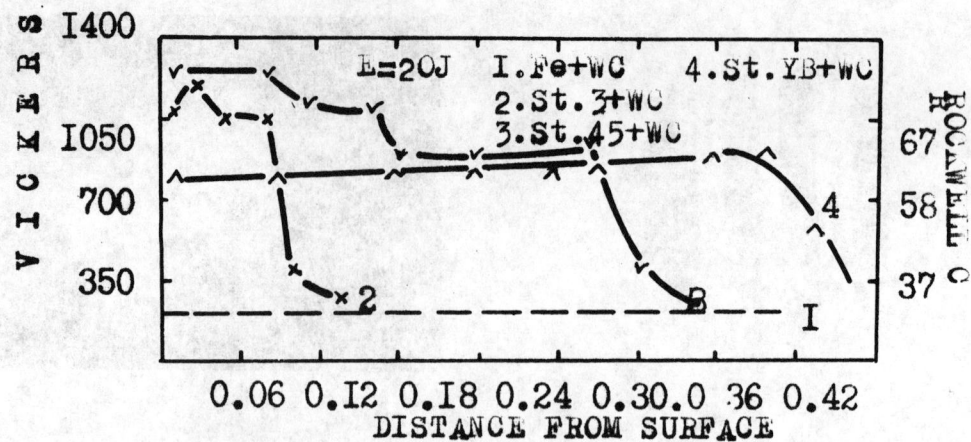

Fig.4. Hardness of alloying zones with CO_2-laser

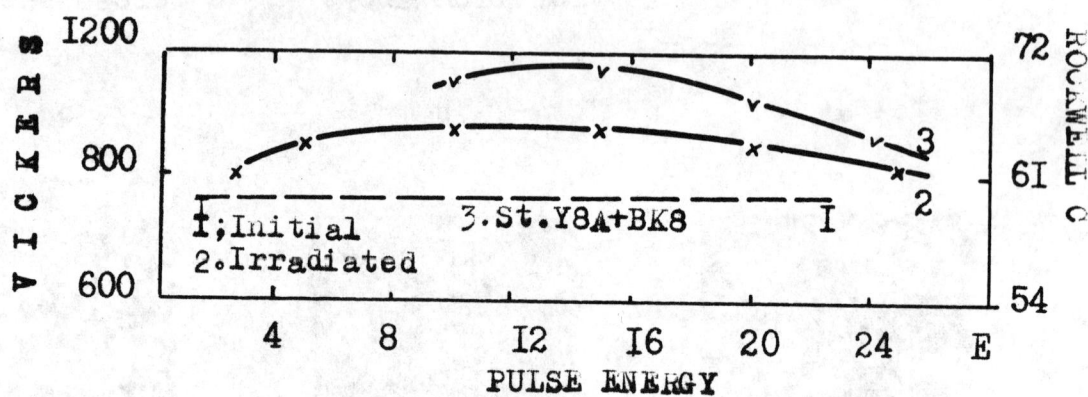

Fig.5. Hardness depending on energy

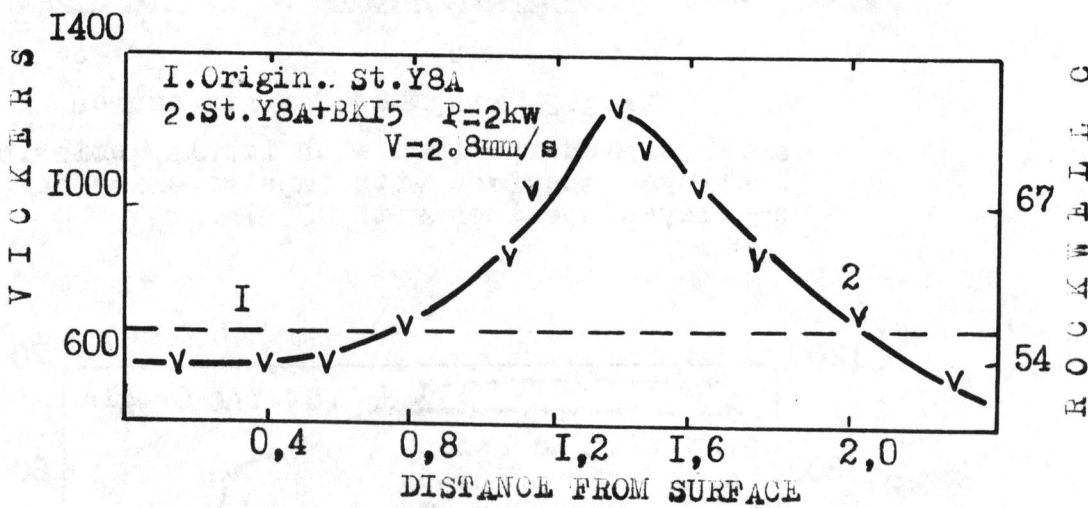

Fig.6. Hardness in zone obtained with CO_2-laser

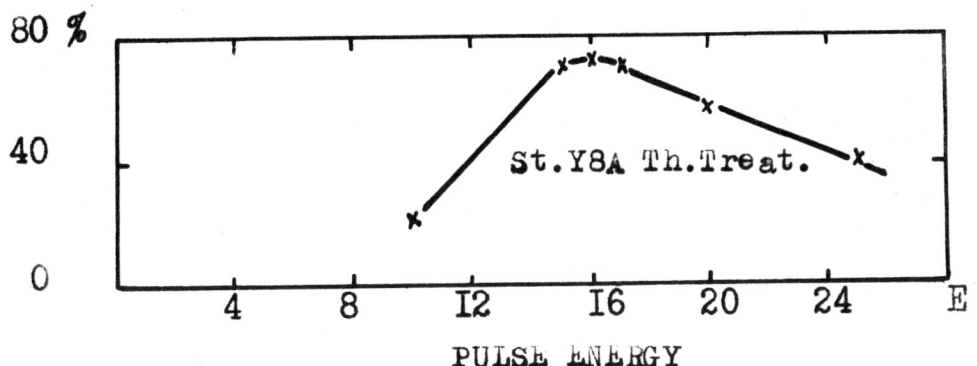

Fig.7　W content in alloying zones

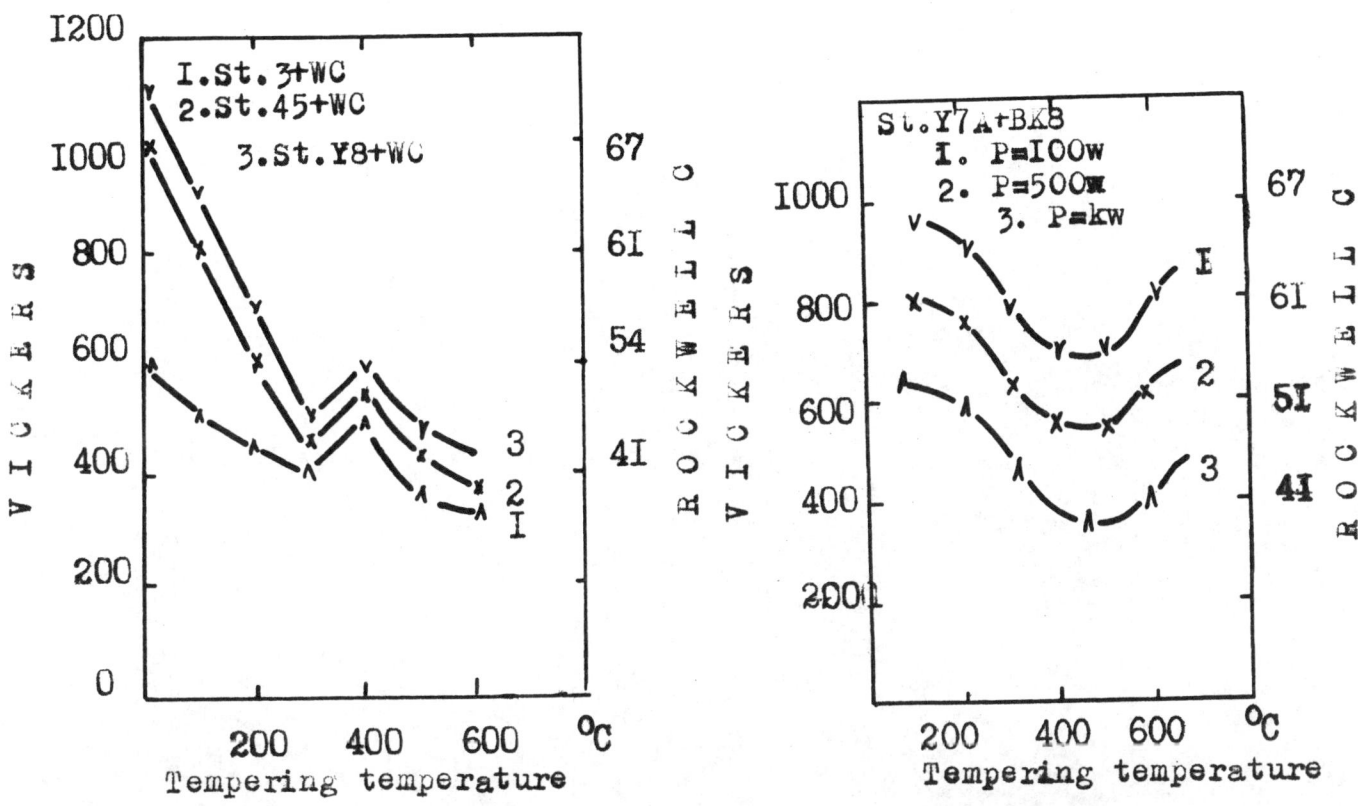

Fig. 8. Heat stability of alloying zones

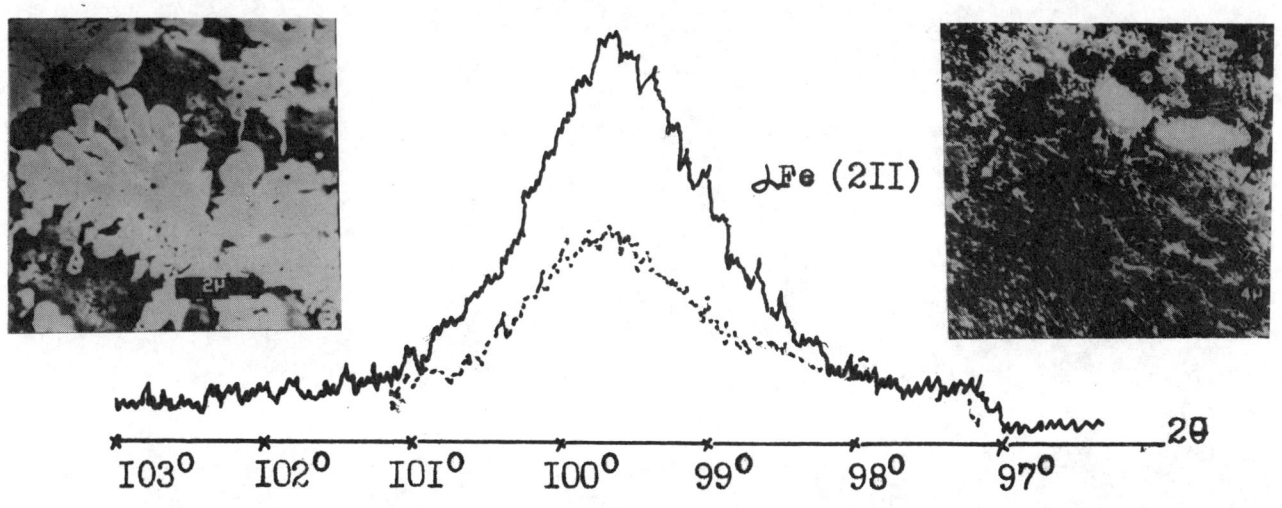

Fig.9.10. Structures and diffractogr. of alloy. zones

Properties of laser melted SG iron

H. W. Bergmann
IWW, Universität Clausthal, West Germany
and
M. Young
University of Birmingham, UK

Abstract

Ductile cast iron is an ideal material for laser surface melting. It is possible to produce economically a large number of components with ideal combination of a ductile tough core and a hard wear resistant surface. The melted depth and translation rate depend on the component size, the application and the dimensional tolerance. As cast components require deep melted layers as sufficient material must be available for the finishing operation. Finished components require only a shallow layer as a finishing operation must be absent or minimal. It is shown in the paper how crack free, smooth (< 1 µm) laser melted surface can be produced. The wear resistance and fatigue of laser surface melted cast components are presented and discussed.

Introduction

Various techniques have been used to modify the surface structure of grey cast iron in order to achieve a white, ledeburitic microstructure whilst retaining the grey core. The most prominent techniques are chill surface casting and Tungsten Inert Gas Welding (TIG). Today, both methods are widely accepted in industry, for instance in camshaft production. The advantage of such treatments is the combination of the core ductility, due to stable Fe-C solidification resulting in an iron-graphite microstructure, and the wear resistance of the surface due to metastable solidification giving a $Fe-Fe_3C$ layer. Recently, the availability of high-power lasers has enabled high speed surface melting and large scale automation[1]. This paper describes the laser melting of SG iron, the microstructures obtained and a selection of resultant properties.

Principles of Rapid Solidification after Laser Melting
of Cast Iron

The solidification process can be characterised by three important para-

meters:
- quenching rate, $\varepsilon = \frac{dT}{dt}$

- solidification velocity, $R = \frac{dx}{dt}$, which is the velocity of the solidification front

- temperature gradient, $G = \frac{dT}{dx}$, across the liquid/solid interface.

These parameters are inter-related by the following equation

$$\varepsilon = R.G. \qquad (3)$$

Depending on the above variables, different microstructures can be achieved on solidifcation[2]. Fig. 1 gives a schematic drawing of the laser melting process and the microstructure obtained in three rectangular sections of the track. Increasing the quenching rate lead to a finer structure of the same morphology, see Fig. 2. If the solidification velocity is increased the morphology of the solidification changes[2,3]

Mechanical Properties of Laser Melted Grey Cast Iron

The properties of laser melted cast irons have not been fully investigated as yet, but the few results available are quite encouraging, especially in the case of S.G. iron. The mechanical properties of laser melted S.G. irons depend on both the area ratio of melt depth to substrate and HAZ to substrate. For S.G. iron the HAZ is small when compared to the melt depth and may be neglected in a first approximation. Fig. 3 shows typical stress-strain curves for untreated and laser melted test samples for ferritic S.G. iron. The diameter of the gauge length was 15 mm and melt depth was 0.3 - 0.4 mm, which results in approximately 10 % of the cross-section being transformed. The stress- strain curves of the lasered tensile test specimens are similar to those of untreated specimens, except that above the yield point, the case cracks circumferentially giving a feature on the curve not dissimilar to discontinuous yielding and this is accompanied by an acoustic emission. At least 1 % elongation occurs before the case cracks. The case is more brittle than the core, though still surprisingly tough. Fractographs show a ductile fracture of the core and a more brittle fracture of the outer layer. The crack propagates through the case mainly in the interdendritic eutectic phase, where microporosity may also be present. A unique feature of this material after tensile testing is the occurrence of equidistant circumferential cracks along the gauge length. These cracks do not penetrate further than the case core interface because of workhardening of the ductile core. This effect is not influenced by metallurgical inhomogeneities between laser tracks as there are none and it occurs with longitudinal tracks as well as for spiral melting. This behaviour occurs for ferritic, pearlitic and bainitic core structures which have a higher toughness than the case. When a martensitic matrix is used, the initial case crack continues through the brittle core.

Tensile test results for different matrix structures show no major differences compared with untreated material providing that the laser melted regions were small when compared with the untreated area. For a deeper case, e.g., ∼1/3 of the total cross section, an increase in strength, a decrease in elongation and a detectable increase in Young's Modulus can be observed. It should be noted that even when the case is that deep, the mechanical properties obtained are still superior to those of the corresponding white castings.

Internal Stresses Introduced by Laser Melting and the Influence of Subsequent Heat Treatment

Laser surface melting results in internal stresses on solidification. Several factors contribute to the final stress situations, for example, the volume difference on solidification, differences in shrinkage in the solid state for core and surface due to both temperature differences and different coefficients of thermal expansion. However, the various transformations in both case and core have an effect. For cast iron with pearlitic matrix and flaky graphite it was found that in the surface, compressive stresses were present which change to tensile stresses at a certain distance below the surface. Depending on the case depth, the sign of the stress changes in either the core or the surface layer. S.G. iron contains compressive stresses in the as lasered surface. Tensile stresses at the surface were found in white cast iron in the as lasered condition, Table 3 [28]. They gradually change to compressive stresses on further heat treatment.

Fatigue Properties

The fatigue properties of ferritic S.G. iron were studied with the pull-pull test. The mean tensile stress was ≅50 % of the yield point, that is ≅150 Nmm^{-2}. A 10 Hz frequency of various amplitudes was then applied to the specimens and the fatigue limit determined. Untreated specimens were ground after machining. As no relevant data is available in the literature, optimised processing parameters for laser treatment had to be defined. This was done by keeping the case depth constant at about 10 % of the cross-section. Compared to the untreated S.G. iron, a decrease in fatigue limit was found in the as lasered condition. This was more pronounced when N_2, CO_2 or Ar was used as the protective gas, but less with He. In addition, a spiral laser track gives better results than a longitudinal series of tracks. Grinding the surface, i.e., smoothing small amounts of roughness produced by the laser treatment leads to a small decrease in fatigue limit as compared with untreated value. The most favourable values, see Fig. 4, were found when the specimens were annealed at 240 °C for 2 hrs after laser treatment with helium and subsequently ground. In a second series the influence of a deeper case was studied. Here the mean stress was chosen as 340 N/mm^2, because of the higher yield point, see Fig. 5.

Wear Properties

The excellent wear properties of ledeburitic surface layers on cast iron has been demonstrated by various authors. Most of the work has been carried out on TIG melted surfaces, while for the new technique of laser melting, only a limited amount of data is available. It is known that surface melted layers produced by the TIG technique can be run under higher applied loads than for hardened steels. The tests were performed employing various lubricants. The optimised combinations are surface melted cast irons running against surface melted and nitrided cast irons. The tests were carried out with camshafts and rocker arms. For lasered grey iron with pearlitic matrix and flaky graphite, Bell and co-workers demonstrated that the wear properties in dry pin on disc tests are improved by laser melting by one order of magnitude. There was still a significant increase compared to fully martensitic microstructures [3].

The advantage of laser melted S.G. iron was demonstrated for rolls which run dry against each other with a fixed relative slip [4], one wheel being driven and the other partially braked. Various stresses were applied (e.g. 500 and 1000 Nmm^{-2}) and the humidity was controlled. For a slip of approximately 3 - 5 % and the higher load, melting occurred after a couple of minutes when steel rolls were used, but the cast irons showed no evidence of any deformation. For comparison, ≅1 % slip was used (Fig. 6,7). It is obvious that the wear properties of lasered cast iron are superior to those of hardened, conventional steels. Combinations of lasered irons and steels must be avoided as massive deformation of the steel occurs. When laser treatments were carried out under He, better results were found than with other gases. Excellent results were obtained when laser melted S.G. irons were used in combination with TiN or TiC coated hardened steels, the TiN/TiC surface suffering the wear[1]. After the tests, no deformation is visible on the ledeburitic surface. Good wear properties are obtained in combination of lasered S.G. iron with nitrided or borided steels. A comparison between lasered and TIG melted ledeburitic surfaces favours the laser process. The finer microstructure may be the reason for this. For extremely high loads ≅4000 N/mm^2 it was found that the ledeburitic case has to be about 1 mm thick to avoid fracture as a result of the Hertzian stresses. If a case of this thickness is required the laser melted or TIG welded microstructures are almost equivalent and so are the wear properties.

Better wear resistance of laser melted S.G. iron compared with steels does not imply a change in the coefficient of friction. This is demonstrated in Fig. 7. The coefficient of friction was determined for various loads and velocities as a function of the slip. In all combinations the same curves were obtained. The wear behaviour, however, does show differences.

Laser melted S.G. iron exhibits also superior wear properties if tested under lubrication. The test results are given in Fig. 8. Tests of real components used in car industry on a testing machine as well as after running in a test car are being undertaken at present.

General Comments and Possible Applications

Apart from the fact that laser melting enhances certain desirable properties mentioned above, there are additional beneficial features obtained with this technique. The melting process can be carried out on an almost finished component as the distortion associated with the process is negligible and surface roughness is in the order of 10 μm, which can easily be improved by grinding if necessary.

There is no need for a sophisticated handling system as a uniform and homogeneous case can be achieved if lasered within the focal distance, i.e., ± 1.5 cm for a 25.4 cm focal length lens.

Laser melting is a hardening process which can be fully automated. It is also a fast and cheap process. In Fig. 9,10 a collection of laser melted components is shown, demonstrating the quality achieved as well as the production time required for each component. The production cost for the equipment used is approximately ₤ 30 - 40 per hour including running costs, maintenance, technical expertise and leasing costs.

Acknowledgements

The fundamental studies were financed by the Stiftung Volkswagenwerk. Investigations correlated with industry were supported by the Bundesministerium für Forschung und Technologie. The author likes to thank the commission of the European Community for financing the colaboration between the University of Birmingham and the University of Clausthal.

References

1. A. V. La Rocca, First European Conference on Optical Systems and Application, Brighton, U. K. (1978)

2. I. Hawkes, W. M. Steen and D.R.F. West in Proc. Electroheat for metals, Cambridge (1982)

3. H.W. Bergmann, Surface Engineering, in print

4. H.W. Bergmann, B.L. Mordike, T. Bell, in print

5. H. Krause, H.W. Bergmann, Ch. Schroelkamp, in preparation

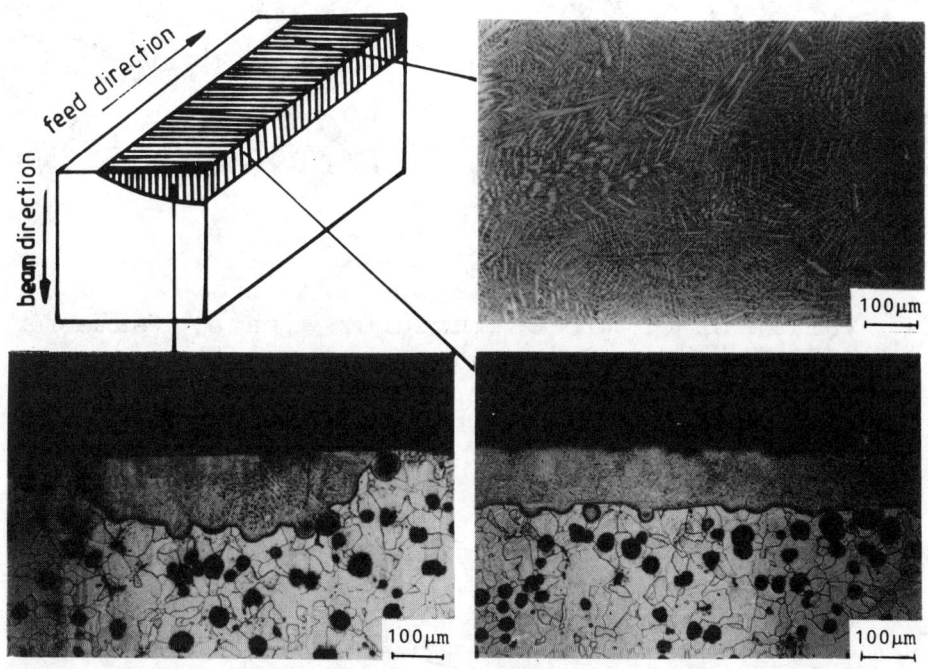

Fig. 1. Laser melted S. G. iron
(a) schematic showing of a single track
(b) micrograph, top view
(c) micrograph, cross section
(d) micrograph, longitudinal section

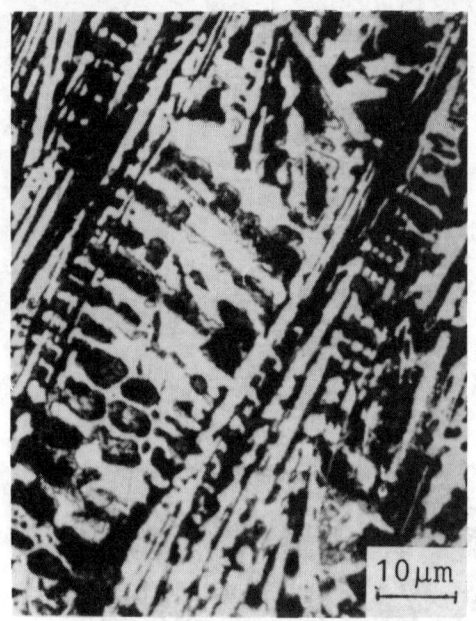 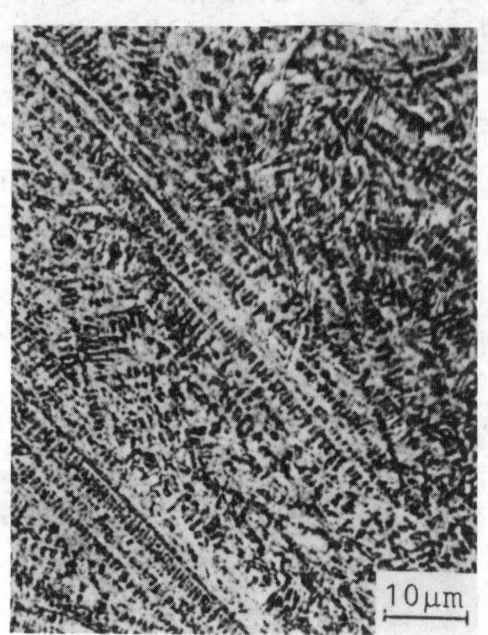

Fig. 2. Variation of microstructures with ε, (same material) (a), (b) increasing quenching rate

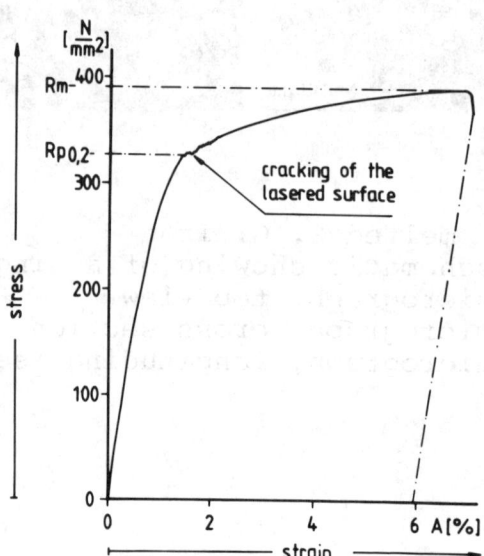

Fig. 3. Typical stress-strain curves for ferritic S.G. iron in the 'as lasered' condition.

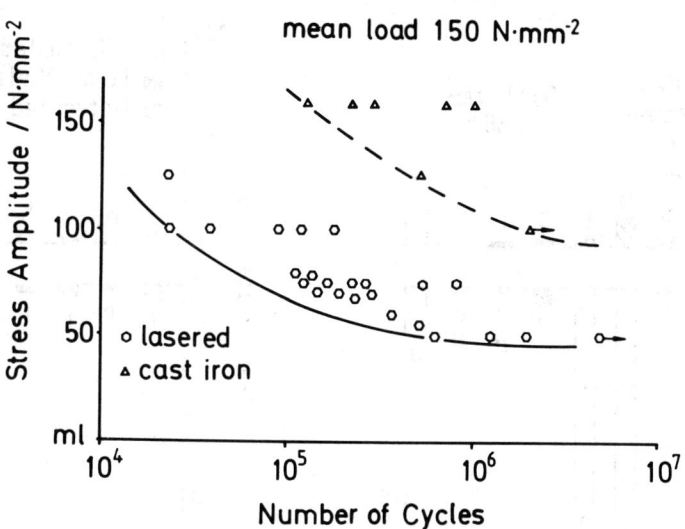

Fig. 4. Pull-pull fatigue behaviour of untreated and laser melted ferritic S.G. iron with a mean stress of 150 N/mm^2 and alternating stress σ_a applied with a frequency of 10Hz. The superimposed alternating stress is plotted against the observed life-time.

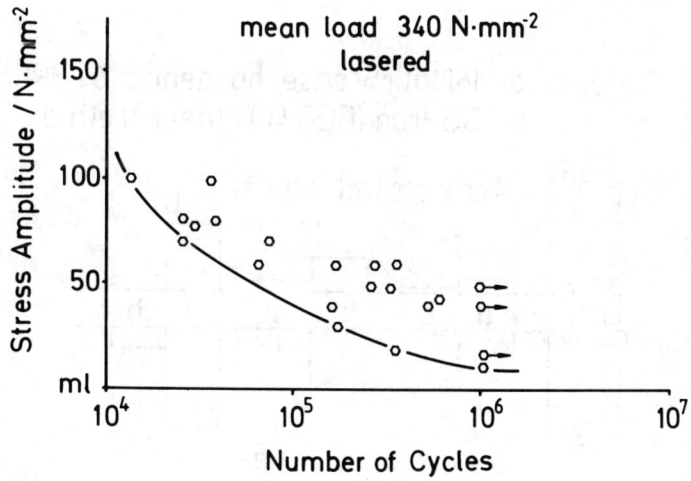

Fig. 5. As Fig. 4 for specimens with ≅30% of cross section laser melted the laser treatment was carried out under a He gas jet and the samples were annealed at 240°C for 2 hrs and then ground. The mean load was 340 N/mm^2 which is approximately the yield stress of the untreated S.G. iron.

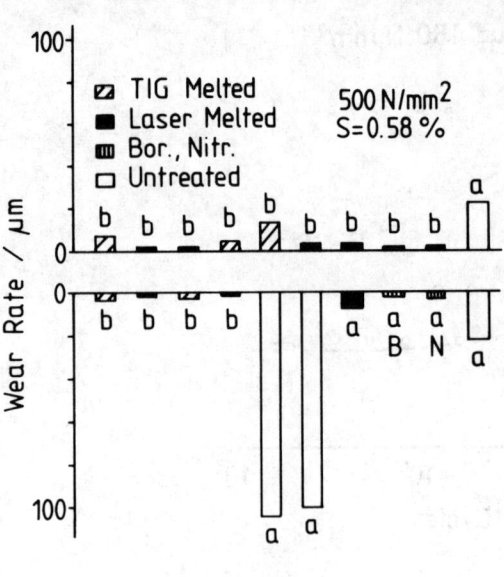
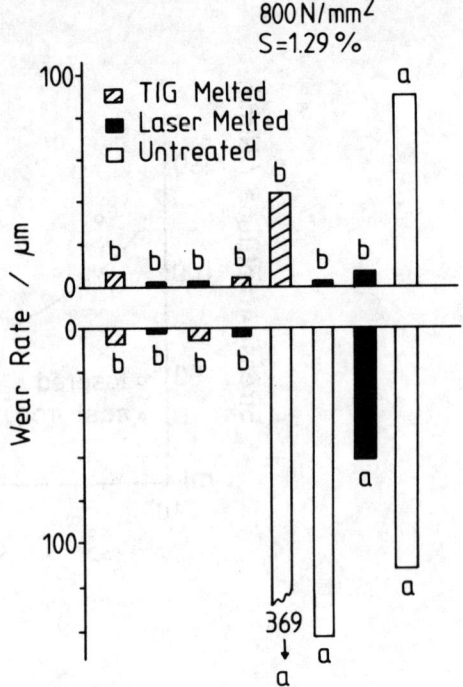

Fig. 6. Wear properties of lasered S.G. iron compared with those of various steels, with and without surface treatments. The applied stress and the slip between the rolls are given in the figures. In Fig. 6 a-c the letter a indicates a carbon steel with 0.6 w% C, the letter b represents surface melted S.G. iron and the letter c is used for case hardened 16MnCr5 (64 HRc).

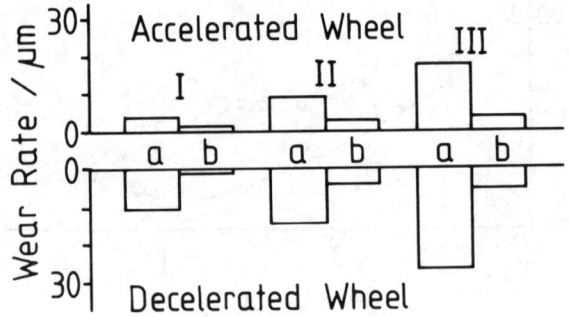

Fig. 7a.

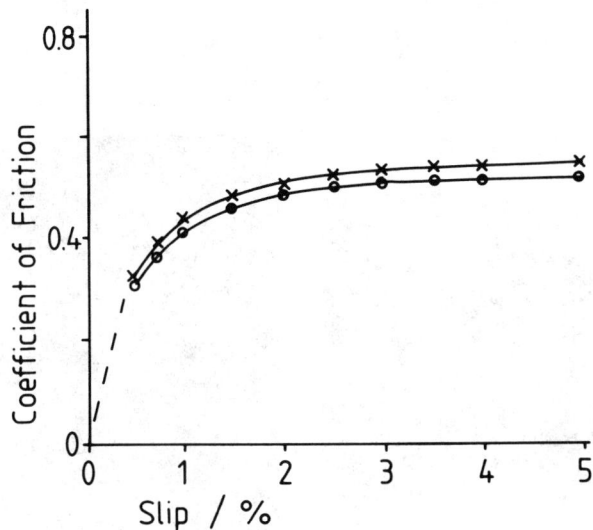

Fig. 7. Wear properties and coefficient of friction
a) wear rates, I = 500 N/mm^2 and 200 rounds/min.
   II = 500 N/mm^2 and 1000 rounds/min.
   III = 1000 N/mm^2 and 200 rounds/min.
b) coefficient of friction
   x stands for lasered S.G. iron
   o stands for case hardened 16MnCr5 (64 HRc)

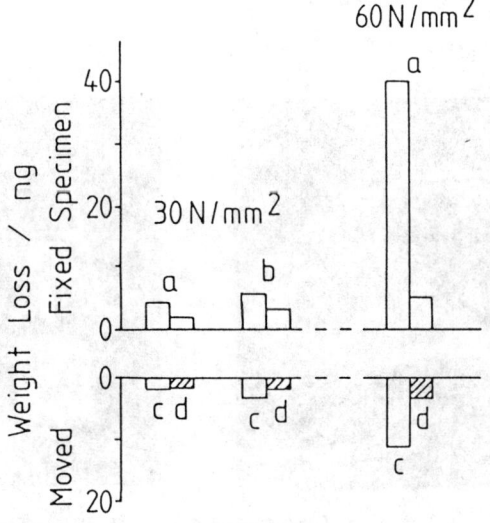

Fig. 8. Wear test under lubrication. The letter c represents hardened 16CrNi6, the letter d the laser melted S.G. iron. The letters a and b represent two different copper alloys. The testing velocity was 80 mm/s.

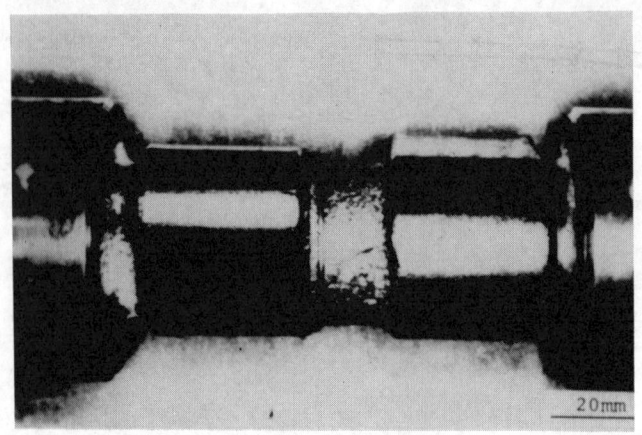 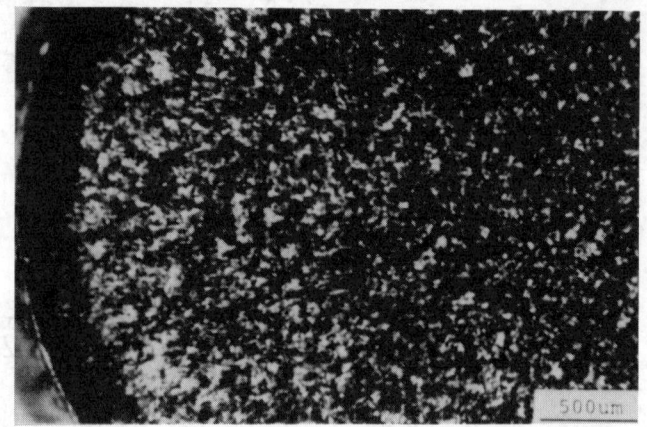

Fig. 9.   Laser melted camshaft
a) macrograph
b) micrograph of cross section

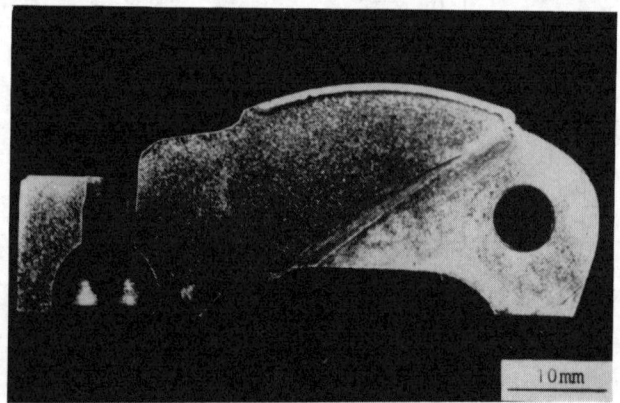 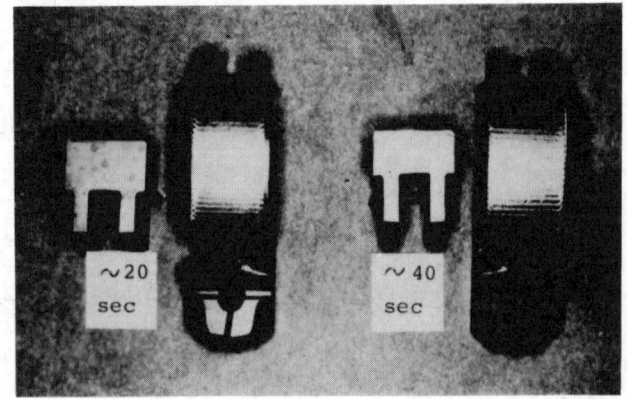

Fig. 10.   Laser melted rockerarm
a) longitudinal section
b) cross sections and top views

Laser hardening of a 12%-Cr steel

M. Roth
Brown Boveri Research Center, Switzerland
and
M. Cantello
RTM Institute, Italy

ABSTRACT

In this work the laser hardening of a turbine blade material, 12 %-Cr steel (X22CrMoV121), was evaluated. The laser treatment was performed with a 15 kW-AVCO Laser at R.T.M. Institute (Italy). Due to variation of the laser processing parameters power and interaction time, different hardness profiles could be obtained. Hardening with multiple passes showed no significant drop of the hardness between adjacent passes. The microstructure of the hardened layers was characterised by light-, scanning- and electron-microscopy. The hardness increase depends on the high carbon concentration in the matrix, the very fine lamellar martensite and the high dislocation density. By melting the surface with the laser, further increase in hardness is possible. In this case, alloying with carbon takes place. The carbon stems from the graphite layer which was applied in order to increase the absorptivity of the material.

A mathematical model for the heat transfer during laser processing was applied in order to evaluate the relevant laser processing parameters, power and interaction time (beam speed), for a desired hardness depth. The experimental check of this method shows that the processing parameters of the laser treatment compare favorably with the theoretically predicted ones.

1. INTRODUCTION

In steam turbines the erosion of last-stage turbine blades by water droplets is well known [1]. Therefore the outer part of the leading edges of these turbine blades made from 12 %-Cr steel is hardened, e.g. by induction heating. In this work the principal possibility of laser hardening a 12 %-Cr steel (X22CrMoV121) was evaluated. Most of the published work on laser hardening has been performed on carbon steels, low alloy steels, tool steels and cast irons [2-4], whereas laser heat treating of martensitic stainless steels has not been investigated so extensively.

2. HIGH POWER LASER PLANT

The laser used for the surface treatment is installed at the R.T.M. Institute in Vico Canavese (Italy).

This laser, made by AVCO (U.S.A.), has a continuous power output of 15 kW, and it is equipped with two working stations (Fig. 1): the first one is equipped with a highly focusing system and it is used for welding and cutting operations, while the second station is equipped with a medium and low focusing system for heat treatment and hard facing operations.

In heat treatment operations, optical systems are used which can uniformely spread the laser energy over the area to be heat treated. A high uniformity of power density is required in order to avoid hot spots which can cause undesired surface remelting or non-uniform depth of treatment.

The most common device used consists of a high frequency scanning (125 and 400 Hz) system for the laser beam. The vibration of the two mirrors is generated electrodynamically. The scanning amplitude can be easily varied. Also, the oscillation, being sinusoidal, causes a longer persistence of the beam along the lateral sides of the treating area which results in a non-uniform and higher power density concentration.

The alternative optical system consists of a composite mirror made by an array of 7 x 7 mm^2 flat mirror units arranged on a spherical surface: each one of them reflects a part of the beam over the area being treated. In this way a beam integration is generated. The energy distribution over the irradiated area is sufficiently constant. However, the system is of limited flexibility because the irradiated surface depends on the dimensions of each mirror composing the beam integrator.

3. EXPERIMENTAL RESULTS

Plates (100 x 50 x 10 mm) of martensitic stainless steel (X22CrMoV121) were hardened by single and by multiple passes using a high frequency scanning system. In order to increase the absorptivity of the metal, an absorption coating (graphite) was applied.

Due to variation of the laser processing parameters, power and interaction time (beam speed), different hardness profiles could be obtained. The hardness of the base material (tempered martensite) was 300 HV. Due to the laser treatment, the hardness was increased to values between 500 and 600 HV. The depth of the hardened layers varied between 0.7 and 1.4 mm (see table 1). Figure 2 gives an example of such a hardness profile; at the top of the diagram the corresponding structure of the heat treated zone is shown. The structure of the base material and of the laser hardened zone was investigated more closely by transmission electron microscopy. In the base material elongated and globular carbides of type $M_{23}C_6$ are present (Fig. 3 a and b), which are characteristic of the tempered martensite structure. The laser treated zone does not contain these carbides at all (Fig. 4 a and b). These pictures show also the extremely high dislocation density and the very fine lamellar martensite (see also the micrograph of Fig. 2). These structures have also been investigated by scanning electron microscopy: numerous carbides are visible in the untreated material (Fig. 5 a) whereas the irradiated layer is completely free of carbides (Fig. 5 b). Thus the hardness increase depends on three different mechanisms:

- high carbon concentration in the matrix
- high dislocation density
- fine lamellar martensite structure

The high dislocation density results from thermal stresses caused by the steep temperature gradient when the laser beam is moved along the surface. However, in spite of these stresses no cracks - either on the surface or in the bulk - could be found. Residual stresses were measured by X-ray diffraction: due to the high dislocation density in the heat treated zone, an exact determination of residual stresses was not possible. However, measurements at the surface and to a depth of 0.2, 0.4 and 0.6 mm, yield values which are predominantly negative.

Laser hardening with multiple passes showed no significant drop of the hardness between adjacent passes (see Fig. 6). A slight tempering effect can be seen when measuring the microhardness at a distance of 0.3 mm parallel to the surface (see Fig. 7).

At a lower critical value of the beam speed the interaction time is long enough to produce melting of the surface. In laser hardening usually this effect should be avoided, but in this case we note that an extremely high hardness depth could be achieved (2.8 mm), and that in the molten layer (0.4 mm) a further increase in hardness took place (see Fig. 8). In the corresponding micrograph a fine network of carbides at grain boundaries is visible, which is only present in the molten zone. The additional carbon obviously stems from the graphite coating. In this way a combination of laser surface alloying and laser hardening is possible.

4. LASER-MATERIAL INTERACTION IN HEAT TREATING PROCESSES

In this chapter a theory [5] is briefly described by which laser processing parameters power and interaction time (beam speed) for a desired hardening depth can be roughly estimated. This method has been checked by comparing the experimental data of the laser hardening of 12 %-Cr steel with the theoretical predicted ones for this material.

During irradiation the metal surface partially absorbs and partially reflects the laser beam. Special coatings increase the absorption coefficient from very low values (0.1 - 0.15), which are typical for ferrous materials, to higher values, e.g. 0.5 to 0.8.

The determination of the absorption coefficient is complex, mainly because laser heat treatment results in combustion phenomena and evaporated gases above the surface. These effects cause variations of the absorption coefficient. Nevertheless, we can rely on the value 0.6 which is the medium absorption coefficient, measured by a calorimeter as results from many experiments on specimens coated by graphite [6].

The absorbed laser energy is converted into thermal energy (entering thermal flux = F_o), which flows within the material according to the normal thermal laws. Within the approximation, such as operating over a very wide area compared to the depth of treatment, a onedimensional mathematical model can be used for the investigation of thermal flow. Under the assumption that the entering flow F_o and thermal parameters of the material are constant during the whole interaction time τ and disregarding the latent heat of the structural changes, the temperature versus time for different thicknesses z is represented by the following Fourier equation [7]:

$$T(z,t) = \frac{F_o}{K} \, 2\sqrt{\alpha\tau} \, \text{ierfc} \, \frac{z}{2\sqrt{\alpha\tau}}$$

where K is the thermal conductivity and α is the thermal diffusivity.

When the radiation ends, at time $t = \tau$, the maximum temperature in the surface layer ($z = 0$) will reach the value:

$$T(0,\tau) = \frac{F_o}{K} \, 2\sqrt{\alpha\tau} \cdot \frac{1}{\sqrt{\pi}}$$

For a correct and practical simple use of these equations the following normalizations are useful [5]:

$T_n = \dfrac{T(z,t)}{T(0,\tau)}$ Temperature related to the maximum one at the surface-layer ($z = 0$)

$t_n = \dfrac{t}{\tau}$ Time related to interaction time τ

$Z_n = \dfrac{z}{D_\tau} = \dfrac{z}{2\sqrt{\alpha\tau}}$ Depth related to the length of diffusivity at the time τ

After such normalizations, the equation governing the heat propagation appears much more compact. It is valid for the various operative conditions (F_o, τ) as well as for various materials:

$$T_n = \sqrt{t_n} \cdot \sqrt{\pi} \, \text{ierfc} \, \frac{Z_n}{\sqrt{t_n}}$$

From this equation, the diagram of Fig. 9 is derived (for the heating phase).

These normalized equations are the basis for easy-to-use plots which can be applied for the heat treatment of specific materials. Such a plot has been established for the material parameters of the 12 %-Cr steel (X22CrMoV121) (see Fig. 10). The following values have been taken:

- thermal conductivity $K = 0.027$ W/mm°C
- thermal diffusivity $\alpha = 5.8$ mm²/s
- transformation temperature $T_{AC} = 900°C$.

In the first quadrant of Fig. 10 the normalized temperature T_n is plotted as a function of various depths Z_n at the interaction time τ. If the ratio between the transformation temperature and the maximum surface temperature (e.g. $T(0,\tau) = 1300°C$) is fixed, we obtain the thickness Z_n in which the transformation temperature is exceeded and, consequently, hardening can take place. In the second quadrant of Fig. 10 for a desired hardening thickness z (e.g. 1.0 mm) the interaction time τ can be determined. Lastly, in the third quadrant, we obtain for the assumed maximum surface temperature the required thermal flux F_o.

This diagram allows an easy determination of the operating parameters (entering thermal flux F_o and interaction time τ) that can be used in order to obtain a treated layer whith a desired thickness, while the surface temperatures are maintained at acceptable values.

This method can be easily checked by using the experimental data of the laser treatment of the 12 %-chromium steel. In Table 1 data for some heat treated specimens are listed. The specific power, measured in W/mm², is defined as the laser power at the specimen surface divided by the area of beam scanning (10 x 10 mm²). The interaction time τ is calculated from the speed of the laser beam and its length along the direction of movement (10 mm). The hardening depth (hardness higher than 400 HV) has been determined from the microhardness profiles which have been taken in the middle of a laser track.

When inserting the hardness depth z and the corresponding interaction time τ in the second quadrant of Fig. 10, one can determine - when going "up" to the first quadrant - the maximum surface temperature out of the relation $T_{AC}/T(0,\tau)$ (for all tests the transformation temperature T_{AC} has been taken as 900°C). Going now - from the second to the third quadrant - to the curve of the above determined maximum surface temperature, we obtain the thermal flux F_o for the different hardening tests (F_o (theoretical) in Table 1). These F_o-values can be compared with the specific power multiplied by the absorption coefficient ($\alpha = 0.6$). We note that the correlation of these values is quite good (see last column in Table 1).

specific power F_o [W/mm²]	interaction time τ [sec]	depth of hardening z [mm]	$F_o \cdot \alpha$ ($\alpha = 0.6$) [W/mm²]	F_o (theoretical) [W/mm²]	$\dfrac{F_o \text{ (theoretical)}}{F_o \cdot \alpha}$
18	1.09	0.7	10.8	10.8	1.0
25	1.00	1.4	15.0	15.7	1.05
22	1.50	1.4	13.2	11.6	0.88
21	1.62	1.2	12.6	10.0	0.79
25	0.63	0.6	15.0	15.0	1.0
24	0.86	0.8	14.4	13.4	0.93
19	1.09	0.9	11.4	12.1	1.06
25	0.63	0.8	15.0	16.6	1.10
29	0.90	1.1	17.4	15.0	0.86
24	0.80	0.9	14.4	15.1	1.05

Table 1: Laser processing parameters and theoretical determined thermal fluxes according to Fig. 10 (F_o (theoretical)).

5. CONCLUSIONS

Laser hardening of a martensitic stainless steel was performed successfully. The heat-treated layers were free of cracks. By variation of the laser parameters, power and interaction time, the hardness depth can be controlled very well. In order to evaluate these relevant laser processing parameters, a heat transfer calculation was applied. We have demonstrated, in particular, that the experimental data of the laser treatment compare favorably with the theoretically predicted ones. However, one has to keep in mind that there are some uncertainties in determining the exact values for the absorption coefficient and the maximum surface temperature. To conclude, it is worthwhile to apply the diagram of Fig. 10 in order to obtain a rough evaluation of the laser processing parameters, power and interaction time, for a desired hardness depth.

This work comprises mainly the study of the materials structure after laser hardening a planar surface. Heat treating a component, e.g. the leading edges of turbine blades, requires further investigations, such as laser processing of nonplanar surfaces.

Another important factor is the susceptibility for stress corrosion cracking (SCC) which usually increases at higher hardnesses. Therefore, with regard to the technical application, it would be of interest to determine if there is a difference in SCC-behaviour between laser hardened and conventionally hardened material.

6. ACKNOWLEDGEMENT

The skilled assistance of Mr. F. Pasquini (R.T.M.) with the laser testing and of Mrs. G. Keser (BBC) with the metallographic investigations is deeply appreciated.

7. LITERATURE

[1] Svoboda, R., Faber, G. "Erosion-Corrosion of Steam Turbine Components", in: Proceedings of the 8th Int. Brown Boveri Symposium on Corrosion in Power Generating Equipment, pp. 269-298, Plenum Press, New York, 1984.

[2] Metzbower, E.A. Source Book on Applications of the Laser in Metalworking. American Society for Metals, Ohio, 1981.

[3] Oakley, P.J. "Laser Heat treatment and surfacing techniques - a review". The Welding Inst. Res. Bulletin, pp. 4-10 (January 1981).

[4] Mazumder, J. "Laser Heat Treatment: The State of the Art". Journal of Metals, pp. 18-26 (May 1983).

[5] La Rocca, A.V. "Material processing by high power laser". Acts of Fifth International Symposium on Gas Flow and Chemical Lasers. Oxford, England, August 1984.

[6] Cantello, M., Pasquini, F., Rudilosso, S. and Canova, P. "Use of the Mathematical Models for Laser Heat Treatment of Thin Materials". Acts of Fifth International Symposium on Gas Flow and Chemical Lasers. Oxford, England, August 1984.

[7] Carslaw, H.S. and Jaeger J.C. Conduction of Heat in Solids. Oxford University Press, London, 1959.

Fig. 1: The 15 kW AVCO Laser installed at R.T.M. Institute.

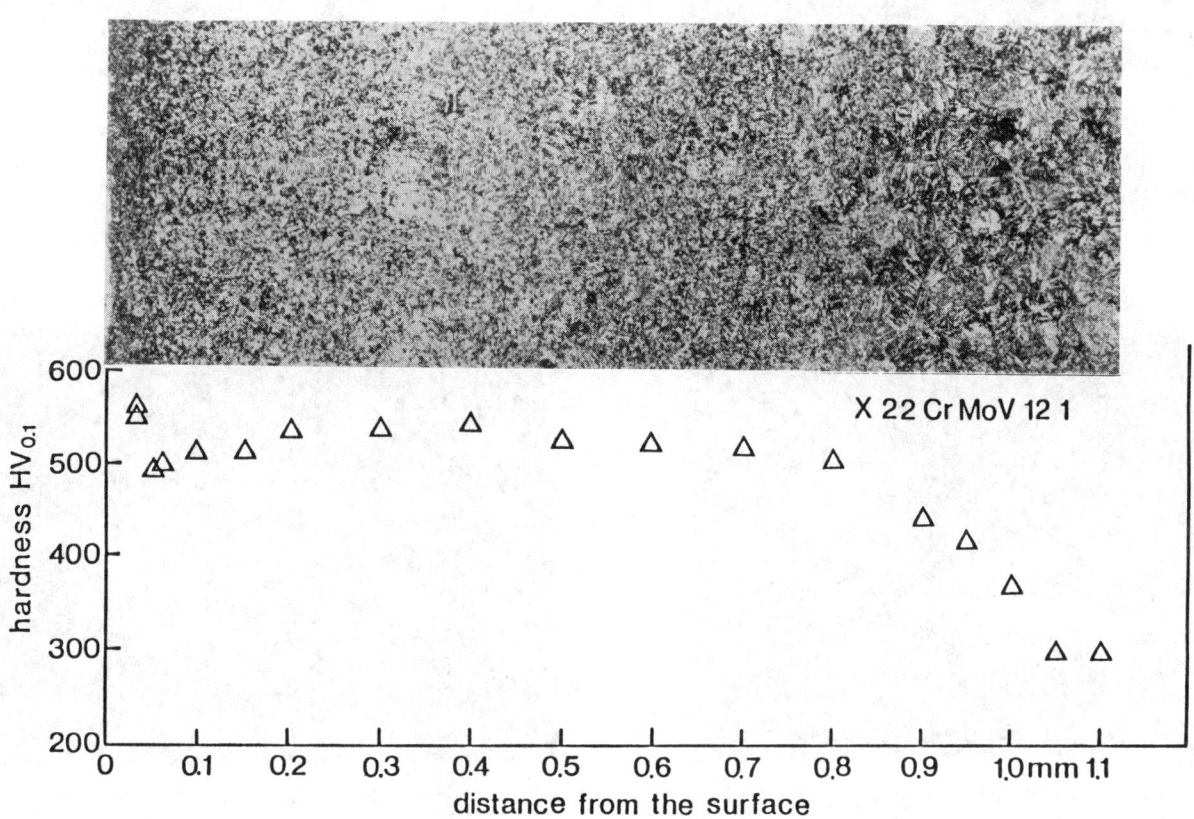

Fig. 2: Laser hardness profile with corresponding microstructure (optical micrograph).

 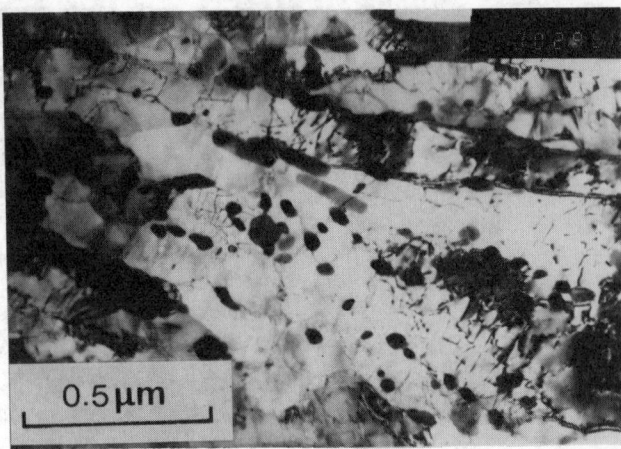

Fig. 3a and b: Microstructure of the base material: the tempered martensite is characterised by elongated and globular carbides of type $M_{23}C_6$ (TEM-micrograph).

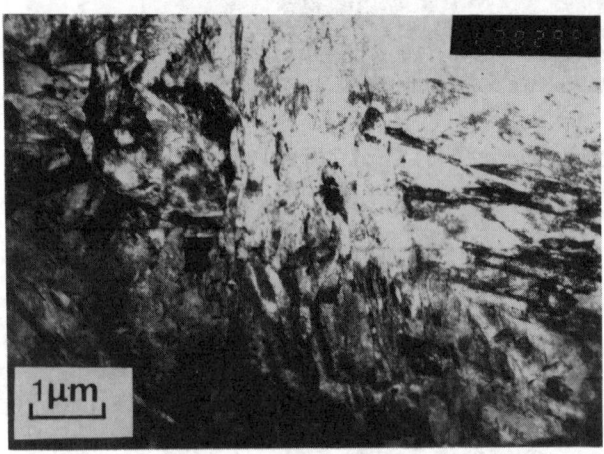 

Fig. 4a and b: Microstructure of the laser hardened material: no carbides are present; the high dislocation density and the very fine lamellar martensite are visible (TEM-micrograph).

 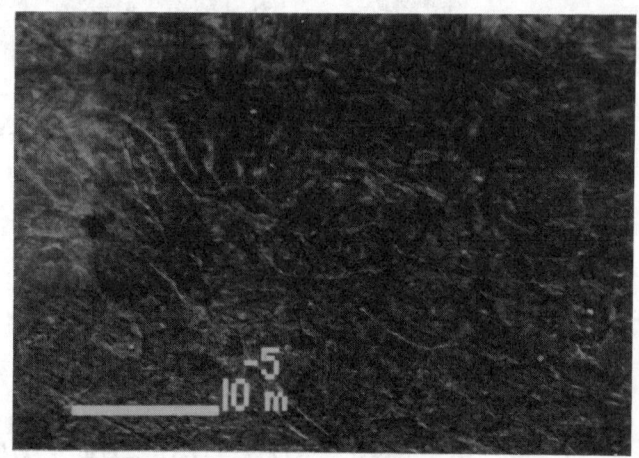

Fig. 5a: SEM-micrograph of the base material, showing numerous carbides.

Fig. 5b: SEM-micrograph of the laser hardened material, showing a structure free of carbides.

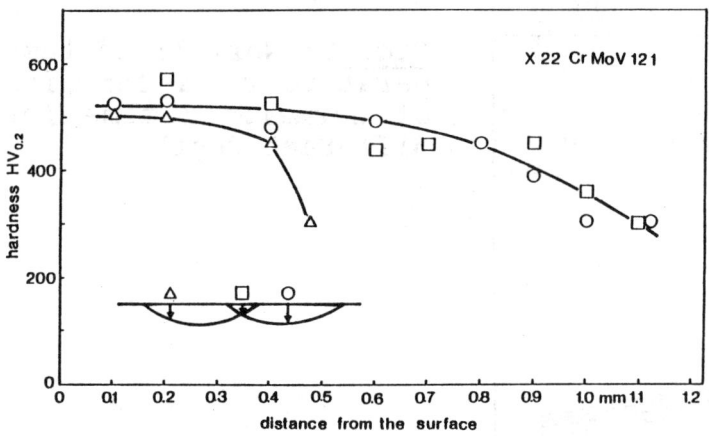

Fig. 6: Laser hardening by multiple passes: no significant drop of the hardness between adjacent passes.

Fig. 7: Slight tempering effect in the left, previously made track.

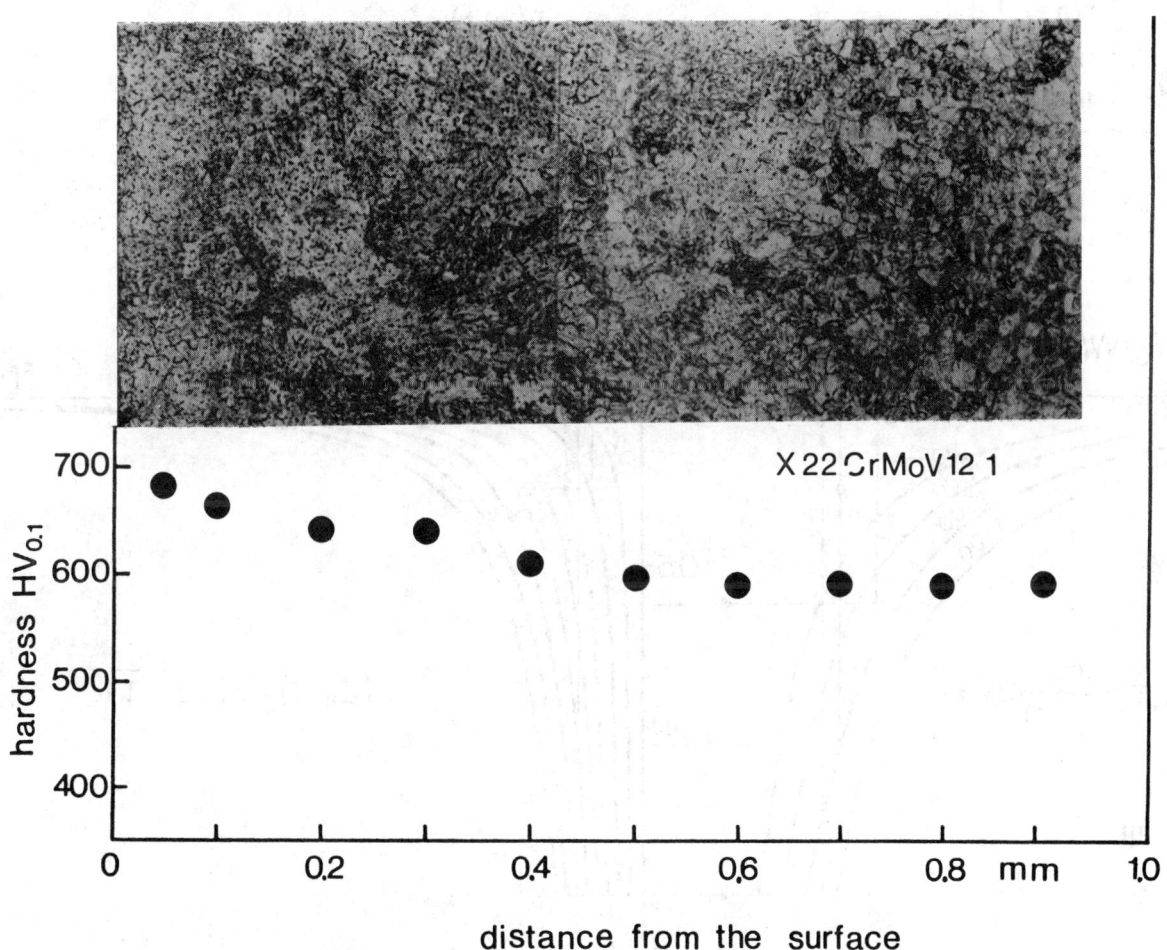

Fig. 8: Hardness profile of a melted and hardened surface: the further increase of the hardness in the outer, molten layer (z ~ 0.45 mm) is due to precipitation of a fine network of carbides (optical micrograph).

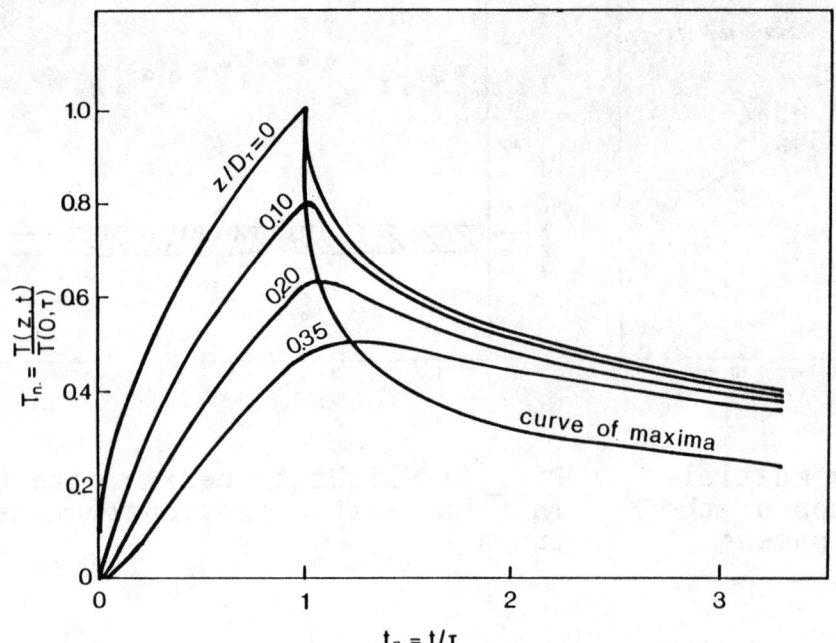

Fig. 9: Normalized temperature as a function of normalized time for different depths.

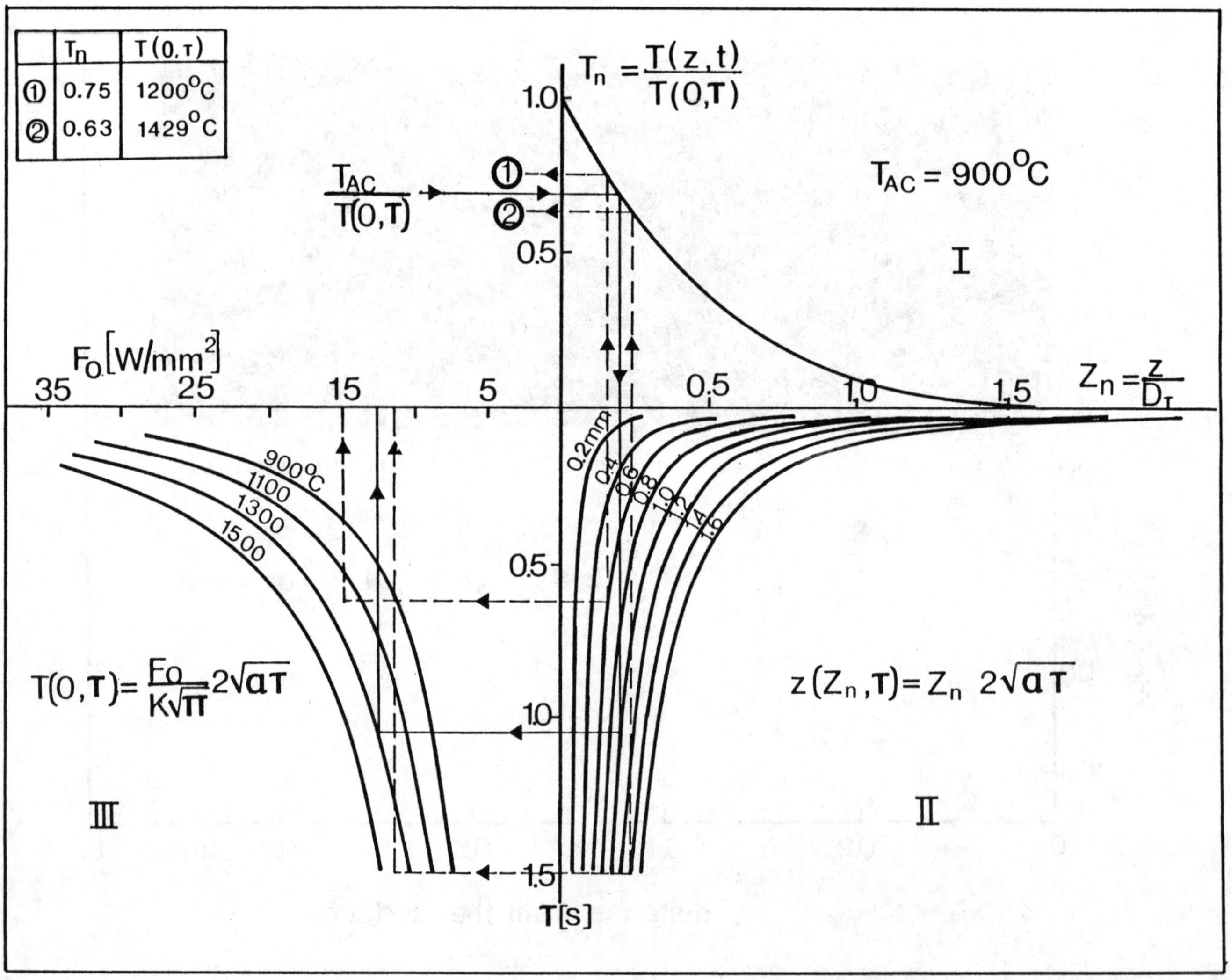

Fig. 10: Diagram for the determination of the laser processing parameters interaction time τ and specific power F_o for a desired hardening depth z (full line). The dotted lines correspond to the experimental check of this method.

NON-DESTRUCTIVE TESTING AND OPTICAL SENSORS

Process monitoring of high power CO2-lasers in manufacturing
W. König, F. U. Meis, H. Willerscheid
and
Cl. Schmitz-Justen
Fraunhofer-Institut für Produktionstechnologie (IPT), West Germany

Essential parameters affecting the process quality of laser beam machining as imposed by workpiece applications are presented and discussed. The method of transformation hardening as an example illustrates the influence of fluctuations in the beam power profile upon the quality result of this operation.

Different techniques to measure and record characteristical properties of the laser beam are to be compared. These are differentiated by the types of detectors used, the procedure of measurement and the equipment design. The prototype of a beam analysing system developed with special regard to the correlation between beam characteristics and quality of heat treatment is to be presented. For a commercially available scanning unit the precedure of continuous processing and evaluation of data during heat treatment of the workpiece is described giving major attention to the implemented process monitoring functions.

1. APPLICATIONS AND OBJECTIVES OF THE LASER AS A MACHINE TOOL

Advanced machining processes are employed within the industrial production to attain a higher product quality at a competitive effort - even for the manufacturing of small batches. Moreover, it is the intention to enhance productivity and economic efficiency. These considerations increasingly include the application of laser beam machining. Ever since its first appearance on the production line in the early seventies the high power CO_2-laser has been utilized for a large variety of machining processes. Among these laser beam cutting - mainly at a power level up to 1.5 kW - presently is the process most widely established in industry. For the time being laser welding is considered to have the highest potential for large-scale production applications in the immediate future. The purpose is to achieve deep-penetration welds of a superior quality, which, however, requires a laser power of more than 10 kW. Several different processes of laser heat treatment, transformation hardening, remelting, and surface alloying are now going through a final development toward the industrial application on suitable parts and structures. This basically implies a laser power of less than 6 kW.

The specific and advantageous characteristics of the laser are substantial for its use as a thermal tool in production engineering: As the energy input can be precisely controlled by power, time and area of interaction, it allows the irradiation of limited zones on the workpiece surface. This provides the instrument of a selective heat treatment at such zones that require a major improvement of material properties due to extreme stress in operation. As for cutting and welding the high power density will cause a small heat affected zone. The lack of process forces in combination with the high intensity ensures a low workpiece distortion in these processes. Optical components to guide and shape the laser beam give the opportunity to cover zones of complex geometry or difficult access as it is put into effect when cutting three-dimensional contours or hardening bore hole walls. The characteristics shown above clearly demonstrate the potential of the laser not exclusively as a substitute for conventional processes but much more as a technologically and economically advantageous complement in certain fields of application.

To fulfill the industrial requirements of high production quality and dependability several tasks have to be accomplished in the development of the next generation of laser sources and processes:

- Optimization of the laser source in terms of higher output power and improved beam quality.
- Further development of beam guiding and shaping components to facilitate flexible laser machining.
- Development of appropriate process analysing and observing devices to assess the beam quality and interaction between laser beam and workpiece.
- Definition and measurement of process specific parameters, compilation of data bases.
- Improvement of peripheral units, enforcement of safety standards.

These assignments can be summed up by the objective of monitoring the laser machining process, based on a beam analysis, a process observation and evaluation of the workpiece quality. Such a system could later be incorporated into a closed-loop control for the application of a laser machine tool in automated manufacturing.

2. INFLUENCE OF PROCESS PARAMETERS UPON THE MANUFACTURING QUALITY IN LASER MATERIAL PROCESSING

If the laser machining process is regarded in its relation to the laser source on one hand and the workpiece, respectively its handling on the other hand, a multitude of variable parameters appears, of which the most significant ones are shown in figure 1. The complexity of the process is particularly characterized by the interdependence and interaction of these parameters. As for the dominant beam properties such as mode structure, power, dimensions, polarization and divergence, they are not only affected by the laser source and the optical elements to guide and shape the beam but also by reactions from the workpiece. Beside those parameters related to the laser and the handling system, changes in the physical and chemical material properties exercise an essential influence upon the machining quality. An example is here the absorptivity affected by wavelength, power density and temperature.

Concerning the martensitic transformation hardening by laser light the depth and width of the hardened zone depend upon power density, mode structure, traverse rate, surface absorptivity, angle of incidence, heat conductivity, and basic temperature of the workpiece. In figure 2 the effect of different parameter irregularities is presented. A detuned cavity and corresponding mode structure, as seen in the upper part of the figure, result in an asymmetrical cross sectional geometry of the hardened track. The comparison with a hardened zone at optimized multimode structure shows the difference in depth and width. Due to local power peaks melting of the surface occurs, thus damaging the workpiece. The hardness values in the martensite zone, however, are approximately equal for both conditions.

A further geometrical aberration of the hardened track, yet at optimized mode structure, shown in a longitudinal section, is caused by the low heat transfer from material to environment at the edge of a workpiece. Such a local accumulation of heat again increases the hazard of surface melting.

Long-term operations of laser systems over some hours also produced differences in width and depth of the hardened zone. These are the results of power and mode instabilities depending on fluctuations of the cooling water temperature, internal heat sources as blowers, pumps, and laser process heat, and external sources like exposure to sun irradiation or heat from adjacent machine tools.

The workpiece characteristics also have a strong influence upon the shape of the hardened track. A 30% increase in hardness depth was obtained for a roughly milled steel surface, compared to a polished surface at use of the same graphite absorption coating. Beyond these microproperties the correct treatment of a real workpiece also depends on macrogeometrical parameters like the angle between the travel direction and the beam axes for a non-circular beam or the angle of incidence between the beam and the normal upon the workpiece surface. This influence is shown in figure 3. The example here is the transformation hardening of a quenched and tempered steel, 42 CrMo 4. It is obvious that incidence at an increasing inclination will result in the growth of the effectively exposed area, and thus in a loss of power density. This reduces the depth of the hardened zone, while the uneven heat dissipation into the base material on both sides causes an unsymmetrical cross section of the track.

3. MEASUREMENT OF LASER MACHINING PROCESS CHARACTERISTICS

Irregularities in the machining quality as discussed in the last chapter which have their correspondents in cutting and welding show the necessity to keep the parameters within a narrow tolerance to ensure a reproduceable manufacturing quality. These comprise the laser beam charactistics as well as such process variables that are specifically attributed to the workpiece. The correlation between results of an online process observation and beam analysis will generate process data which allow the monitoring of the laser machining quality.

Workpiece testing includes the conventional geometrical and metallographic inspection. The intention of this development in the sense of an improvement in manufacturing dependability also is an online assessment, e.g. by the interpretation of magnetic properties in terms of residual stress or microstructure.

Other quality characteristics - the forming of dross and striation for cutting, the surface constitution and temperature distribution for hardening and welding - are observed and recorded by high speed cinematography and thermography. Highly dynamic events which are typical for laser machining due to the rapid generation of heat and quenching require special recording devices. Either the film strip is rotating on a wheel (35.000 frames/s) or a rotating mirror is used for exposure (up to 25 million frames/s). Only these high-speed methods allow the optical analysis of the interaction between the laser beam and the workpiece during machining processes.

A thermography system gives information on the position and geometry of a heat radiator and especially the temperature distribution on a surface. This is qualitatively shown for a hardened track in figure 4. The colorized, respectively isothermal reproductions were generated by computer aided processing of a grey-scale image. The quantitative measurement of surface temperatures without mechanical contact implies the knowledge of those physical coefficients that characterize the radiation properties also at high temperatures. Moreover, errors of measurement have to be compensated which are caused by external heat sources, atmospherical transmission, and nonlinearities within the measuring unit. A qualitative or quantitative assessment of the temperature distribution below the workpiece surface could be possible only by means of a mathematical model using measured surface temperatures and workpiece geometry as preconditions. Such a method, however, is presently not yet known for laser beam machining.

As shown in figures 2 and 3, an insufficient knowledge of the laser beam characteristics and the beam performance is leading to qualitatively nonacceptable workpieces. Assessment of the beam properties is usually done by "burn-ins" into acrylic glass and by methods based upon integral power measurements respectively measurement of a splitted or selected beam portion. While the burn-in is a basically optical check of the mode structure (also see figure 6), all the other methods, of which some are depicted in figure 5, are using detectors which supply an electrical output signal proportional to the amount of power absorbed. The relevant detector types are differentiated by function, noise equivalent power level, sensitivity and rise time. At the 10.6 μm wavelength of the CO_2-laser thermal or quantum detectors are suitable.

In thermal detectors the rise of temperature caused by absorption of laser radiation is transferred into a thermovoltage (thermocouple), a change of resistivity (bolometer) or a charge displacement (pyroelectric detectors). The function of quantum detectors is based upon the internal photo effect. Compared to thermal detectors their rise time is significantly shorter. Due to the high thermal noise level, however, they have to be cooled by liquid nitrogen or helium.

The continuous integral power measurement of a cw-beam up to the multikilowatt-range is achieved by the use of beam choppers or partially transmissive mirrors. As shown before, however, process monitoring requires the online registration not only of the integral laser power but also of the local intensity distribution without any significant loss. The presently known diagnostic systems incorporating local and temporal resolution are based upon splitting of the beam in combination with a cross sectional scanning procedure or the selective deflection of a beam portion by highly reflective elements of small size (e.g. needles, mirror strips). The latter system is used in some commercially available units. The "sensor head" of the laser beam diagnostic system shown in figure 5 basically consists of a spoke wheel. The 8 reflecting spokes are set up axially shifted and inclined by 45°. If 3 detectors are used this gives a 3 x 8 scanned lines over the beam cross section for a sufficiently large beam diameter.

The quasi-continuous measurement of the Laser Beam Analyser is put into effect by the deflection of a beam section upon detectors by a thin reflecting needle rotating normally to the beam axis. The special configuration of the detectors as depicted in figure 6 allows a simultaneous registration of the beam intensity profile in two Cartesian axes. Figure 6 represents a comparison between an x/y intensity profile averaged over 5 individual measurements and a burn-in into acrylic glass. A transversal flow CO_2-laser was used at 2.000 W integral output power. The scanning procedures in commercial analysing systems are fixed in terms of local and temporal resolution. The design of both devices discussed above does not allow the registration of rapid mode fluctuations - as caused by mechanical or electrical instabilities - at any given position in the cross section of the beam. Even modifications of these units such as additional detectors, needles or spokes, and extended speed ranges will not give a significant improvement due to the error or measurement.

Thus a method based on selective deflection was developed to obtain an online registration of the laser beam intensity in a wide range of local and temporal resolution. The function of this "Beam Measurement System" is illustrated in figure 7 and figure 8. 5 highly reflective needles which pass through the beam provide an approximately straight scanning in a plane normal to the beam axis. Figure 7 shows the setup for measurements of high local resolution. The deflected parallel beam segment is focused upon detector "A" by a concave mirror. A rapidly rotating aperture wheel ($n_{aperture\ wheel} > n_{needle\ carrier\ wheel}$) generates straight profile sections of the beam which are presented in a computer simulated illustration for a Gaussian mode. Corresponding profiles of a multimode beam are shown in figure 8. The number of scanned tracks and thus the grade of local resolution is determined by the variable rotating speed of the aperture and the needle carrier wheel. Additionally the intensity profile in travel direction of the needle is measured by detector "D". This is also used as a trigger signal to facilitate computer aided data processing.

Figure 8 shows the setup for high temporal resolution measurements incorporated in the same device. As the needle passes through the beam on the upper side a portion of the beam is reflected into 2 detectors which may travel parallel to the needle axis. If both detectors are positioned opposite to each other two intensity profiles along one axis can be scanned at a time difference of $\Delta t = d_N/v_N$ where d_N gives the needle diameter and v_N the circumferential velocity. This configuration allows registration of mode instabilities up to the kHz range. If both detectors have different axial positions, the independent measurement of two separate longitudinal beam profiles is possible. A measurement of the beam divergence within the same unit is obtained by an appropriate arrangement of detectors C and D at an identical axial position. Precondition, however, is a relatively stable mode structure.
The general alignment of this device is achieved optically by use of a He-Ne aiming laser which is reflected upon alignment plates by the individual needles.

4. EVALUATION OF PROCESS PARAMETERS FOR MONITORING OF LASER BEAM MACHINING

The detector output signals proportional to the intensity distribution are amplified for analogue, or digital display or transfer to a computer for further processing. A computer aided laser beam analysis comprises the calculation and documentation of parameters affecting the machining quality such as mode structure and its difference to a reference mode, integral laser power, and the beam dimensions. Moreover, derivative variables like the power density and position of local peaks are included. Figure 9 exemplarily illustrates the evaluation of the interaction between beam and workpiece affecting the beam intensity. The rotating needle method (Laser Beam Analyser) was used for measurement in the unfocused beam while the workpiece was positioned at different distances to the focus. The output signal of the pyroelectric detectors were digitized and stored by means of a personal computer. The reactive mode modulations caused by interaction with the workpiece are made obvious by the differences between the corresponding intensity signal and a reference mode which was taken without any workpiece. Significant aberrations of the beam characteristics are identified by variably preset difference limits. A complete process monitoring, however, does not only include the determination and checking of beam properties but also of those parameters typical for the individual processes, e.g. surface treatment, cutting of welding. This is summarized in figure 10.

Those input values relevant to control a process, like current, voltage, cooling water temperature and mode structure concerning the laser source as well as focal distance, process gas and traverse rate concerning the handling of the workpiece are presently compiled in process data sheets for individual materials and workpiece geometries. These parameter values are generated and optimized in practical tests by correlation between results of geometrical and metallographic investigations, process observation and beam analysis. These process data sheets allow an assessment of aberrations to preset values in industrial applications.

With computerized data processing monitoring functions are carried out in the way of crt or printed messages, alarm or interruption of the machining process depending on the grade of parameter deviation.

Based upon the information of process data sheets and the online monitoring the ultimate step of development will be an adaptive control of laser machining processes as it is envisaged in figure 11. In such a closed loop parameters are adjusted as actuating signals corresponding to the set point deviations measured for the process characteristics. To achieve the goal of high manufacturing quality and dependability it is again important to include laser, workpiece, and process specific characteristics.

5. SUMMARY

The high power CO_2-laser as a machining tool has gained an increasing importance within the thermal processes of cutting, welding, and surface heat treatment. The precondition of continuously high quality and reliability for large-scale industrial applications has to be fulfilled by an appropriate process management. This requires the development of online measurement devices covering laser beam and process characteristics of which the major concepts have been discussed. The correlation of beam and process data with the results of workpiece testing constitutes the core of a process monitoring system. This implies a computerized handling, processing and evaluation of data in anticipation of a further development toward a closed loop process control.

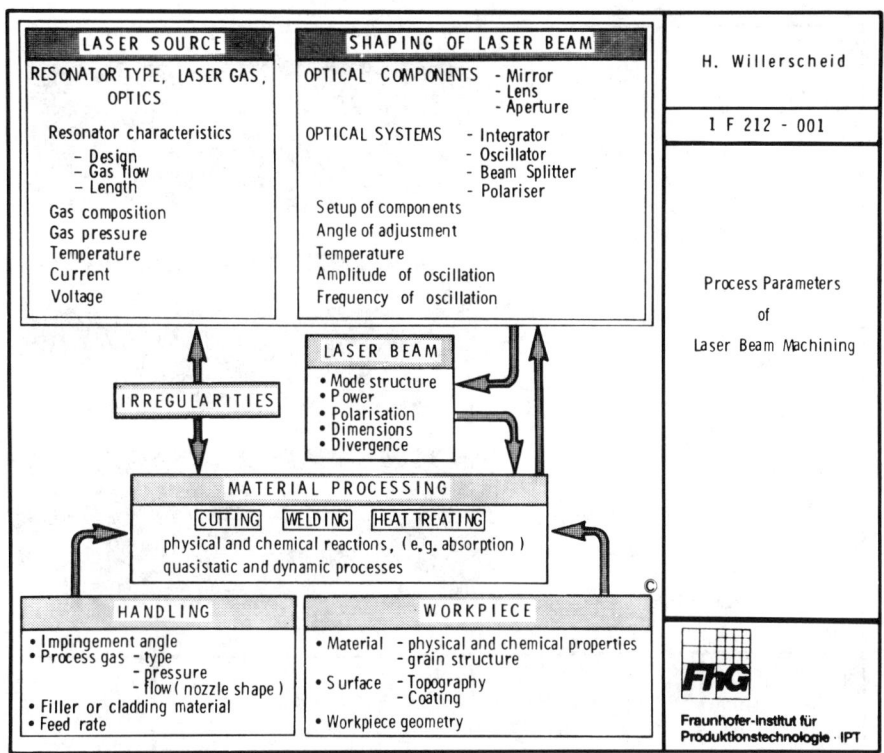

Figure 1

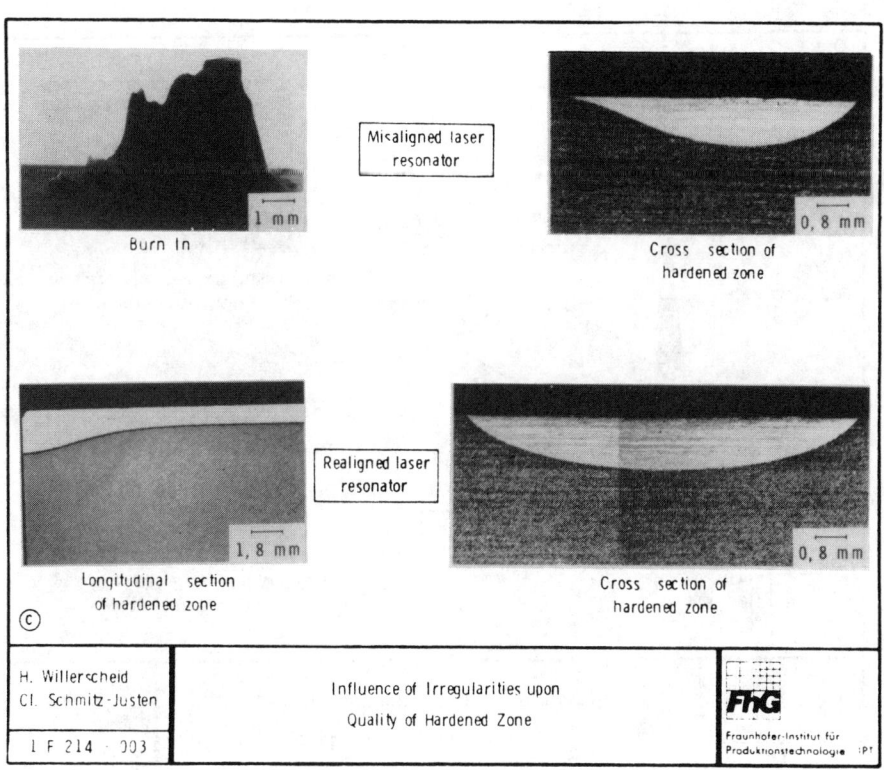

Figure 2

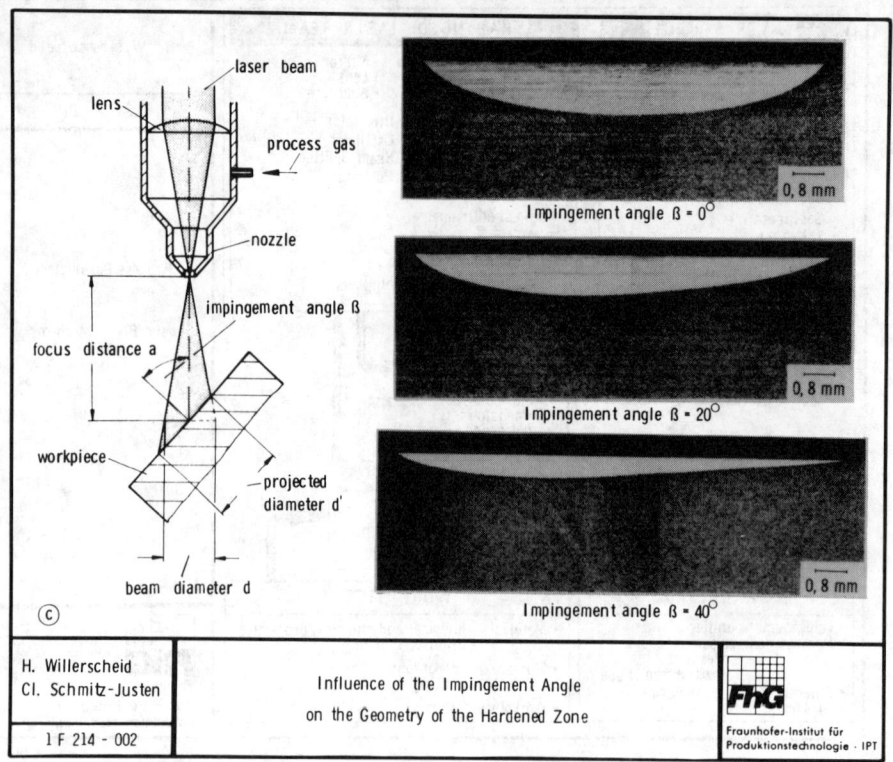

Figure 3

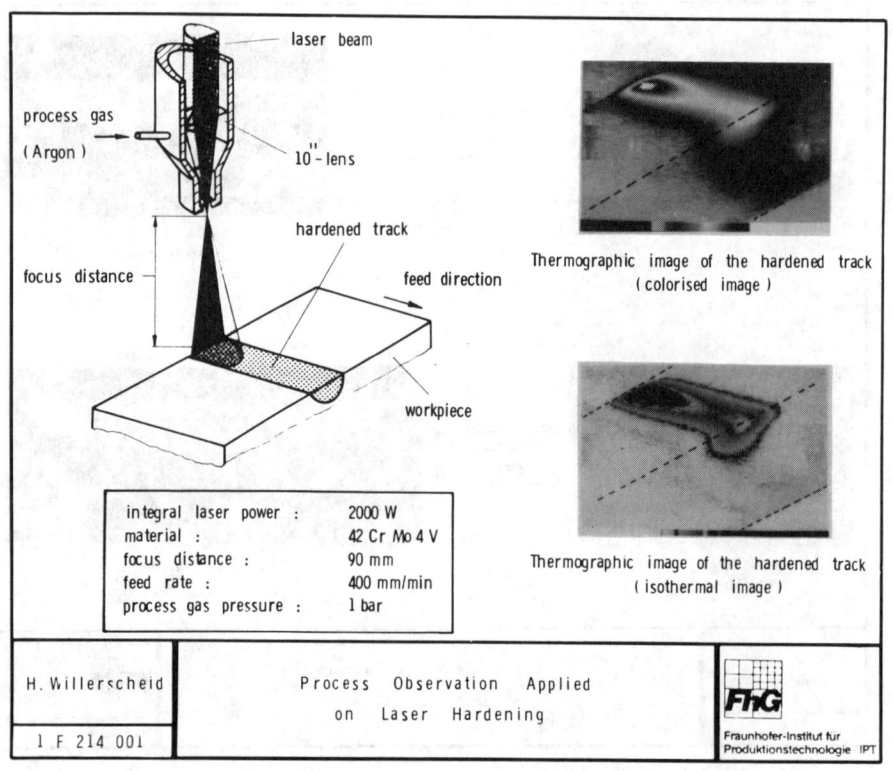

Figure 4

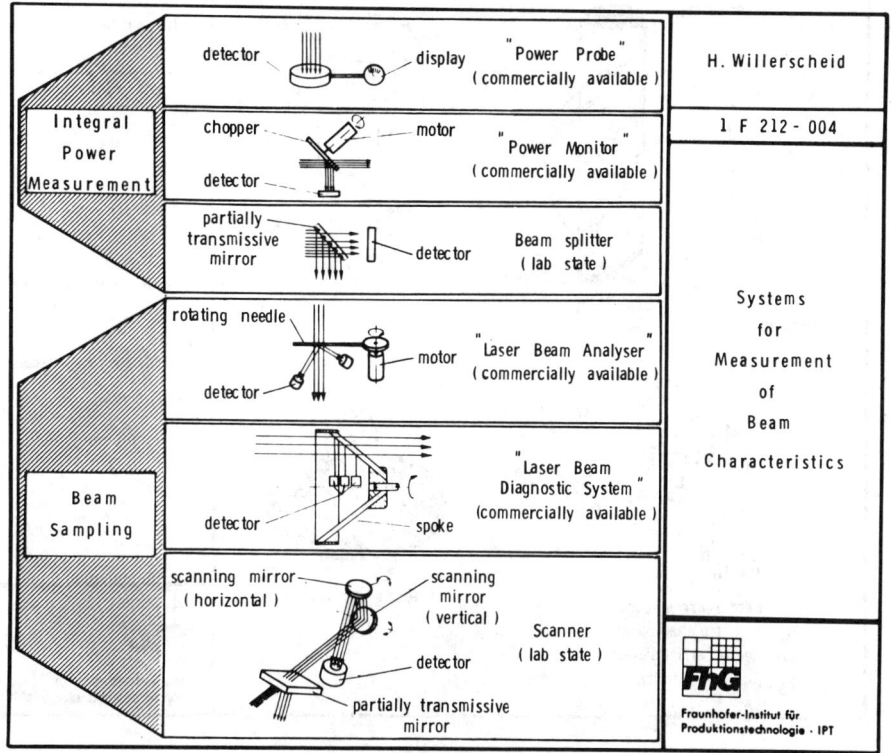

Figure 5

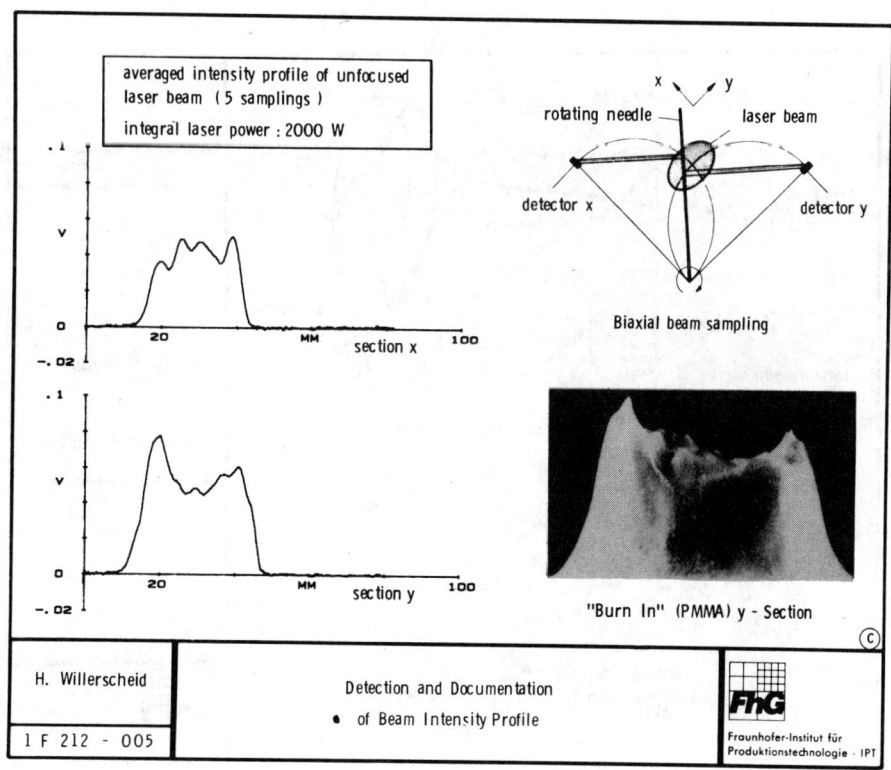

Figure 6

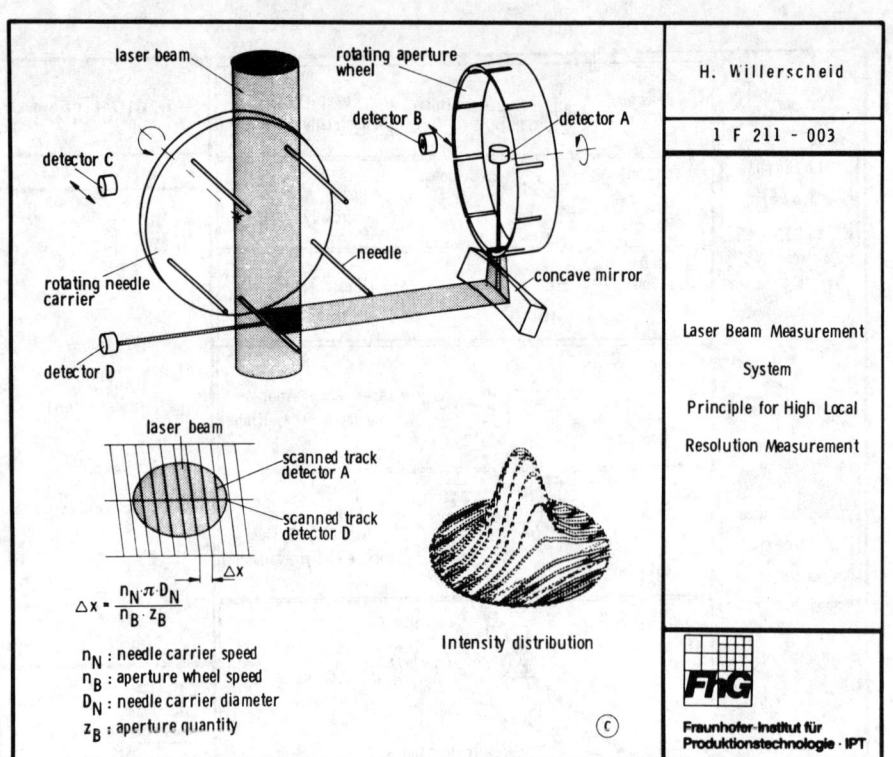

Figure 7

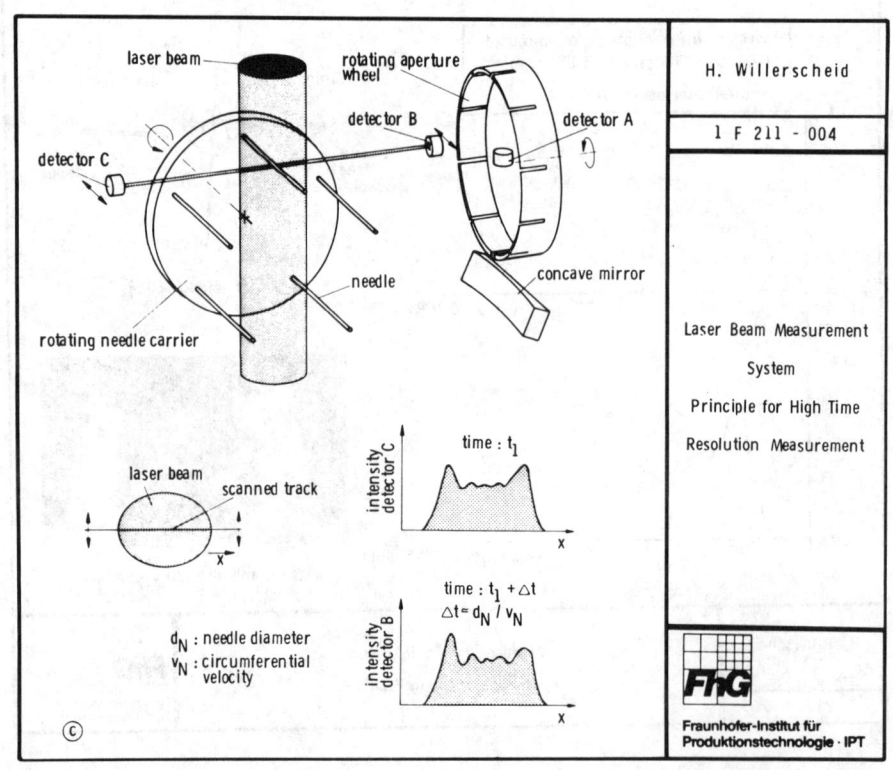

Figure 8

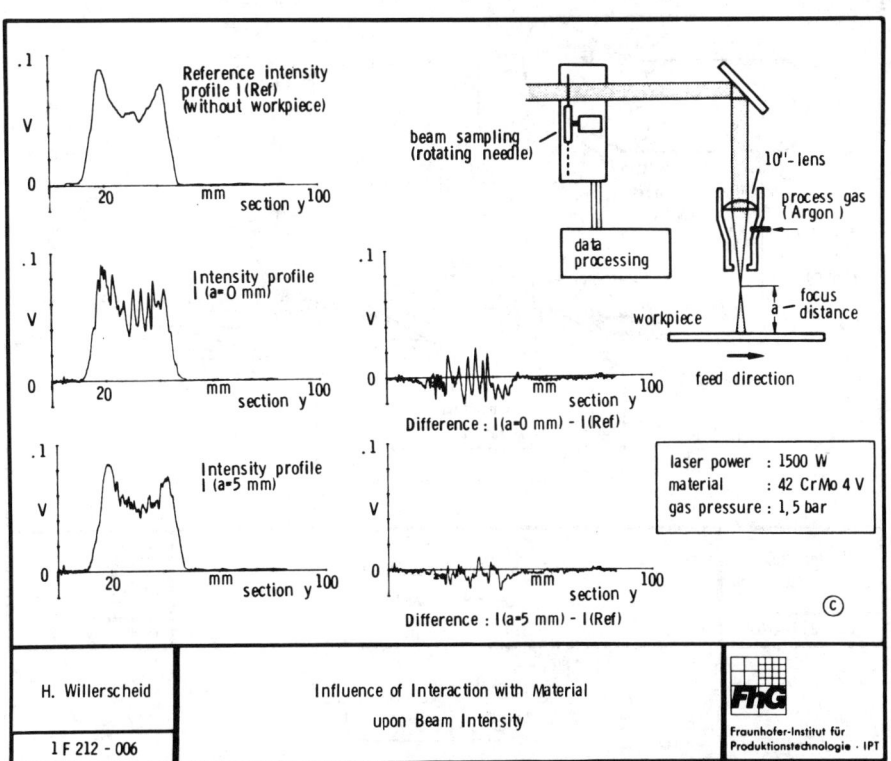

Figure 9

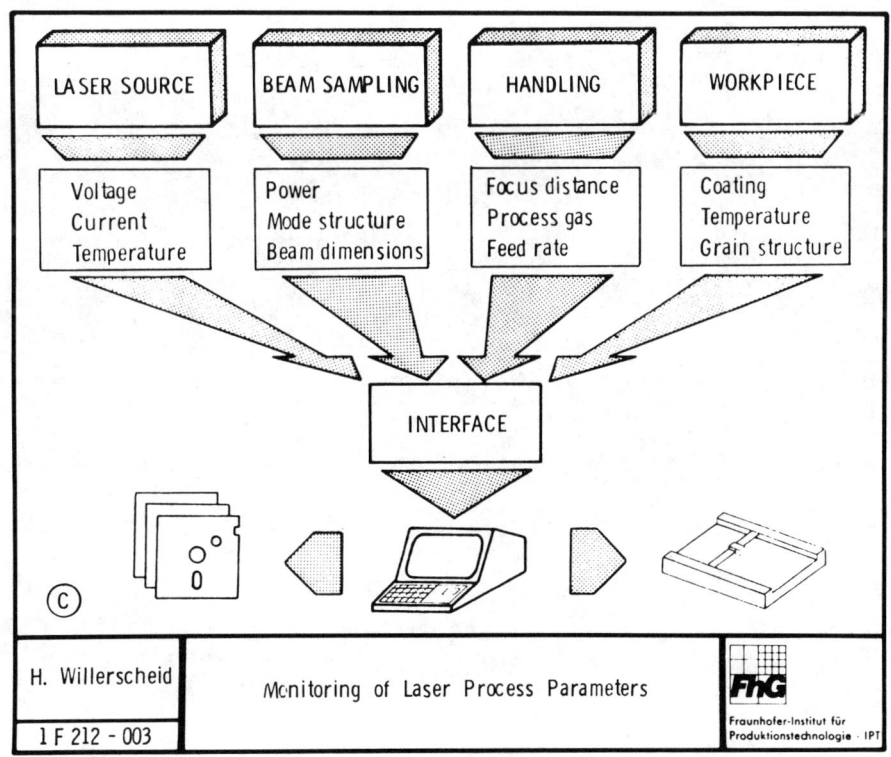

Figure 10

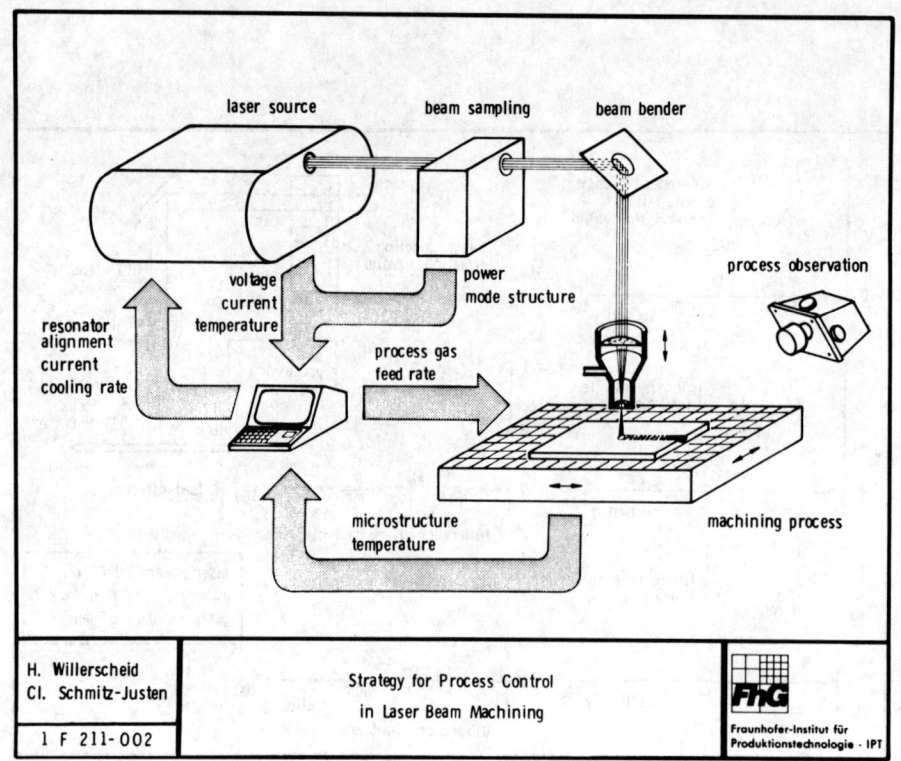

Figure 11

The use of electronic speckle pattern interferometry (ESPI) as an inspection tool

P. C. Montgomery
and
J. Tyrer
Loughborough University of Technology, UK

ABSTRACT

Electronic Speckle Pattern Interferometry (ESPI) has become a useful laser technique for observing surface deformations of less than 1 µm over a component area ranging from 1 mm^2 to 1 m^2. Results, displayed as a contour map on a T.V. monitor, provide information helpful to design, quality control and in-service testing. This paper presents typical applications of the technique to component testing in manufacturing. The authors also discuss current and future advances for improving performance and reducing the cost of ESPI systems.

INSPECTION IN MANUFACTURING

Inspection is a vital part of the manufacturing process and is often dependant, to a great extent, on the skilful eye of trained operators. To extend the capabilities beyond that of human detection there are many tools enabling precise measurements to be made and the accruel of normally inaccessible information.

Inspection Techniques

There are many approaches available, for example; shape comparators, coordinate measuring machines, photographic or video systems, photoelasticity, strain guages and brittle lacquer techniques which make use of visible light, infra red, X rays, ultrasonics and various types of sensors. The choice of technique for a particular problem depends on the parameter being measured, the accuracy required, the cost effectiveness and environmental suitability.

ESPI

Electronic Speckle Pattern Interferometry is a non-contact video technique for detecting vibrations and deformations in the surface of a component. Using laser light, it gives measurements down to a fraction of a micron in real time as the component is subjected to static or dynamic loading. The place of ESPI as an inspection tool is seen perhaps as a companion to photoelasticity, strain gauges and accelerometers, yielding valuable data over the complete area of the actual component.

This paper presents some applications of ESPI and discusses its viability as an inspection tool for use in design, quality control, and in-service testing.

THE ELECTRONIC SPECKLE PATTERN INTERFEROMETER

How it Works

A block diagram of a basic ESPI system is shown in Fig. 1 and a photograph in Fig.2. The instrument consists of two main components; the optical head for illuminating and viewing the component and the electronics for processing the viewed image (Butters et al, Jones et al).

Optical Head - The technique is very similar to focussed image holography, in that a Helium-Neon laser beam is divided into two, with one beam illuminating the object. The other beam forms the reference beam which illuminates the faceplate of a video camera. This interferes with a focussed image of the object to form a phase referenced speckle pattern, unique to the object surface at rest.

Electronic Processing - The Video signal of the pattern is digitised and recorded in a digital frame store, to be continuously substracted point by point from the live image. While the object is at rest, the speckle pattern remains the same, resulting in a black picture on the monitor. As the object surface moves, each speckle changes in intensity with variations in the object beam path length in relation to the fixed reference beam. By processing the whole image, fringes of loci of equal phase change are mapped out on the screen.

Fig. 3 shows the effect of twisting a steel plate. A phase difference of 2π between fringes represents a path length change of λ. For illumination within 10° of the viewing direction using He Ne laser light of 0.633 µm wavelength, these contours show out-of-plane movement of 0.3 µm per line.

Capabilities and Limitations of ESPI

Measurement Range - ESPI is useful over areas from less than 1 mm^2 to greater than 1 m^2 for measuring out-of-plane displacements of a fraction of a micron upto 12 µm in one fringe pattern. Using phase modulation it is possible to detect as small a movement as 3 nm and, by re-referencing the speckle pattern, to monitor deformations under continuous loading to destruction of the component. For most applications a Helium-neon laser of 5 mW to 25 mW power is sufficient, but for large or difficult components, an Argon-Ion laser of 1 w to 3 w may be used.

Fringe Interpretation - Contours such as those in Fig. 3 are interpreted by manual tracing and subsequent point by point calculation using the fringe order and geometry of the optical arrangement. Alternatively, by digitising the video signal of the contour pattern, this analysis may be done automatically using a computer, (Nakadate et al).

Fig. 4 shows an example of an image processing system used for smoothing, production of a binary image and tracking of the fringes. The ease and speed with which automatic measurement is made depends upon the quality of the original pattern, the complexity of the object and the type of load applied.

Operational Modes - Modification of the illuminating optics makes it possible to detect purely in-plane strain with ESPI over the whole of the surface studied, (Jones). Because of the high sensitivity of the technique to both in-plane and out-of-plane movement, ESPI is of great use in NDT for detecting hidden faults or weaknesses under thermal or mechanical stress.

ESPI is also sensitive to vibrations on an object, for example the cone of a loudspeaker shown in Fig. 5. (a) vibrating at 4.0 kHz. The white areas signify stationary points whilst the contours represent maximum out-of-plane vibration. By a fringe integration technique (Lokberg et al) higher contrast fringes can be obtained either to assist fringe analysis or simply improve the output quality for presentation, (Fig. 5. (b)).

Component Surface - Although fringes are obtained on non-reflective surfaces such as machined parts, the highest contrast fringes are obtained on matt white surfaces. This may be a temporary surface which can readily be removed.

Stability - Being sensitive to a fraction of the wavelength of light, ESPI has certain stability requirements. The component must be stable in relation to the optical head and both must be isolated from excessive floor vibrations. Fig. 2. shows the instrument operating on a wooden table nearby to a running I.C. engine, showing the possibilities of it being used on the shop floor. The engine is vibration-isolated from the concrete floor and the optical bench is mounted on anti-vibration feet. Warm air currents from the engine cooling fan produced a drift in the pattern but even under these conditions valuable results were obtained (see Fig. 3). A protective hood over the instrument could be used to eliminate this problem.

Safety - The expanded laser beam illuminating the object is no more harmful to the eye than a torchlight. Because all unexpanded beams are enclosed, the instrument is a class 1 laser system that can be used without other safety equipment under normal lighting conditions.

ESPI IN DESIGN

Visualisation of vibration modes and deformation of complex structures under load is an important aspect of design. ESPI is particularly useful in this respect in providing data for use in making modifications or in supporting Finite Element studies.

Using the basic ESPI system described, an optical rig is designed for a particular application, depending on the size and accessibility of the component. Initially the whole structure is viewed and subjected to loading, followed by close up studies, wth a zoom lens, of areas of interest.

Study of Engine Resonance Modes

In collaboration with British Leyland Technology, work is being carried out at Loughborough to investigate the use of ESPI in studying engine resonance modes. Two areas of interest are the siting of piezo-electric 'knock' detectors and finding points of noise emission.

'Knock' Detection - Sophisticated electronic ignition systems now make use of automatic sensing of unwanted combustion chamber detonation, or 'knock' to help run the engine more efficiently, (Scarlett). This is achieved by detecting the pulses emitted when a cylinder fires prematurely with a piezo-electric accelerometer fixed to the block. The ignition timing is then retarded until the condition ceases. Background noise in an engine environment is quite high, so detection is improved by using a 'tuned location' for the detector. Such a location consists of a point that resonates in the range of 'knock' frequencies characteristic of the engine.

Fig. 6. (a) shows typical results in the search for a good resonant site on an engine block. The engine is mounted on a large optics table and excited with a transducer to induce different resonant modes which are seen immediately on the monitor screen. Scanning through a range of frequencies, it is not too difficult to find relevant sites on the block, the position of these being further optimised with accelerometers on a running engine.

Noise Emission - The results in Fig. 6. (b) show vibrations in the wall of the sump of an engine showing possible sources of noise emission. Future work correlating noise emission with resonant modes is yet to be carried out.

Validation of Finite Element Studies of a Welded Structure

British Gas are interested in studying the response of welded structures to loads simulating offshore structures in the North Sea. They have already carried out Finite Element studies on models of typical structures such as the tandem brace, where two parallel steel tubes are welded onto a larger diameter tube. ESPI is being used to study the displacement of the joints under load to investigate the correlation with F.E. analysis. Fig. 7. shows typical results of such studies one one of the joints.

Other Applications

Being able to visualise small displacements in a component under load is useful in many other design situations. For example, ESPI is being used successfully in Norway for measuring resonant modes of turbine blades to ensure dangerous resonant frequencies are not induced in running gas turbines (Lokberg et al). It is also being used in vehicle component noise suppression, and crank shaft bearing cap studies. Many problems requiring empirically deduced component modifications can make use of ESPI.

ESPI IN QUALITY CONTROL

Because ESPI is operational under moderately noisy conditions on the shop floor, it is a valuable quality control technique for vibration, static loading, and thermal testing of components. The component is clamped in place in the machine, stressed, and the fringe pattern compared with a control response. With a set of known pass/fail patterns, the system can be used by an unskilled operator.

Loudspeaker Testing

The frequency response of the cone of a loudspeaker very much determines the final quality of the sound from a speaker pair. Fig. 8. shows the results of comparing two B. & W. soft dome tweeters vibrating at 7.0 kHz. Speaker (a) is moving in a uniform manner, indicated by the circular fringe pattern. The broken pattern in (b) indicates non-uniform behaviour. At this frequency this may be unacceptable, indicating a sub-standard speaker. An immediate result is therefore obtained with the simple procedure of clamping the speakers in the instrument and connecting them to a variable frequency generator.

Phase mapping of the speaker can be carried out if required (Lokberg et al), as can computer processing of the results (Hurden).

Other Applications

ESPI is widely applicable in NDT and quality control (Butters) for testing laminated and honeycomb section panels, reinforced plastic components, glued joints, plastic compression welds etc. for detecting unbonded regions, fractures and other discontinuities.

IN-SERVICE TESTING

Once a component or structure has been commissioned, it is sometimes necessary to study its load response under working conditions or after part of its expected working life. This may be carried out on site or in the laboratory using a dedicated ESPI system.

Safety Testing of a Gas Main Syphon

Syphons are used throughout the gas main network in Britain to remove water that seeps in through pipe joints sitting in the ground below the local water table. These rectangular cast iron sumps undergo varying degrees of corrosion depending on the surrounding soil conditions.

A 50 year old syphon was prepared with a matt white surface and tested in the laboratory with an ESPI system using an Ar-Ion laser at a wavelength of 514 nm. The tank was hydraulically pressurised to thirty times its normal working pressure and an incremental pressure was added to produce the contour pattern shown in Fig. 9. Counting the fringes, the side of the tank was been displaced outwards by approximately 4.0 μm.

Studying the symmetry of the pattern by eye, there appears to be no significant weak points due to corrosion nor any detectable cracks induced at this particular pressure. These would have caused sharp discontinuities in the fringe pattern if they had been present.

Further work will establish a test procedure for measuring safety factors in such components.

Other Applications

The testing of thermoplastic welds connecting sections of plastic gas piping is also being investigated in the laboratory, with the possibility of producing an instrument for use on-site. NDT tests could then be carried out on each weld as the piping was laid in the ground.

Such applications of ESPI as an in-service diagnostic tool are obviously unlimited.

FUTURE DEVELOPMENTS IN ESPI

The development of ESPI is clearly related to the rapidly advancing technologies of electronics and optics.

Two major advances will be in automatic fringe analysis and in the use of new electro-optic devices in the optical head.

Camera Resolution

The present video system uses a 625 line camera and monitor with a 512 x 512 pixel frame store. High resolution T.V. systems using 1200 lines are now being investigated for use with store arrays of 1024 x 1024 pixels. Presently this is non-standard T.V. equipment, but changes to broadcast standards shortly to be introduced should make these systems more widely available. The result will be higher quality fringe patterns.

Optical Head

A present limitation relies on the component being visible. The use of coherent fibre bundles will allow inaccessible locations to be inspected.

Semi-conductor lasers and solid state camers will help to produce a more compact optical head. Combining them with integrated optics will result in a more stable and versatile inspection tool.

Pulsed Laser System

To enable on-site analysis of components and structures, a double pulsed laser removes the problem of instability. Pulsed ESPI using a ruby laser, (Cookson et al) allows only low repitition rates because of the inherent lasing efficiency of the laser. It also requires a cumbersome power supply. Using improved frequency doubling crystals with a Nd: YAG system would enable the matching of pulsing and T.V. frame rate. The result would be a portable double pulsing system that could update displacement data every 1/25 second for fast NDT analysis.

Fringe Analysis

Improvements in automatic fringe analysis will result from the growing field of image processing. The development of suitable algorithms will be followed by incorporating them in array processors for high speed operation.

CONCLUSIONS

ESPI is a useful laser technique for detecting small surface displacements in a component which can be used throughout industry in design, production and in-service testing. It has numerous applications in both dynamic and static structural analysis With a range of optical configurations it can be used in the laboratory, on the shop floor and out on site. The results are readily digitised for analysis by computer.

Present limitations are low resolution and the need for some surface preparation. Being an interferometric system, the instrument requires moderate vibration isolation.

Future developments in electro-optics and automatic fringe analysis all point towards a versatile remotely operating inspection tool.

ACKNOWLEDGEMENTS

The authors would like to thank the Science and Engineering Research Council for their financial support of this work, Mr. B.D. Bergquist for his help, Mr. P. Henry for his work on image processing amd Mr. K. Topley for the photographs.

REFERENCES

Butters, J.N. "Applications of ESPI to NDT". Optics & Laser Technology, Vol. 9, pp 117 - 123 (June 1977).

Butters, J.N. & Leendertz, J.A. "Holographic and Video Techniques Applied to Engineering Measurements". Transactions of the Institute of Measurement and Control, Vol. 4, No. 12, pp 349 - 354 (December 1971).

Cookson, T.J., Butters, J.N. & Pollard, H.C. "Pulsed Lasers in Electronic Speckle Pattern Interferometry". Optics and Laser Technology, Vol. 10, pp 119 - 124 (June 1978).

Hurden, A.P.M. "Vibration Mode Analysis using Electronic Speckle Pattern Interferometry". Optics and Laser Technology, Vol. 14, pp 21 - 25 (February 1982).

Jones, R. "The Design and Application of a Speckle Pattern Interferometer for Total Plane Strain Field Measurement". Optics and Laser Technology, Vol. 8, pp 215 - 219 (October 1976).

Jones R. & Wykes, C. "Holographic and Speckle Interferometry". Cambridge University Press (Cambridge 1983).

Lokberg, O.J. & Hogmoen, K. "Use of Modulated Reference Wave in Electronic Speckle Pattern Interferometry". Journal of Physics E: Scientific Instruments, Vol. 9 pp 847 - 851. (1976).

Lokberg, O.J. & Slettemeon, G.A. "Improved Fringe Definition by Speckle Averaging in ESPI". ICO-13 Conference Digest, Japan, pp 116 - 117 (August 1984).

Lokberg, O.J. & Svenke, P. "Design and Use of an Electronic Speckle Pattern Interferometer for Testing of Turbine Parts". Optics and Lasers in Engineering, Vol. 2, pp 1 - 12 (1981).

Nakadate, S., Yatagai, T. & Saito, H. "Computer-Aided Speckle Pattern Interferometry". Applied Optics, Vol. 22, No. 2 (January 1983).

Scarlett, M. "Montego!" Autocar, Vol. 161, No. 4557, pp 20-29 (28th April 1984).

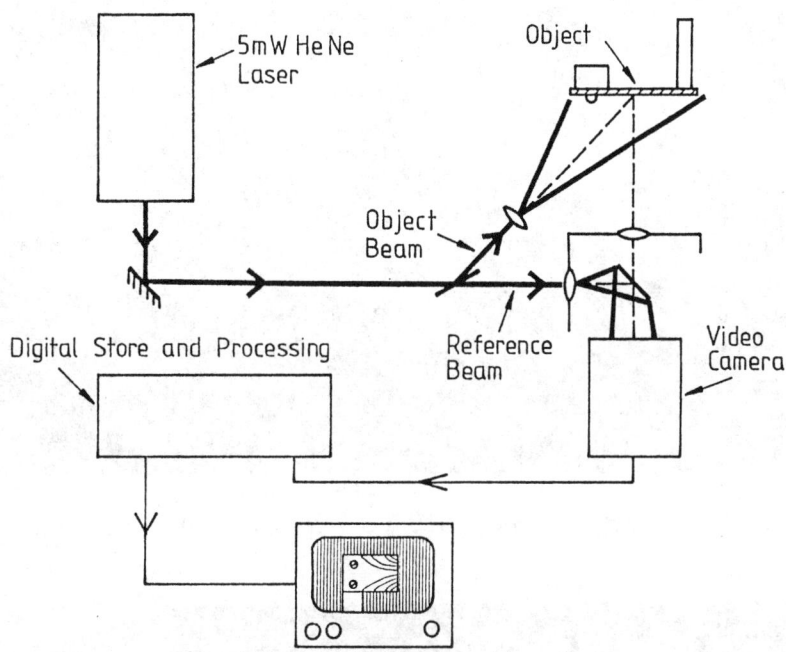

Fig 1. Block diagram of the Electronic Speckle Pattern Interferometer.

Fig. 2. Optical bench and electronics for ESPI showing how it can be used in a noisy environment (with a running I.C. engine in the background).

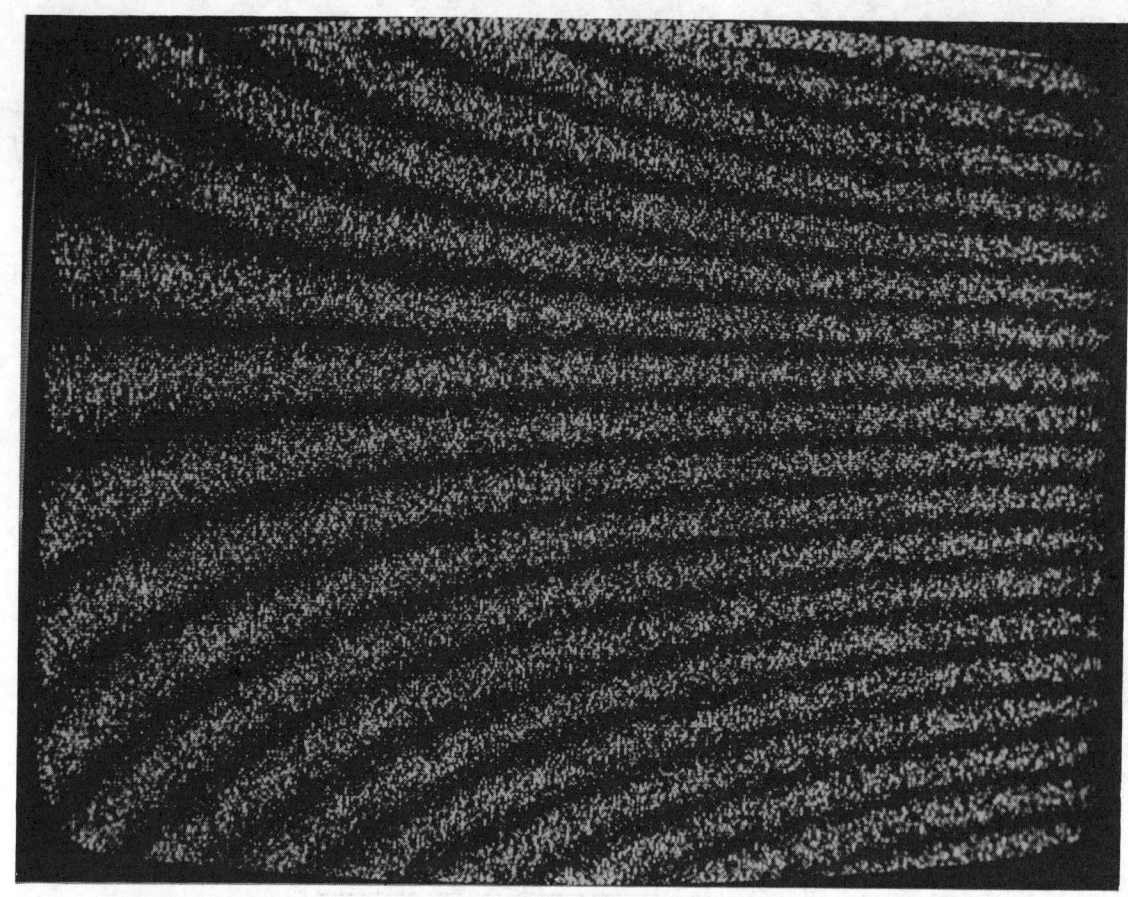

Fig. 3   Out-of-plane contour fringes (0.3 µm separation) from Fig. 2 showing torsion in a steel plate.

Fig. 4   Automatic processing of an ESPI fringe pattern.

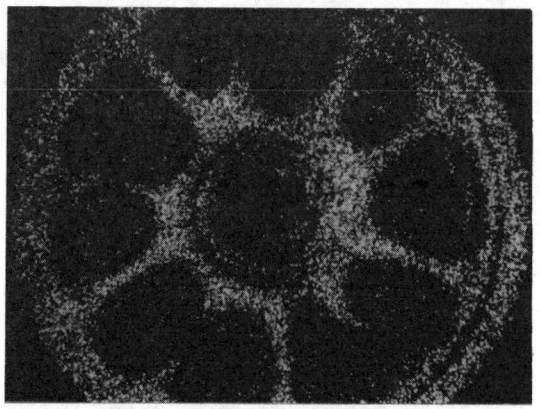 

(a) Ordinary time average speckle pattern.

(b) Smoothed speckle pattern by integration technique.

Fig. 5 ESPI analysis of a 50 mm diameter loudspeaker cone vibrating at 4.0 kHz.

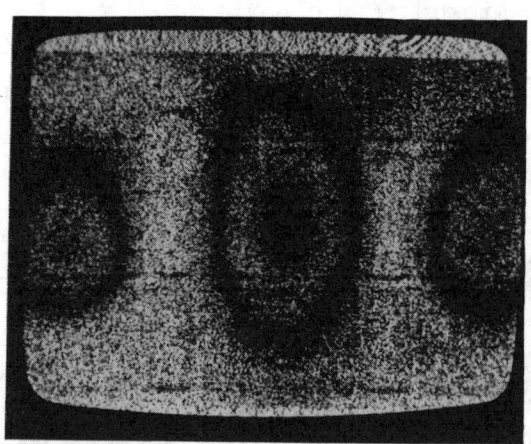 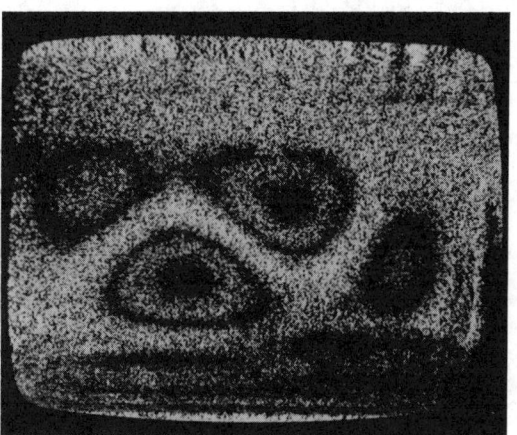

(a) Resonance in the wall of an engine block excited at 3.9 kHz.

(b) Resonance in the wall of an engine sump excited at 3.3 kHz.

Fig. 6 Using ESPI in engine design for detecting resonant modes.

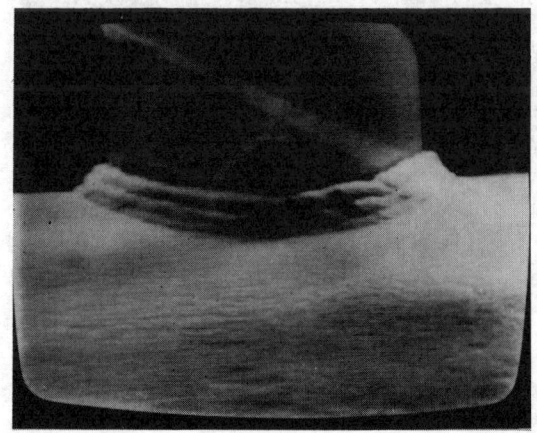 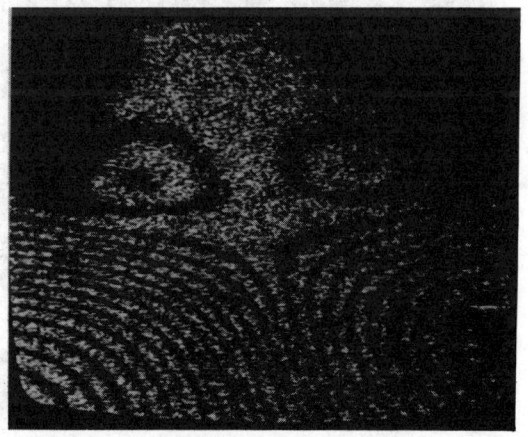

(a) View of welded steel tubing.

(b) Vertical tube under lateral bending moment.

Fig. 7 Verification of F.E. analysis of bending in a welded steel structure.

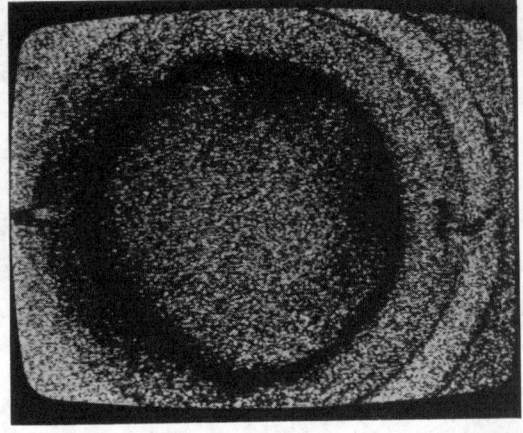 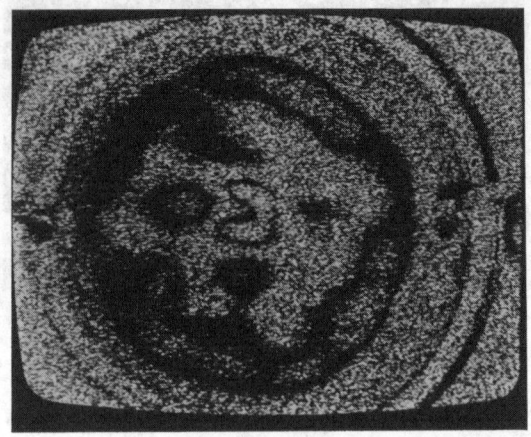

(a) Good speaker indicated by circular fringe pattern.

(b) Sub-standard speaker indicated by broken fringe pattern.

Fig. 8 Comparison of two B. & W. soft dome tweeters vibrating at 7.0 kHz.

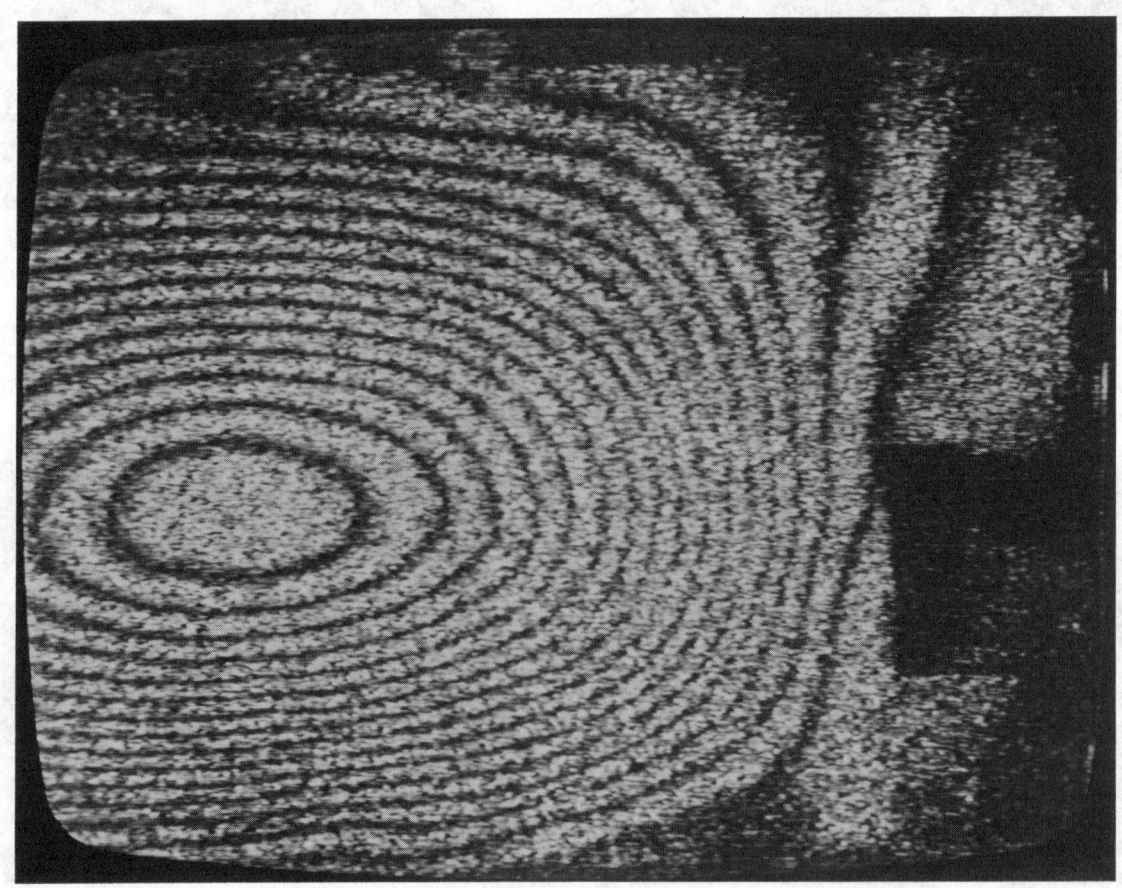

Fig. 9 Expansion of a gas main syphon under hydraulic pressure - the uniform pattern indicates no significant weaknesses in the corroded cast iron wall (0.5 m screen width).

Non-destructive testing of adhesive joints
B. P. Holownia
Loughborough University of Technology, UK

ABSTRACT

Electronic speckle pattern interferometry (ESPI)[1] which was used is capable of detecting debonded areas with certain limitations. These depend on the size of the defect and the thickness of metal used. Parepared specimens with debonded areas in the middle of adhesive joints and on the periphery of an overlap shear joints, were tested using various loading techniques. It was found that mechanical loading and vacuum chamber gave the best results when using ESPI system for detecting debonded areas. Some overlap shear specimens using copper and steel strips were subjected to prolonged environmental testing in fresh water at 40°C. The results obtained looked encouraging.

1.0 INTRODUCTION

Adhesive joints offer an efficient alternative to mechanical fasteners in many applications. In aircraft applications they can be used to produce laminated metal structures which, because of their good fracture toughness characteristics and low crack propagation rates, replace components normally machined out of solid. Many workers therefore have been examining large honeycomb panels[2] to detect debonded areas using various methods.

Adhesive quality control requires the non-destructive testing of intermolecular forces between adhered surface and the cured adhesive layer, after the bond has been created[3]. This is similar to the non-destructive determination of intermolecular forces within a single material. As yet no single successful non-destructive solution is known for this problem today and hence there is no method which gives reliable indication of adhesive strength of the bonded joints after the joint has been made.

Most adhesives suffer from moisture ingression where a gradual weakening of adhesive joint occurs due to moisture penetration with a complete break of the bond around the periphery making the effective bonded area smaller and therefore weaker.

A number of different methods have been used by other workers such as thermal inspection capacitance measurements, infrared methods acoustic inspection [4,5,6], holographic interference [7,8,9,10] and laser speckle photography [11].

A review of non-destructive testing of composite materials is given in reference 12.

For many years now Admiralty Research Establishment (Holton Heath) have been testing metal-to-metal adhesive joints using different types of adhesives with different types of adherends. The current research is concerned with the deterioration of the joints by water penetration. This is particularly difficult since to monitor the deterioration of an adhesive joint in hostile environment, hundreds of samples have to be made. These are then placed in water under various conditions and a representative number of samples are taken out periodically and tested to destruction to examine the joints visually.

A non-destructive testing offers a number of advantages by reducing the number of samples to be made and also more accurate study can be monitored enabling the study of the progress of water ingression into the joint, thus reducing the statistical scatter.

The following loading methods were tried using prepared specimens with various degrees of success:-

(i)   Pressure loading,
(ii) Vacuum loading.
(iii) Thermal loading.
(iv) Vibrational loading.
(v)   Mechanical loading.

It was found that the most effective way of detecting debonded areas in the middle of a joint was when the joints were subjected to a vacuum loading. The best way of detecting debonded area around the periphery of the specimen was using mechanical loading.

Thermal loading was particularly sensitive when joints were made up with two different materials. A thin metal strip (high conductivity) bonded to thick non-metal adherend (low conductivity such as perspex).

A similar result can be obtained using thick glue line (2 mm) between two metal adherends.

The present study was concentrated on the degradation of the adhesive when the joints are subjected to ageing in fresh water at 40oC.

EXPERIMENTAL SET-UP.

Electronic Speckle Pattern Interferometry (ESPI). When laser light, which is optically coherent, is scattered from a diffuse object and viewed through an imaging system, such as a TV camera or the human eye, a phenomenon known as speckling is observed. Each speckle contains information within its relative phase, about the position on the object from which it originates, the pattern is a fingerprint of the static object.

By combining this reflected pattern with a reference beam, taken directly from the laser, the phase information is extracted from the speckles. Each speckle undergoes constructive or destructive interference with the reference, depending on its phase, producing a speckle pattern of varying intensity. If the object is moved very slightly so as the speckles themselves do not move but their phase changes, the speckle pattern changes in intensity. A combination of two such speckle patterns will lead to larger changes in intensity, or fringes, across the object. Work was first reported on this by Leendertz [14], in 1970.

In the ESPI technique, the speckle pattern and reference are combined on a video camera and the signal then processed in several ways (see Fig 1), depending on the type of study being carried out. The resultant processed image being observed on a TV monitor and recorded on a video tape.

Theoretical concepts of ESPI are outlined by Jones & Wykes,[15], and some applications of ESPI by Butters,[1].

PREPARATION OF SPECIMENS.

(i) <u>Specimens With Perspex Backing.</u> A thin (0.14 mm) brass metal plate was glued with araldite on top of a 60 mm x 60 mm x 12 mm thick perspex plate containing drill voids of various diameters, Fig (2). The specimen was then examined using the ESPI system. By heating the metal plate very slightly, using a hand-held hair dryer, all the voids and any accidental de-bonded areas were visible on a TV screen, Fig (3).

Different thicknesses of brass, aluminium and mild steel plates were used on perspex backing. These were examined by the ESPI system using loading modes of heat, pressure and vacuum. The metal plates varied in thickness from 0.1 mm to 1.7 mm.

It was found that for thin plates (0.1 mm) all voids down to 3 mm diameter or less, could easily be detected using any of the loading methods. As the thickness of the metal plates increased, higher pressure, higher vacuum and higher temperature had to be used to detect the faults.

For brass plates 1.7 mm thick, a pressure of up to 50 lb/in^2 is needed to identify faults of about 6 mm in diameter.

(ii) <u>Overlap Shear, Thick Glue Line.</u> A number of specimens were prepared with thick glue line (2 mm) using PTFE tape to simulate debonded areas. These were then tested using different modes of loading.

The fringe spacing indicates the amount of out of plane movement hence it is easy to differentiate between glued areas (with fringes wide apart in the centre as in Fig. 4), and debonded areas (with fringes close together as in Fig. 5, showing debonded central area).

Some difficulties were experienced when forming an overlap shear joint using metal strips. When applying adhesive, the adhesive spread past the overlap onto the metal strip. The excess adhesive, when set, was difficult to clean from the thin metal strips, thus it was difficult to define the edge of the glue line.

(iii) <u>Overlap Shear, Thin Glue Line.</u> A total of 12 overlap shear joint specimens with thin glue line were made by M.O.D. Six of the specimens were made with 1" wide x 1/16" thick copper strip glued to a thin (.007") steel strip and the other six were the same size with the thick strip being steel, glued to thin (.007") steel strip. The adhesive used was epoxy-polyamide which gives a high rate of degradation in water. To cater for mechanical loading the thin strip was made to overlap the thick strip so that slight force can be applied on the overlap of the thin strip around the pheriphery. (Fig. 6).

ENVIRONMENTAL TESTS.

All thin glue line specimens were checked both, in vacuum and using mechanical loading to reveal any debonded areas before ageing in fresh water at 40°C. No faults in this adhesive were visible.

The examination by ESPI system requires a white reflective surface and therefore a lot of time was spent in masking off the unwanted areas and spraying white the exact area of the adhesive. After ageing for four weeks in fresh water at 40°C the white area on each specimen had to be resprayed before examination by ESPI.

A special framework had to be made for mechanical loading, (Fig. 7). The thick metal strip was firmly held in the framework and very fine threaded (70 t.p.i.) screws were used to apply slight bending on the thin strip around the periphery of the joint. Figs. 8 and 9 shows progressive degradation of the same specimen after 10 weeks ageing in fresh water at 40°C.

Photographs taken from a TV screen are not very clear. The speckle appearance over the whole photograph is due to the inherent speckle associated with ESPI system. However the photographs show fairly defined edge of the bonded area.

CONCLUSIONS.

The present work shows that ESPI system can be used successfully for non-destructive testing of adhesive joints. It demonstrates that mechanical loading can be used to measure the progressive degradation of the adhesive around the pheriphery of an overlap shear joints when they are subjected to ageing in water at 40°C.

For internal debonded areas, vacuum loading gives the best results.

ACKNOWLEDGEMENTS.

This work forms part of a programme of research supported by the Procurement Executive, Ministry of Defence.

REFERENCES.

1. Butters, J.N. "Application of ESPI to NDT". Optics and Laser Tech. Vol. 9, p. 117, (1977).

2. Hagamaier, D.J. "Bonded Joints and Non-Destructive Testing, Bonded Homeycomb Structures". N.D.T. International p.401-406. (December 1971).

3. Schliekelmann, R.J. "Non-Destructive Testing of Adhesive Bonded Metal-to Metal Joints 1". N.D.T. International p.79-86. (April 1972).

4. Batsien, P. "Difficulties in the Ultrasonic Evaluation of Defect Size". N.D.T. International p.147-151. (February 1968).

5. Ding, W.Z. and Chen, J.M. "Principles of an Acoustic Impedance Method for Detection and Location of Non-Bonds in Adhesive-bonded Multi-Layered Joints". N.D.T. International p.137-142. (June 1982).

6. Schliekelmann, R.J. "Non-Destructive Testing of Adhesive Bonded Metal-to Metal Joints 2". N.D.T. International p.144-153. (June 1972).

7. Grant, R.M. and Brown, G.M. "Holographic Non-Destructive Testing (H.N.D.T.)" Materials Evaluation p.79-84, Vol. 27,(1969)

8. Hockley, B.S. "Non-Destructive Testing by Holography". British Journal of N.D.T. p.115-121. (July 1972).

9. Wells, D.R. "N.D.T. of Sandwich Structures by Holographic Interferometry". Materials Evaluation p.225-231, Vol. 27,(1969).

10. Schliekelmann, R.J. "Non-Destructive Testing of Bonded Joints, Recent Development in Testing Systems". N.D.T. International p.100-103, (April 1975).

11. Gregory, D.A. "Laser Speckle Photography and the Submicron Measurement of Surface Deformations on Engineering Scructures". N.D.T. International p.61-70. (April 1979).

12. Scott, I.G. "A Review of Non-Destructive Testing of Composite Materials". N.D.T. International p.75-85. (April 1982).

13. Holownia, B.P. "Non-Destructive Testing of Overlap Shear Joints Using Electronic Speckle Pattern Interferometry", Optics and Lasers in Engineering. (In Press).

14. Leendertz, J.A. "Interferometric Displacement Measurement on Scattering Surfaces Utilizing Speckle Effect". J. Sci. Instr. "J.Phys.E", Vol. 3, No. 3, p.214-218 (March 1970).

15. Jones, R. and Wykes, C. "General Parameters for The Design and Optimization of ESPI". Optica Acta, Vo. 28, No. 7, p.949-972 (1981).

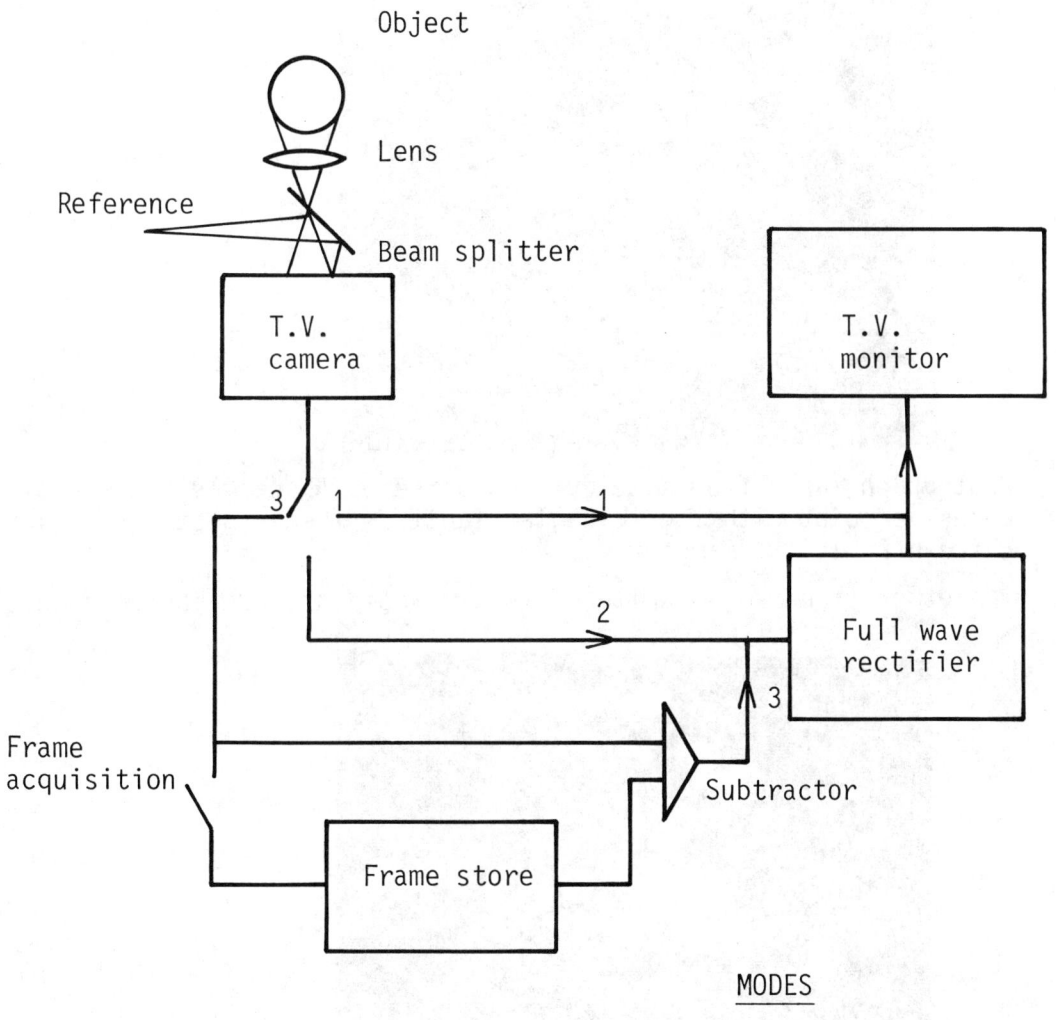

MODES

1. Camera Direct
2. Filtered Camera Signal
3. Subtracted Signal Filtered

Fig. 1 - E.S.P.I. System.

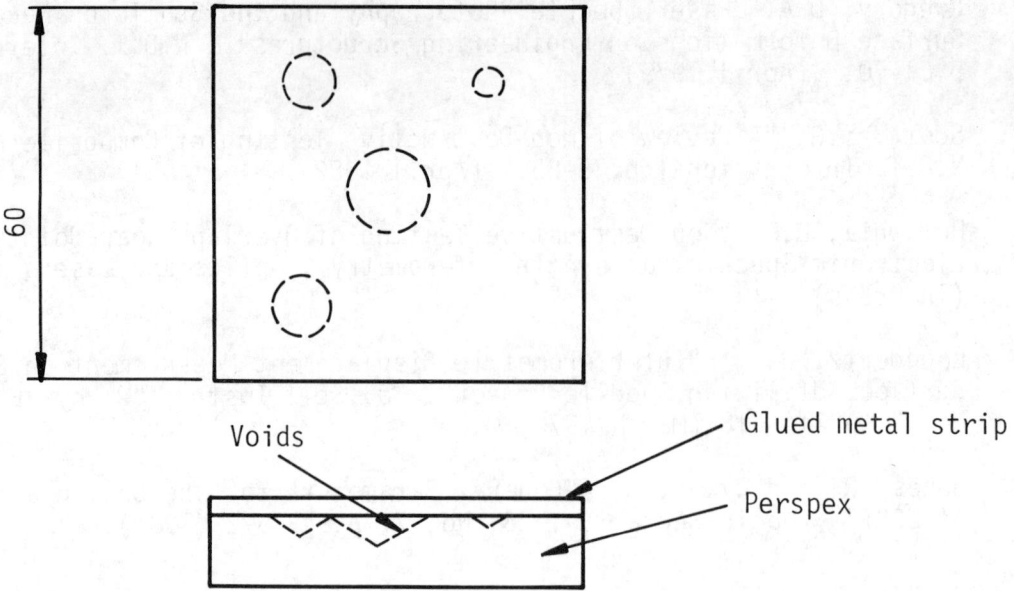

Fig. 2 - Specimen with perspex backing.

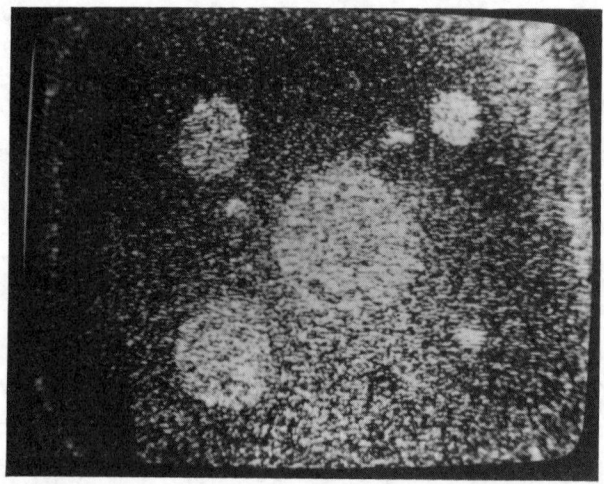

Fig. 3 - Photograph taken from a television screen. A speckle pattern of adhesive joint with four circular faults. The largest circle in the centre is ½" diameter.

The other three small white spots are accidental de-bonded points.

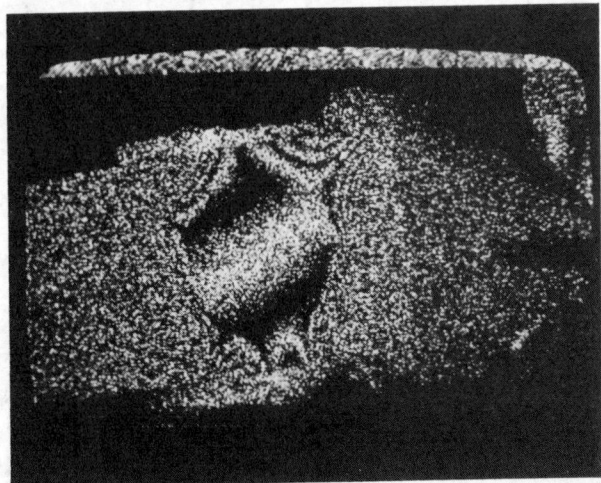

Fig. 4 - Mechanically loaded specimen on four sides showing glued central area.

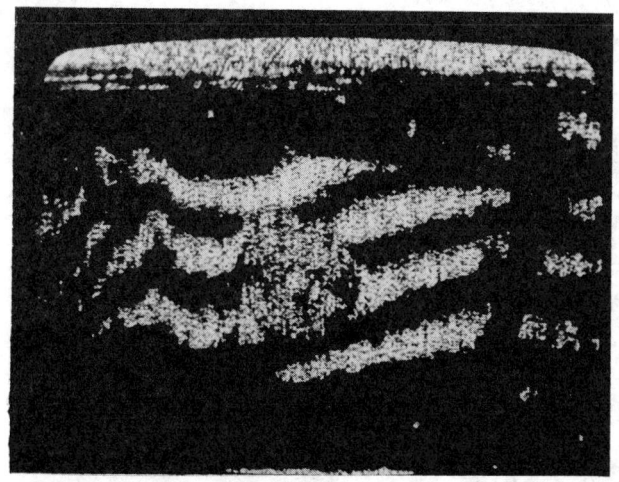

Fig. 5 - Steel specimen in a vacuum chamber showing de-bonded area in the middle of the adhesive joint.

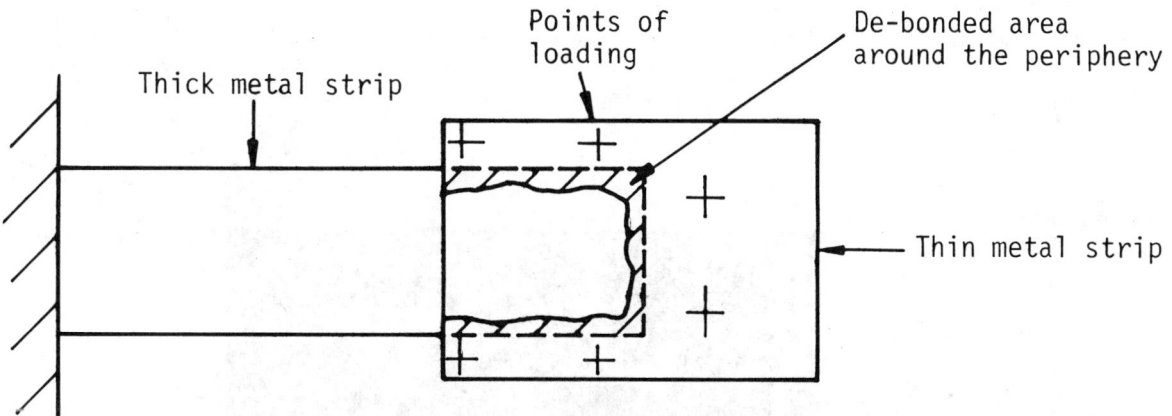

Fig. 6 - Overlap shear joint specimen for mechanical loading.

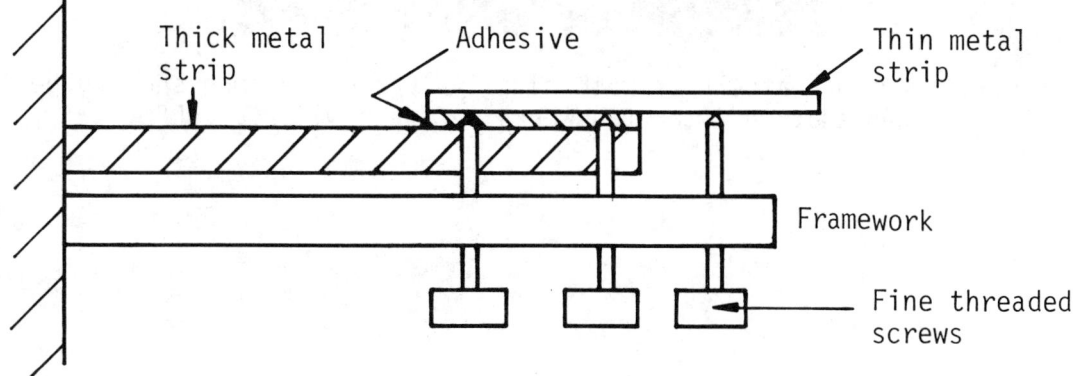

Fig. 7 - Mechanical loading of metal to metal overlap shear joint.

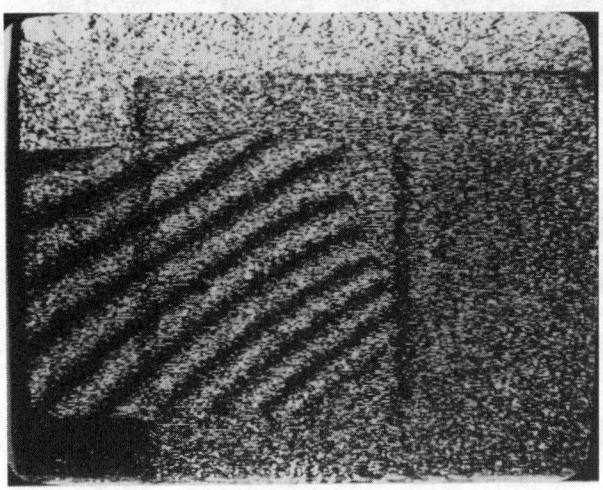

Fig. 8 - Fringe pattern of mechanically-loaded specimen showing peripheral degredation after four weeks in fresh water at 40°C.

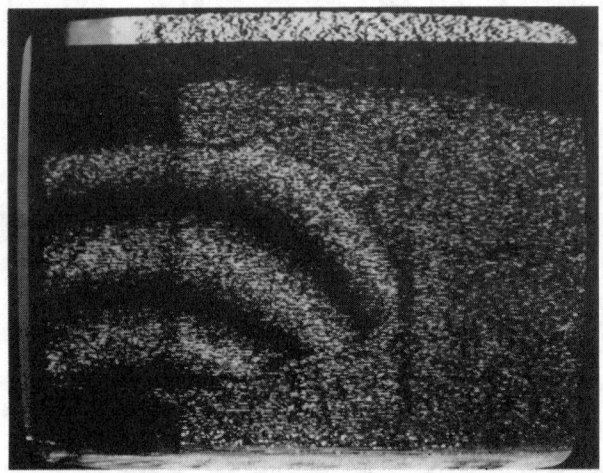

Fig. 9 - Fringe pattern of mechanically-loaded specimen showing peripheral degredation after ten weeks in fresh water at 40°C.

Application of immersion technique on the measurement of some engineering products

M. M. El Sayed
Helwan University
and
M. M. Koura
Ain Shams University, Egypt

M.M.EL-SAYED[*] & M.M.KOURA[**]

[*] Assistant Professor , Mech. Design Dept. , Faculty of Engineering & Tech., Mataria , Helwan University , Cairo , Egypt.
[**] Professor , Design & Production Eng. Dept. , Faculty of Engineering , Ain Shams University , Cairo , Egypt .

ABSTRACT

The recently developed techniques of holographic interferometry had shed new lights on many problems of experimental mechanics. For example holographic interferometry has been applied in immersion technique which has been employed to demonstrate the generation of surface elevation contours of any product. This paper shows the feasibility of the holographic immersion technique in mapping the contour of some engineering products which faces the inspection personnels in industry with various difficulties. The paper also, includes holographic interferometry patterns which were taken for some objects together with the qualitative and quantitative analysis that illustrate the used technique.

INTRODUCTION

Holographic interferometry has many applications in our modern life. One of these applications is mapping of surface contours for any object with high accuracy. As in all methods of holographic interferometry , interference between the monochromatic light must take place. In this case there are two wave fields in which the difference between their phases indicates the difference in the optical path lengthes .This difference gives the exact information about the distance between each point on the tested object and any reference plane.

Two methods may be used in realising the above mentioned condition . The first , depends upon illuminating the object with two laser beams , consequently,having different wave-lengthes; such as He-Ne laser and Argon laser , where the first beam has a wavelength of 0.6328 μm. and the second has a different wavelength of 0.3243 μm .The second method depends on the use of immersion method while using only one type of the laser beam and changing the refractive index of the used liquid .The first method needs special and expensive equipments , while the second method is easy to be applied in an ordinary holographic laboratory .Therefore , immersion method was used in the present investigation .

IMMERSION METHOD AND ITS MATHEMATICAL VERIFICATION

In fact the immersion method technique depends upon making use of the phenomenon which states that , the optical path length is directly affected by the refractive index of the used liquid (in which the laser beam is passing through). This refractive index can be changed in a suitable time . The contour map of the object can be obtained by using any holographic arrangement . For simple interpretation it is prefered to use parallel illumination as shown in fig. 1 . The tested object is to be dived in a water tank , so that one side of this tank is made of high quality glass . The object is to be illuminated by a laser beam through this glass side .

The lapsed time (Frozen holographic interferomtry) technique was used in carrying out the laboratory testing /1-4/ . As shown in figs. 1& 2 , the laser beam is coming from the laser tube (1) and is reflected from the mirror (2) , then passes via two lenses (4) and through the beam splitter (5) . The object beam is directed from the beam splitter to illuminate the tested object which is dived in the water tank (6) and is reflected from the object (7) .

The reflected beam comes back through the beam splitter to the holographic plate (8) . The reference beam comes through the beam splitter (5) to the mirror (6) , and then is reflected from there to the holographic plate (8) .Camera (9) is used to photograph the holograms .

In order to get a good quality of the hologram /5,6 &9/, the difference between the object beam length and the reference beam length must be withen the coherence length of the used laser light .The optical path length of the object beam consists of two parts , the first is running in the air while the other part is passing through the liquid which has certain value of the refractive index , i.e.

$$l_{opt_1}(x,y) = l_g(x,y) n_1 + p(x,y) \tag{1}$$

where : $l_g(x,y)$ the geometric path of the object beam in the used liquid.
$l_{opt_1}(x,y)$ the optical path length of any point (x,y) on the object.
$p(x,y)$ the first part of the optical path length which is passing through the air.
n_1 the refractive index of the used liquid.

If parallel illumination is to be used figs.1&2 and the glass side of the water tank is at right angle to the object beam and by using the coordinates shown in fig.1, then the following equation may be used

$$Z(x,y) = 1/2 \cdot l_g(x,y) \tag{2}$$

At this stage the refractive index (n_1) of the liquid can be changed to $l_{opt(2)}(x,y)$, and it may be calculated from the following equation

$$l_{opt(2)}(x,y) = l_g(x,y) \cdot n_2 + p(x,y) \tag{3}$$

The difference in the optical path length can be calculated by using eq.(1) and eq.(3) as well as from the following equation

$$\Delta l_{opt}(x,y) = l_g(x,y)(n_1 - n_2) \tag{4}$$

By reconstruction such double exposure holographic interferometry, some points can be found on the object where the difference in the optical path length equals an odd number of the half wave length of the used laser light. Then black fringes will appear on the tested object. Each black fringe is the locus of these points which have constant distance to the reference plane (the glass side). This means that fringes appear if the following condition is realised.

$$\Delta l_{opt}(x,y) = (2i+1) \frac{\lambda}{2} \tag{5}$$

where Δl_{opt} the difference in the optical path length ,
λ the wave length of the used laser &
i integral number and equals to 1,2,3,.....

From equation (4) we get the following equation

$$l_g(x,y) = -\frac{2i+1}{n_1 - n_2} \cdot -\frac{\lambda}{2} \qquad (6)$$

If eq.(2) is used also, then the following equation may be obtaind.

$$Z(x,y) = -\frac{2i+1}{n_1 - n_2} \cdot -\frac{\lambda}{4} \qquad (7)$$

The right side of this equation is independent on (x,y). Then it can be said that the quantity (Z) describes all points on the surface of the object which lay on (i) fringe order with respect to the reference plane. Then the following equation may be used.

$$Z_i = -\frac{2i+1}{\Delta n} \cdot -\frac{\lambda}{4} \qquad (8)$$

where $\Delta n = n_1 - n_2$

Equation (8) is used for quantitative analysis, but only for the frozen (double exposure) holographic interferometry. If real time method is used, then the factor $(2i+1)$ in eq.(8) must be replaced by $(2i)$, and eq.(9) may be used for such quantitative analysis.

$$Z_i = -\frac{2i\lambda}{4\Delta n} = i - \frac{\lambda}{2\Delta n} \qquad (9)$$

The topographic map of the object may be obtained by using eq.(8) for lapsed method or eq.(9) for real time method in calculating the location of each point on the object related to the reference plane (glass side).
The density of the obtained fringes depends directly on (Δn) i.e. on the choice of the used liquid and on the change of its refractive index which can be calculated from eq.(10)

$$\Delta n = \frac{\lambda}{2\Delta Z} \qquad (10)$$

where $\Delta Z = Z_{i+1} - Z_i = \frac{1}{\Delta Z} \cdot \frac{\lambda}{2} \qquad (11)$

As clearly shown the density of the fringes for the obtained hologram (which describes the tested object) is dependant on the difference between the refractive index (Δn) for the used solution /7,8/.

While carrying out the experimental tests in the holographic laboratory, we must take care of the temperature /9/, because the refractive index (n) depends upon the temperature value. For increasing the temperature by 1°c, (n) may be changed within the limits of 3.5×10^{-4}. The refractive index (n) increases with the decrease of the wave length value of the used light. Its value must be calculated and corrected according to the used laser light. It may be mentioned here that some gases can be used instead of liquids such as freon /7/.

THE EXPERIMENTAL WORK

Lapsed time method was used to map the surface contours for some engineering objects. Water was the first liquid to be used, and the object was dived in it. The holographic plate was illuminated firstly with the half of the exposure time. Then two drops of alcohol was added to the water for obtaining good value of Δn (1 cm^3 alcohol per two liters of water and mixing for two minutes). So that (n_1) is the refractive index of the water and (n_2) is the refractive index of the new mixture (water + alcohol). The obtained holographic interferometry patterns are shown in figures (3,4,5 & 6). Figures (3 &4) are made for the same object but with another position with respect to the reference plane (the glass side of the water tank). Figs.5&6, show the obtained holographic pattern of a piston and of a turbine blade.
For quantitative analysis, a reference cone may be used /8/, its holographic pattern is shown in fig.7. The holographic pattern of the

tested piston while mixing the alcohol in the water is shown in Fig.8. Figure 9, shows the holographic pattern of another piston while adding one drop of alcohol to the water.

Qualitative analysis

In this part qualitative analysis will be discussed for some of the obtained patterns. The holographic interferometry pattern shown in Fig.3, will be taken into consideration firstly. These fringes which cover the object looks like the fringes which may be obtained for the object if it is turned a small angle about the vertical axis. But in reality there is no rotation, and the obtained fringes state that the object is tilted vertically with a small angle with respect to the reference plane. This angle can be calculated as given in the quantitative analysis. If these fringes were in the horizontal direction, that means that the object is tilted about horizontal axis with respect to the reference plane.

The obtained holographic interferometry of the piston Fig. 5, states that the object is either concave or convex about the vertical axis. The density of the fringes increases in both sides due to the bend of the object with respect to the holographic plate. This is proved to be convex since exactly similar pattern can be obtained from another illumination angle.

Quantitative analysis

The quantitative analysis for holographic interferometry pattern is made for Fig. 3. to calculate the tilted angle (θ) of the object with respect to the reference plane. e.g.

$$\theta = \Delta Z / 1$$

where $\Delta Z = Z_i - Z_1$ and (1) is the actual distance between fringes (Z_i) and (Z_1).

Using eq. (8), and taking into consideration $\lambda = 0.6328 \ \mu m$, $\Delta n = 0.0006$, then the value of ($\Delta Z = 475 \ \mu m$) is obtained and the tilted angle $\theta = \tan^{-1}(0.0475/40)$, where the actual distance between fringes order (21) and (1) equal to (40 mm).

Also, the quantitative analysis can be made for circular parts as the piston in Fig. 5, e.g. referring to Fig. 10. If δ is the out of roundness where $\delta = r - m$ and $Z = r - \sqrt{m^2 - x^2}$

$$\therefore m^2 - x^2 = (r - z)^2 \text{ since } \delta^2 \ll 2r\delta$$

$\therefore \delta^2$ can be neglected and the above equation reduces to

$$\delta = (z^2 + x^2 - 2rZ)/2r \quad (12)$$

Again if the out of roundness (δ) is considered to be zero, the diameter can be found from eq. (13)

$$2r = \frac{z^2 + x^2}{z} \quad (13)$$

The value of (Z) can be obtained from equation (8) while the value of (X) are measured from Fig. 5, taking the magnification factor into consideration. The results are shown in Fig. 11 which shows that the calculated piston diameter (D) is equal to 110 mm.

CONCLUSION

Holographic technique using immersion method proved to be a powerful tool when inspecting the geometrical form and dimensional error of complicated components. Also from ecconomical point of view only one laser unit is used with immersion method which recommend its application.

FUTURE WORK

Suitability of inspection and quality control of components using immersion technique in industry is the subject under study at the moment by the authors is Faculty of Engineering, Ain Shams University.

REFERENCES

1. Stroke G.W. , "An Introduction to Cocherent Optics and Holography" New York - London. Academic Press (1969).
2. E.R.Roberston and J. Haruey, " The Engineering Uses of Holography" Cambridge (1970) .
3. J.N. Butters,"Holography and its Technology" IEE Monograph Series 8 , (1971) .
4. Collier R.J. and Others, "Optical Holography" New York and London, Academic Press, (1971).
5. R.K.Erf, "Holographic Nondestructive Testing" Academic Press, New York and London, (1974).
6. Cathey W.T., "Optical Information Processing and Holography", New York , London , Sydney , Toronto , (1974).
7. Feuer T. and Others, "The Immersion Method", in Polish Warszawa, WAT, (1976).
8. Feuer T.,"The Technical Condition for Holographic Registration" in Polish Warszawa , WAT, (1980).

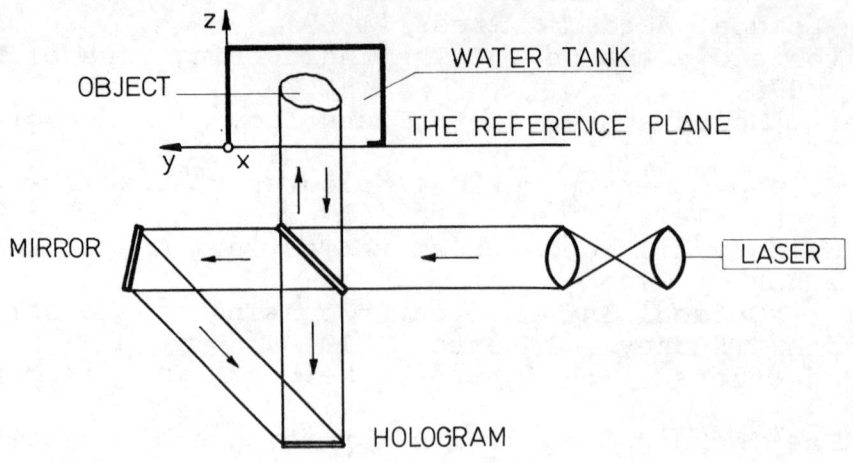

Fig.1. Determination of the contour map using immersion method.

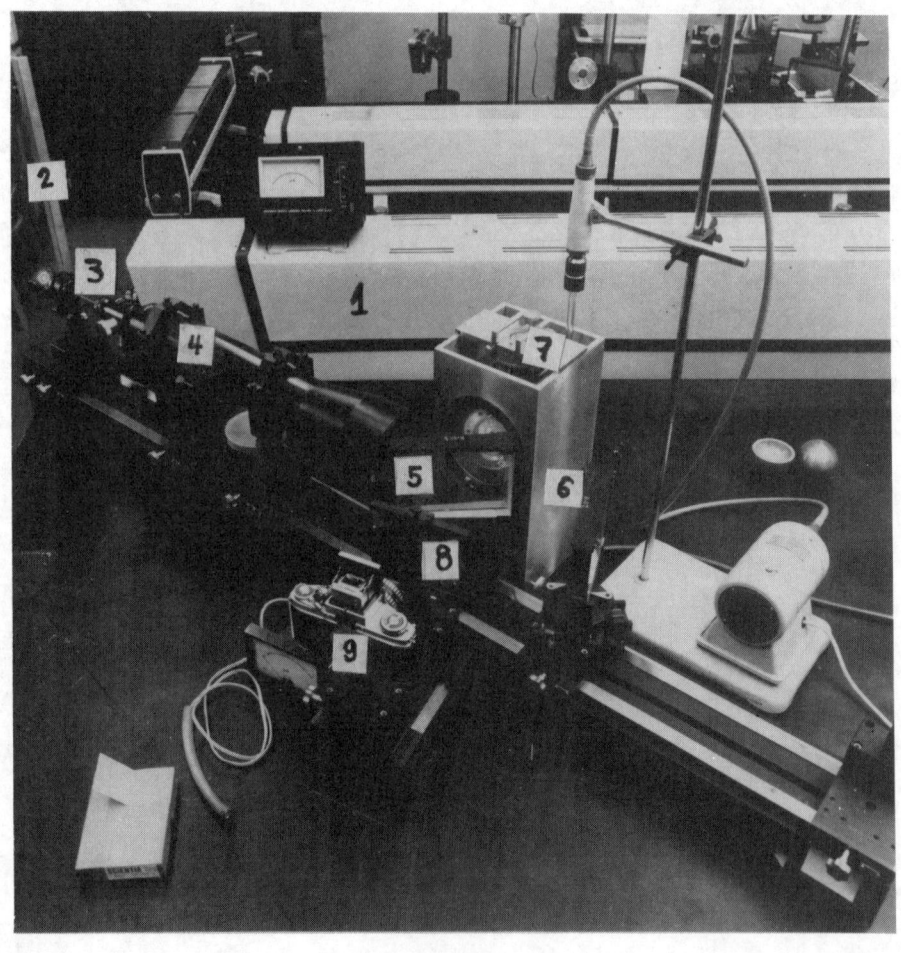

Fig.2. The used holographic arrangement.

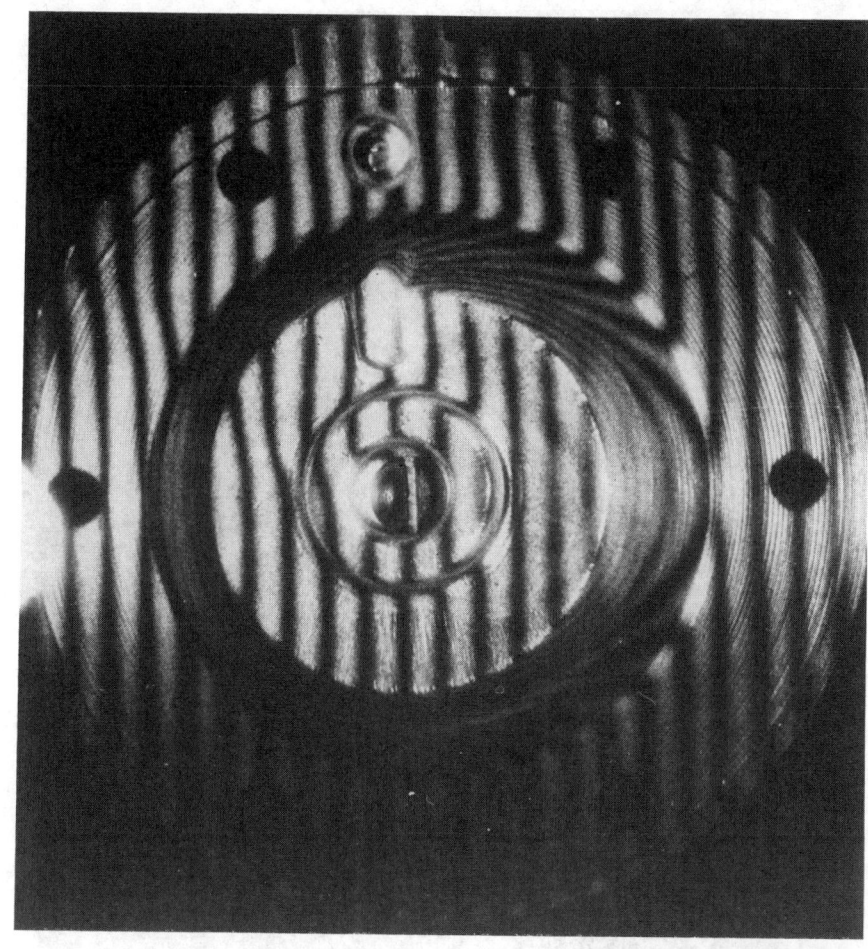

Fig.3. The holographic interferometry for certain object.

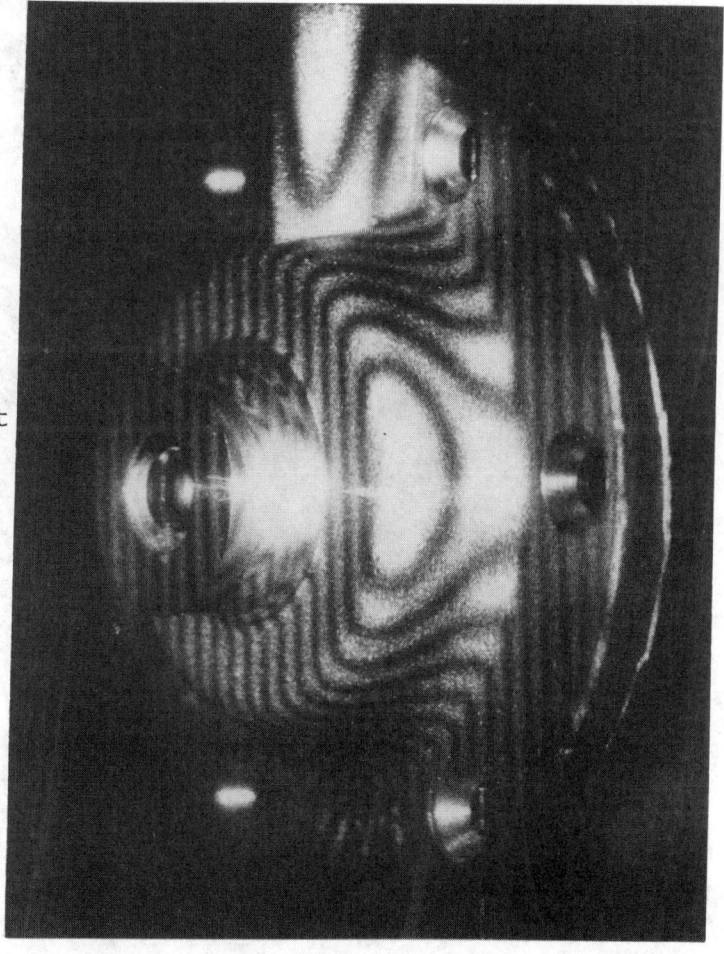

Fig.4. The holographic interferometry for the same object in another position.

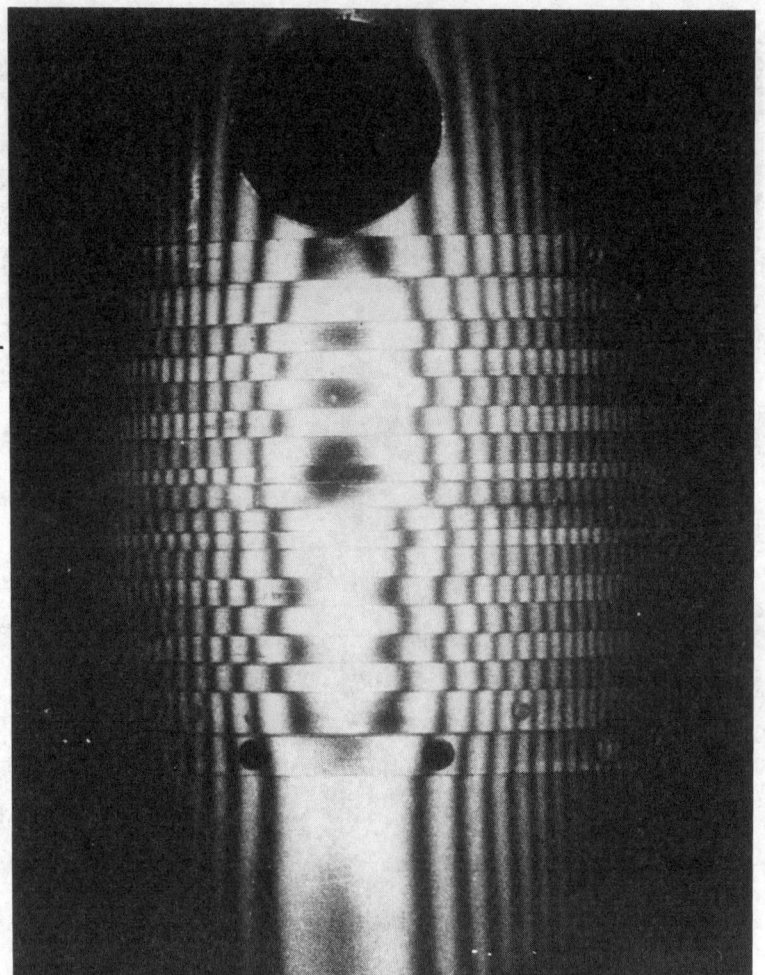

Fig.5. The holographic interferometry for the tested piston.

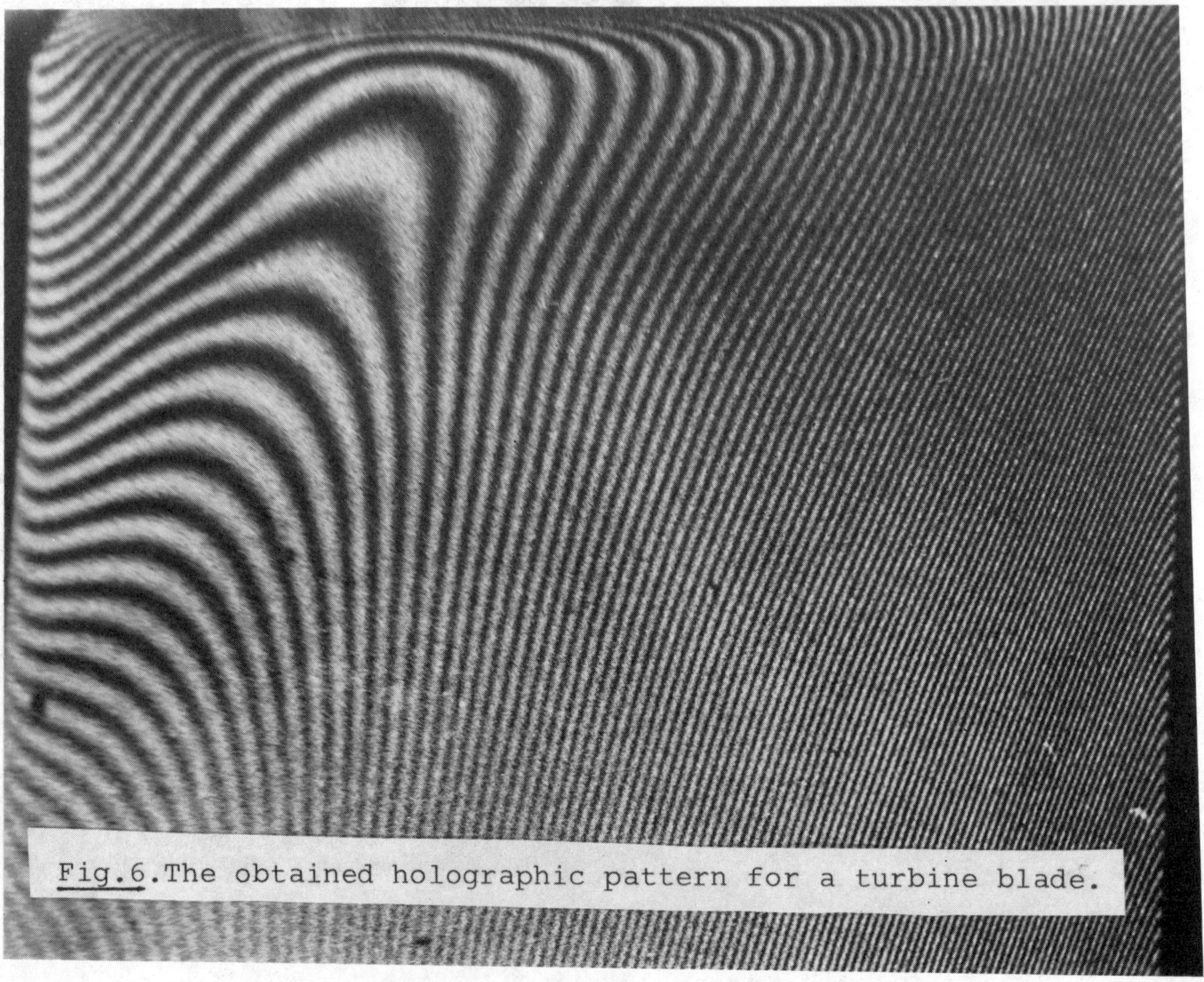

Fig.6. The obtained holographic pattern for a turbine blade.

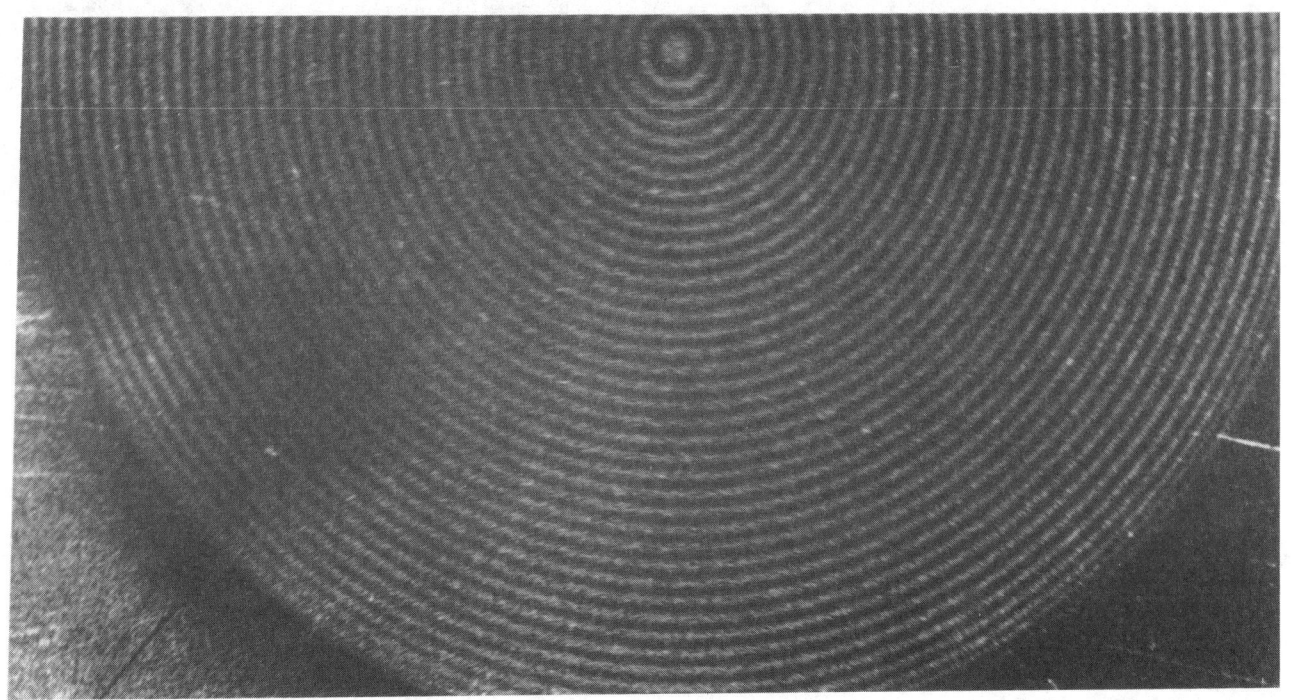

Fig.7.The holographic pattern for the used reference cone.

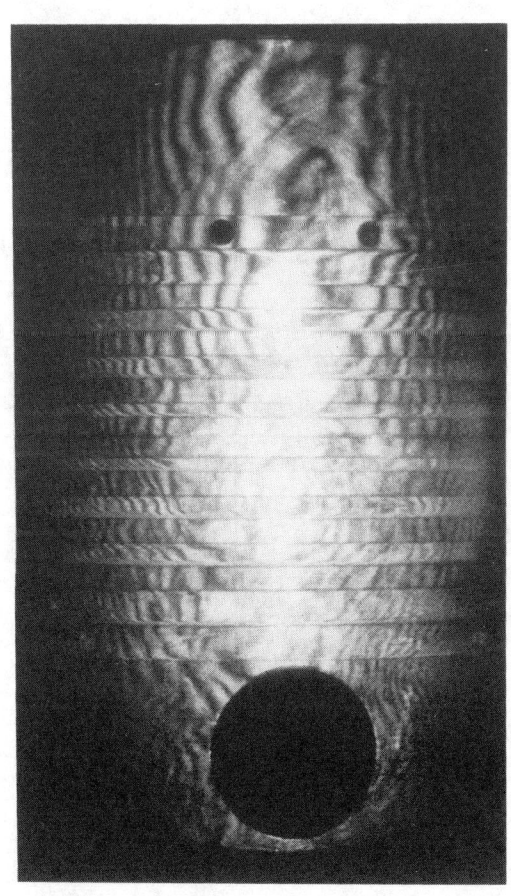

Fig.8.The holographic pattern of the tested piston while mixing the alcohol in the water.

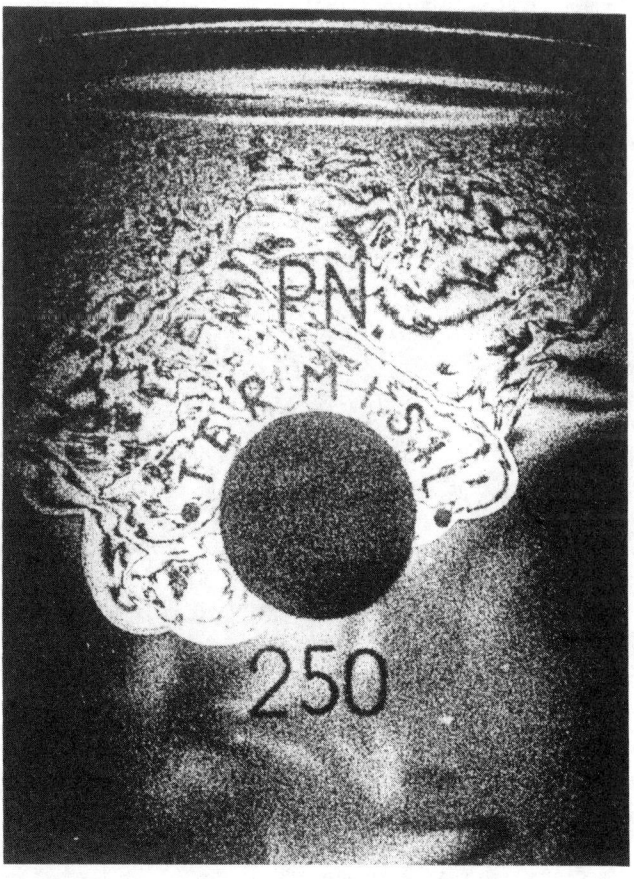

Fig.9.The holographic pattern of another piston while adding one drop of alcohol to the water.

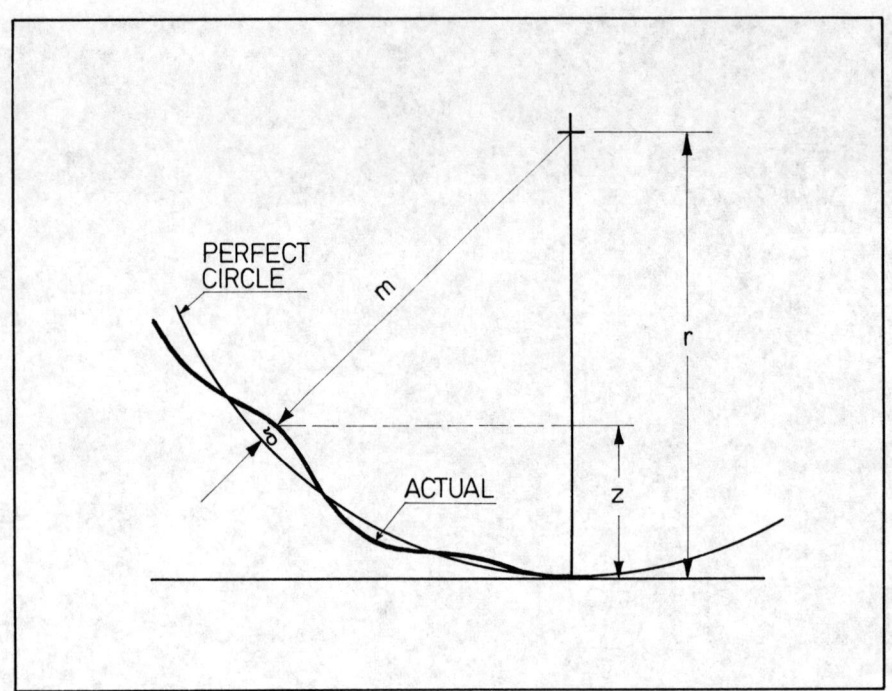

Fig.10.Theoretical analysis of circular arc.

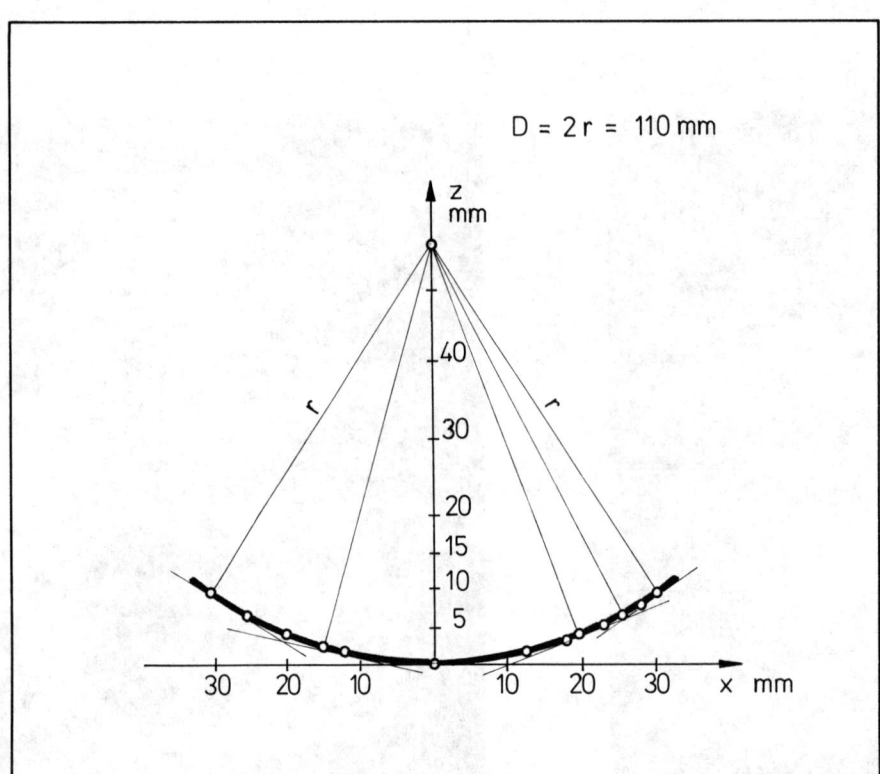

Fig.11.Graphical represéntation for the piston.

Optical polar profilometer: a new method for analysis of surfaces with circular symmetry

G. Laufer
and
E. Lenz
Israel Institute of Technology, Israel
and
Y. Fainman
University of California, USA

ABSTRACT

A noncontact optical measuring method is presented. This technique allows to record polar curves of surfaces with circular symmetry at a high accuracy which is obtained by using differential detection. It was demonstrated that surfaces may be analysed at an accuracy which exceeds 2.5μm.

1. INTRODUCTION

In a recent paper [1] a new profilometer has been discussed. This instrument, by using differential detection, allowed to obtain a longitudinal resolution of 0.1μm and transverse resolution of 2μm. A similar instrument which combines the optical profilometers of ref. [1] and [2] has been also constructed [3]. These instruments [1-3] can be used to generate projections of the boundaries of 2-D intersections in cartesian coordinate system. The construction of a 3-D surface requires cartesian projections of several judiciously spaced cuts. However, in many applications it is desirable to determine one parameter, such as radius of curvature or a cone angle and, therefore, one projection suffices.

On the other hand there exist many engineering applications where the surfaces of an axially symmetric 3-D circular body must be studied. An example for such an application is the drill point geometry. The geometrical parameters of drill points are most important in determining the material removal rate, tool life etc. A drill analyzer using a reciprocating stylus was developed by Forgacs and Ber [4]. Their instrument generates polar plots of the drill point geometry. Nevertheless, the application of this instrument was limited to hard surfaces.

Sensitive surfaces such as lenses and mirrors can not be tested by a stylus, therefore, it is desirable to apply a noncontact optical technique for the measurement of axially symmetric bodies. A new instrument based on the principle of differential detections [1] is presented. This instrument, the Optical Polar Profilometer (OPP), was de-

---

* Yeshaiahu Fainman is with the department of EE&CS, University of California, San Diego, La Jolla, CA 92093.

signed for the analysis of circularly symmetric surfaces. It is expected that the differential detection, in comparison to position detection using photodetector arrays [5,6], will introduce such advantages as high resolution and wide dynamic range.

2. THE MEASUREMENT PRINCIPLE.

The objective of the Optical Polar Profilometer is to describe graphically the boundaries of transverse cross sections of the sample, in contrast to the longitudinal profilometers [1-3] which describes profiles of longitudinal cuts. Since the bodies to be tested have axial symmetry, cylindrical coordinate system will be used. The OPP (see Fig. 1) is designed to follow the locii of points which are at a preset distance, h, from a reference point on the surface of the sample (e.g. the drill point in our experiments) and record their θ and r coordinates, while the sample is rotating.

A collimated laser beam parallel to the z-axis is focused on the sample surface by the lens L1. The focal point A is at a distance h from the tip. The sample is mounted on a stage with controlled r, z and θ-drives. The diffusively scattered light is collected by lens L2 which through the beamsplitter, BS, images the point A onto points A' and A" on detectors D1 and D2 respectively. Two knife edges are placed in front of the detectors and are adjusted to block half of the incident light power. Both detectors are callibrated to yield at this point identical signals. Thus, the output of the differential amplifier (DA), to which these signals are fed, is zero.

If the sample is not a perfect cone, a θ-rotation will cause the incidence point to shift, e.g. to point B. The imaged spots will then be shifted to points B' and B". As the imaged spots move, the light power transmitted through this knife edge on D1 decreases, while the power on D2 increases. The output signal of the DA changes now from zero. This new signal activates the r-drive of the sample mounting stage and moves it to the left until point C which is on the sample surface at a distance h from the tip coincides with the stationary focal point A. At this point the imaged spots coincide with points A' and A" on the edges of the knives: the signals of the detectors are balanced, the output of the DA vanishes and the r-drive is stopped. The θ-drive may be activated when the output of the DA vanishes or, alternatively, it can be left continuously scanning at a sufficiently slow rate allowing the r-drive to follow the surface pattern. Both θ and r coordinates may be obtained by monitoring the overall θ and r translations.

The sign of the DA serves to determine the scanning direction of the r-drive. Therefore, if during a scan, point B crosses the beam in front of the focal point A, i.e. the local surface is convex, the sign of the DA output is reversed and the r-drive is moved to the right. The measuring system is a closed feedback system similar to the system of ref. 1.

Unlike the longitudinal profilometer which uses the quadratic phase detection, the present profilometer responds to the transverse shifts of the illuminated spot. Therefore, an incoherent light source may also be used. The application of a laser in this instrument is primarily justified by the narrow, diffraction limited, focal diameter and the high fluence of the beam.

The resolution of this instrument is limited by the spot size at the focal point and by the slope of the measured surface. The highest sensitivity is for a detection at 90° to the illuminating leg, however, sometimes the angle between the illumination leg and detection leg must be selected such that collection of specular reflection is avoided. This is necessary for eliminating detector saturation and erratic behaviour.

If the maximum signal power detected by either detector is 2P(0), the maximum difference obtainable from the DA can not exceed this value. Assuming that the knives were initially adjusted such that both detectors read P(0), a shift of half the spot size transverseley to the detection line, results in a maximum reading, 2P(0), at the DA. We consider this as the longitudinal on-off resolution limit. In reality, the sensitivity of the instrument to translations of the surface, transversely and longitudinally to the detection line exceeds this limit and is limited by the signal noise level. In our experiments the limit was set by the resolutions of the mounts stepper motors which was 2.5μm for the r transalotion.

3. EXPERIMENTAL SETUP AND MEASUREMENTS

Since the system resolution is limited by the spot size at the sample surface, a lens with small F# was selected. In order to insure high collection efficiency the detection was performed also through a small F# lens.

The experimental setup consisted of a 50 mW He-Ne laser focused by a microscope objective (N.A.=0.12) onto the sample surface, which was mounted on r-θ stage. Due to the size of the laser beam the working F# in this experiment was 23. The diffusively scattered light was collected by a similar microscope objective and focused through a beam splitter, on two simple solar cells equipped with knife edges. The differentially amplified output of the detectors was used to drive the stepping motor of the r-positioning stage. The stepping motor of the θ-drive was run continuously at a rate of 0.6 deg/min, this allowed the system to follow fluctuations in the r-direction. The resolution of the θ-drive was 0.175mrad (1800 steps/rev), while the resolution of the r-drive was 2.5μm (400 steps/mm). The on-off resolution of the optical system was 9.3 μm for the lens F#=23.

In all our experiments the position of r-drive was monitored by a Tesa indicator with a resolution of 0.2μm while the θ-position was monitored by a counter. Both readings were used to form r-θ plots.

In order to demonstrate the use of the profilometer for industrial applications the cutting surfaces of a 20mm drill were measured. Three contours at 2mm, 3mm and 4.6mm from the web have been recorded. These contours were then superimposed on the projection obtained from a profile projector as it is presented in Fig. 2. Only the contours which lie on the clearance surfaces were recorded. The flute region could not be maped, because the cutting edge intersected the detection line when the laser beam was propagating within the flute. This figure compares favorably with polar plots of a 9.5mm drill previously investigated by a microprocessor controled stylus scanner [4]. The sensitivity of the system was limited only by the step size of the stepper motors which for the r transalation was 2.5 μm.

REFERENCES

1. Fainman Y., Lenz E. and Shamir J. "Optical Profilometer: A New Method for High Sensitivity and Wide Dynamic Range". Appl. Opt., Vol. 21, pp. 3200, 1982.

2. Arecchi F.T., Bertani D. and Cilberto S.: "A New Versatile Optical Profilometer", Opt. Commun., Vol. 31, pp. 263, (1979).

3. Dobosz M.: "Optical Profilometer: A Practical Approximate Methos of Analysis. App. Opt., Vol. 22, pp. 3983, (1983).

4. Forgacs R.L. and Ber A.: "Microprocessor - Controled Drill Analyzer", IEEE Trans. on Inst. and Meas., Vol. IM-30, pp. 258, (1981).

5. Parker L.: "Measuring Height with a Position Sensing Detector", Laser Focus, p. 40, (July 1980).

6. Waters J.P. "Gaging by Remote Image Tracking" Opt. Eng., Vol. 18, pp. 473, (1979).

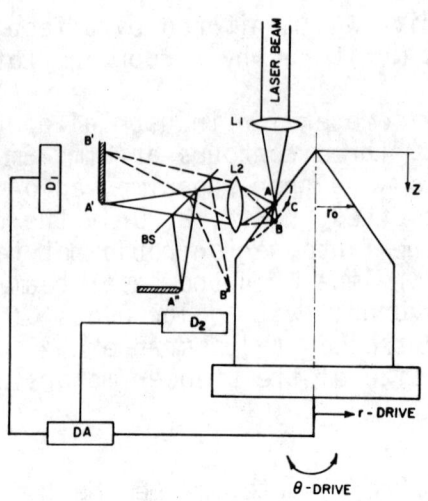

Fig. 1. Schematic representation of the Optical Polar Profilometer.

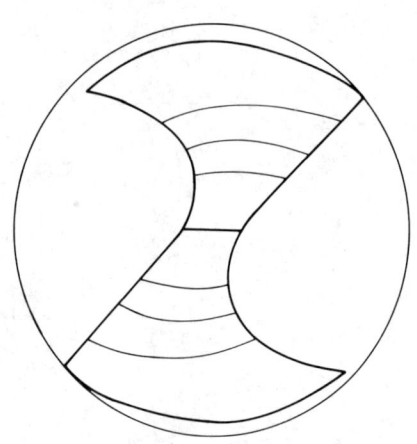

Fig. 2. A polar r-plot of a 20mm drill: heavy lines are projections obtained from profile projector while the light lines are contours of equal height at 2mm, 3mm, and 4.6mm from the web as measured by the Optical Polar Profilometer.

Optical fibre sensors in industrial process control
N. Macfadyen
Barr & Stroud Ltd, UK

Optical fibre sensors are expected to take ten per cent of the market for sensors in industrial process control in the 1990's. The presentation will review the current state of development of optical sensors and identify those application areas where the power and coherence of a laser make it the preferrable, or only possible, light source.

The exploitation of coherence in a sensor requires the use of single mode optical fibre, and once the information is transmitted in a single mode format integrated optical circuits are prime contenders for the pre detector signal processing which is often required. Developments in integrated optics will be described, and relevent circuit architectures discussed.

The total sensor market in 1983 was $5 Billion, and is expected to grow at 30 percent a year into the mid 1990's. This high rate of growth is explained by the demand for sensory inputs by the increasing numbers of microprocessor controllers, and the economic benefits derived from tighter control of automated processes.

Among the initial applications for these sensors, process control is an example where signal transmission distances can be quite long (a few Km's) parameters must be sensed with a high degree of accuracy (0.1%) and in many applications intrinsic safety is required. Optical signal transmission offers the benefits of intrinsic safety, immunity from interference, a dielectric transmission path and a high bandwidth with low loss, unavailable from any other technology.

Optical Sensor technology encompasses as wide a range of performance as is available from electronic or pneumatic equipment, and in their simplest form optical sensors have been in use for many years. The change in emphasis in the last ten years which has attracted the attention of industrial and military research labs in addition to vast numbers of academics, has been in the application of optical techniques to making accurate, calibrated measurements and in addition exploiting the low loss of optical fibre to transmit the optical interrogation beam to the passive sensor, and return the encoded optical signal to the detection electronics.

An example of the first generation of optical sensors is the photonic sensor illustrated in figure 1. An object is illuminated by a fibre or bundle of fibres, and the reflected light is collected by a second bundle and transmitted to a detector. This is an example of an 'extrinsic' fibre sensor, because the fibre is used purely as a means of transmitting light to and from the sensing point. In most applications such a device is used purely as a proximity sensor, and does not offer a calibrated output. The fibres are incorporated to reduce the number of light sources/detectors required by a sensor array, and to provide design flexibility.

A second example of a simple monitoring (uncalibrated) sensor is the fibre optic cryogenic temperature alarm sensor. This device is an example of an 'intrinsic' fibre optic sensor, because the fibre itself is the sensor, and in addition it is an example of a distributed sensor, as the sensitive fibre can be laid out in a grid or along a pipeline and sense sudden reductions in temperature at any point along its length.

The principle of operation is illustrated in figure 2, and also serves to illustrate the principle of low loss light propagation in an optical fibre, for those who are not familiar with it. Light is injected into the core region of the fibre, and because the core is of a higher refractive index than the surrounding optical cladding, the light is confined by total internal reflection, and propagates along the core. In normal operation the propagation process is so efficient that the light is only attenuated a few dB's (50%) per kilometer. However, if the fibre is constructed out of materials whose thermal coefficients of refractive index differ, such that the index of the cladding increases faster than that of the core as the temperature drops, then there will be a temperature at which the core index is lower than that of the cladding, the light guiding conditions will no longer exist and the light will leak out of the fibre, detected as a loss of signal at the receiver.

Both these fibre optic sensors are production items, the first from Pilkington Fibre Optic technologies, where it is configured to monitor the quality of cigarettes during production. The second is available from Pilkington Security Equipment, the application being the detection of leakage of low temperature liquids from storage tanks. Barr and Stroud, who are also a member of the Pilkington Electro Optical Division, are undertaking a development programme targeted with the production of a range of fibre optic transducers for use in the measurement of physical parameters such as temperature, Pressure, Flow, Level Current and Electric Field.

In the remainder of my presentation I shall review some of the techniques that are being developed to sense these parameters, and in particular I shall describe those techniques which exploit the advantages of semiconductor lasers, namely high power (compared to LED's) large modulation bandwidth and coherence.

The available sensing techniques may be grouped under four headings.

1. Intensity
2. Wavelength/Frequency
3. Polarisation
4. Phase

1. INTENSITY

The simple photonic sensor described in the introduction to this talk is an example of an intensity modulated sensor. It cannot be used for calibrated measurements at the end of an arbitrary length of fibre cable because the signal attenuation induced at the measurement point by the measurand cannot be distinguished from the attenuation which occurs in the fibre optic cable and at the various connectors which might be installed along its length. One method of overcoming this problem is to transmit light at two different wavelengths, narrowly separated, such that the cable and connectors affect both wavelengths in substantially the same way but a filtering arrangement is used at the sensor head itself so that, for example, one wavelength signal experiences a constant reflectivity whilst the other is modulated by the measurand. Taking the ratio of the two wavelength signals gives an output which is compensated for variations in the transmission system. Further refinement of the technique as illustrated in figure 3, includes monitoring the intensities of the two input beams and multiplexing the two light sources onto the two detectors, so that detector degradation is also compensated for.

Because the two wavelengths are not identical, errors are introduced by cable length, connectors, detector sensitivity and also optical coupler stability.

Despite these problems the technique has attracted considerable attention and has been adapted for many sensors, for example the temperature sensor based on the absorption of edge shift in a coloured glass filter with temperature, as illustrated in figure 4. One wavelength is chosen to the side of the absorption edge and is unaffected by temperature, whilst the second wavelength is chosen to lie on the absorption edge and the transmission varies with temperature.

The semiconductor laser finds application in these sensors because of its narrow linewidth and because its operating wavelength can be selected.

An alternative intensity modulation scheme overcomes uncertainties over signal level by using digital transmission. As illustrated in figure 5, a simple coding plate placed between the input and output ends of the fibre cable link will modulate the transmitted beam into a series of pulses, giving an incremental measure of displacement or a measure of velocity if the pulse repetition frequency is monitored. In this example the detector trigger levels can be set to accept optical signals varying in power by as much as 30 dB, and as the incoming signal is binary, variations within that range are of no concern.

The technique can be extended to provide an absolute measurement of displacement by incorporating a set of optical delay lines. If a pulse of light is launched into ten fibres simultaneously and (for example) the fibre lengths range from 5m to 50m in 5 m steps, then the light pulse emerging from each fibre will be delayed to an extent determined by the fibre length. In glass light propagates approximately 1 metre in five nanoseconds, therefore 5 metre delay lines will give a train of pulses separated by 25 ns. ie. a 40 MHz data rate. If each fibre interrogates a channel of an absolute coding plate, then we can produce a serial coded indication of displacement.

In this application, a laser light source provides a short high power pulse of light that will still only be a few nanoseconds wide after propagation through several kilometers of fibre.

2. WAVELENGTH/FREQUENCY

Wavelength and frequency are directly related but have been named individually because in optical terms they tend to be used to describe different magnitudes of bandwidth. However for optical sensors, both offer the possibility of a readout independant of optical intensity.

Broadband sources (incoherent or low coherence) are described in terms of their bandwidth, eg. an 850 nm LED with a linewidth of 50 nm, or a multimode laser diode with a linewidth of 5 nm. In terms of frequency a 5 nm linewidth is 2 THz. By comparison a single longitudinal mode semiconductor laser can have a linewidth of only 3 MHz, or 10^{-5} nm, which can be tuned over a range of 100 GHz or more (2 nm). Coherence is also related to linewidth and determines, for example, the number of fringes with an acceptable degree of contrast that can be observed in an interferometer. An LED has a coherence length of 15 um, a multimode laser diode 150 um, and a 3 MHz, SLM diode 100m.

I shall give two examples of this form of sensing, one using wavelength (over a vast frequency range) and the second using frequency, over a very small wavelength range.

The output of the coloured glass temperature sensor described in the previous section could, for example, be monitored using a spectrometer, allowing operation over a wider temperature range with a suitably broadband light source than can be achieved with a single sensing wavelength where the range is limited by the absorption edge moving away from that wavelength. In a ruggedised form the spectrometer would use a CCD or photodiode array for the output, and the broadband light source might be provided by two LED's of overlapping spectral response. Other more complex output spectra may be monitored in the same way.

The main errors in this system approach come from distortion of the output spectrum due to variation in the input spectrum and differential spectral absorption in the fibre. The technique uses broadband light sources and therefore lasers are not suitable.

The most common form of sensing using frequency shifting techniques in Laser Doppler Anemometry; light scattered from a moving object has its frequency altered due to the linear phase shift induced on reflection from a body moving at a constant velocity w.r.t. the light source. If the scattered light is mixed with coherent unscattered light the two frequencies beat together and the beat frequency can be detected to measure particular velocity. The wavelength shift induced in this process is extremely small in most industrial applications and therefore cannot be detected by a spectrometer. However high velocity objects such as stars can have their velocity measured without the need for a coherent reference, because the observed shift in their emmitted spectra is large.

A similar technique can provide range information for stationary objects. As illustrated in figure 6, if the frequency of the laser light source is chirped, and returned reflected signals are mixed with the transmitted signal then the beat frequency is a function of the range of the object from transmitter.

Both these frequency monitoring techniques can be implemented remote from the laser source by using transmission via single mode optical fibres, in which the coherence of the light is largely retained. A laser light source is obligatory and its degree of coherence determines the dynamic range of instrument.

3. POLARISATION

Two examples of the sensing of polarisation effects will be given. Polarisation is not sensed directly, but by monitoring the intensity passed by an analyser. This can be done at the sensor head, in which case subsequent polarisation changes during fibre transmission are irrelevant, or immediately before detection.

The first example I will take in that of the conversion of light from linear polarisation to an eliptical polarisation in a birefringent crystal. Figure 7 illustrates the mechanism, where linearly polarised light is propagated through a birefringent crystal at 45 degrees to the ordinary and extraordinary axes. The polarisation that emerges is a function of the retardation experienced by the extraordinary rays, and the intensity that passes the analyser is a COS^2 function of that retardation.

Taking the example of lithium niobate, the ordinary and extraordinary indices as a function of temperature and wavelength are shown in figure 8. The transmitted intensity at a particular wavelength is a function of temperature but equally the transmitted spectrum from a broad band source can be monitored to give a intensity independant temperature output.

The second example of polarisation sensing I will describe involves Faraday Rotation. In the presence of a magnetic field a linearly polarised beam is rotated, and the phenomenon can be used to sense current flowing in a conductor by non-contact measurement of a line integral of the magnetic flux (a single loop of optical fibre placed round a conductor carrying 12Ka will rotate the propagating polarisation by 7°. The analysis scheme is described in figure 9. When no current is flowing the analyser delivers equal signals to the detectors. When a current is flowing the two signals will be unequal and the ratio will give a measure of the current.

A semiconductor laser is a reasonably well polarised source, giving up to 23 dB extinction, but for a high performance instrument an external polariser is needed, so the main advantage of the laser is once again high input power.

4. PHASE

Optical phase detection is undertaken in an interferometer, the coherent mixing of the two beams shifting the 300 THz carrier frequency down to baseband where phase shifts are detected by fringe counting or intensity monitoring. An excellent and extremely simple version of this is the optical fibre hydrophone, or pressure sensor. As illustrated in figure 10, light from a semiconductor laser is coupled into an optical fibre and then split into two paths by a fibre coupler (beamsplitter). One arm of this interferometer is exposed to the parameter to be sensed and the second arm is isolated from it, but hopefully exposed to identical unwanted influences such such as temperature. Hydrostatic pressure applied to the sensing fibre induces a strain along the fibre length, which is equivalent to an increase in the optical path in the sensing arm of the interferometer, and will cause a fringe movement at the output. Where the sensing and reference arms are mixed in a second fibre coupler, the light in the two output ports (ends) will be intensity modulated 180° out of phase.

The efficiency of such a pressure sensor is enhanced by biasing the output such that it operates at the point of maximum sensitivity (a bias). This can be introduced by a controlled stretching of the reference fibre using a piezoelectric cylinder or by the use of an integrated optical phase controller. Using this technique phase changes as small as 10^{-6} radians can be detected ie. a displacement of 1×10^{-13} M, or less than 1/1000th of an atomic radius

The hydrophone is a successful fibre optic sensor because it is extremely sensitive and, due to its distributed nature, can be formed into complex arrays providing high directional sensitivity. It is also successful because it is an AC sensor with limited bandwidth, and therefore low frequency drift and instabilities can be filtered out electronically. The laser provides high power and a reasonable degree of coherence so that the fibre lengths in the interferometer need not be identical.

CONCLUSION

To include, I have illustrated with examples some of the techniques available for the realisation of optical fibre transducers, and I hope this brief review has

indicated the potential for the technology in progess control. There is
considerable UK interest in the development of optical sensors, from both users and
manufacturers, and the majority of companies with a serious interest are members
of OSCA, the Optical Sensor Collabrative Association, which is sponsored by the
DTI, and funds R & D projects to its members mutual benefit.

BIBLIOGRAPHY

1. D.H.S.Jones "Fibre Optics and their Applications to Miniature Flow Transducers"
    CME, February 1984, pp 35-37.
2. A.Rodgers "Measurement Using Fibre Optics
    New Electronics, October 27th 1981, pp 29-36.
3. First International Conference on Optical Fibre Sensors. London
    26-28th April 1983 IEE Publications, 1983.
4. Second International Conference on Optical Fibre Sensors. Stuttgart
    5-7th September 1984, VDE-VERLAG
5. Third International Conference on Optical Fibre Sensors. San Diego
    13-14th February 1985, IEEE/OSA Publication.
6. OSCA is managed by SIRA
    Mrs.P.West, SIRA Ltd, South Hill, Chislehurst, Kent BR7 5EH.

FIBRE BUNDLE

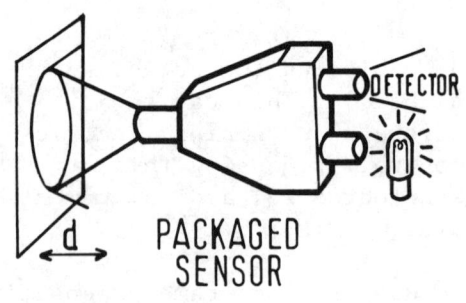

PACKAGED SENSOR

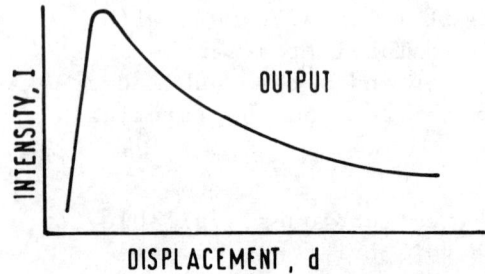

Figure 1   THE PHOTONIC SENSOR.

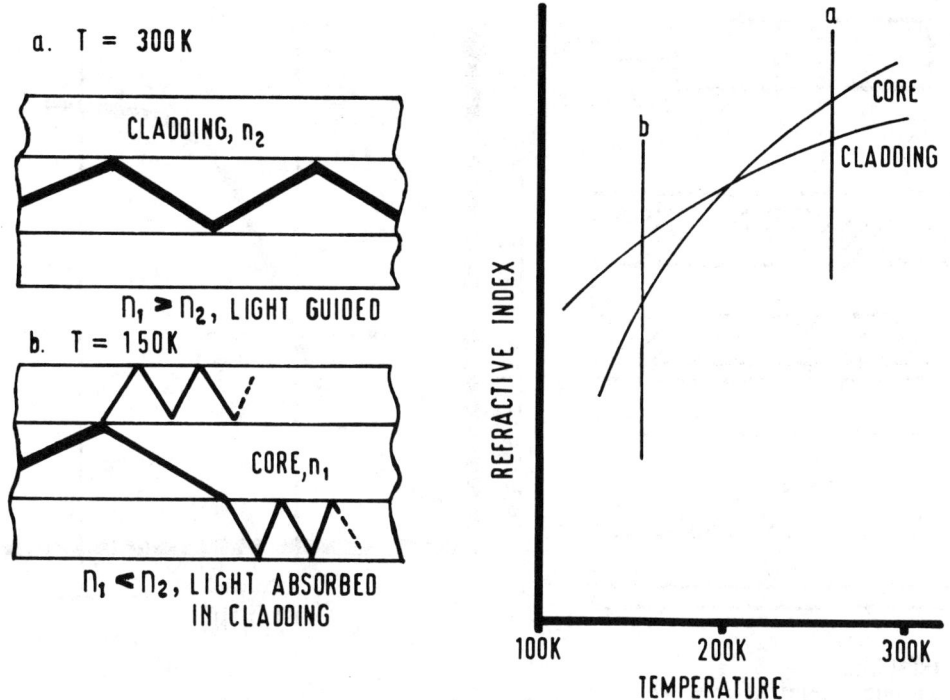

Figure 2   TEMPERATURE ALARM SENSOR
(a) Normal Operation of the Fibre
(b) 'Alarm' State

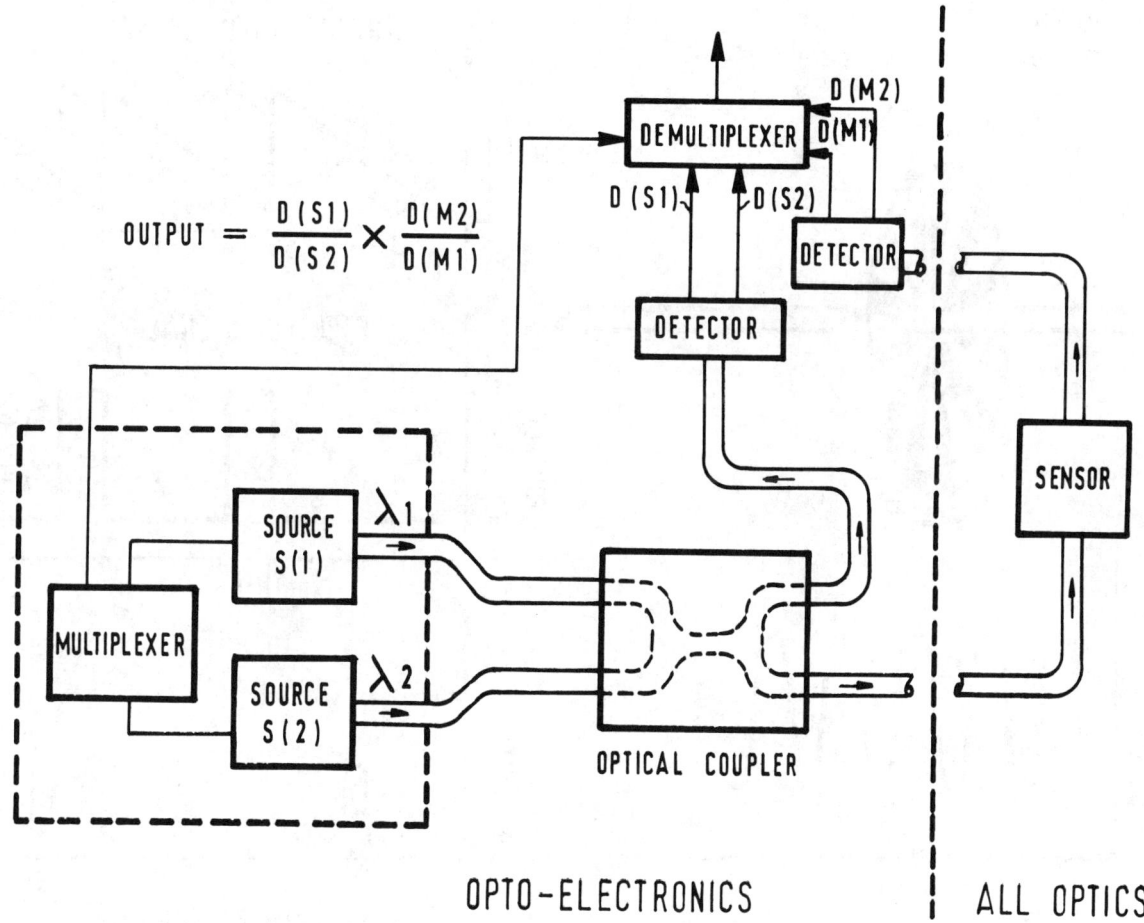

Figure 3   TWO WAVELENGTH REFERENCING

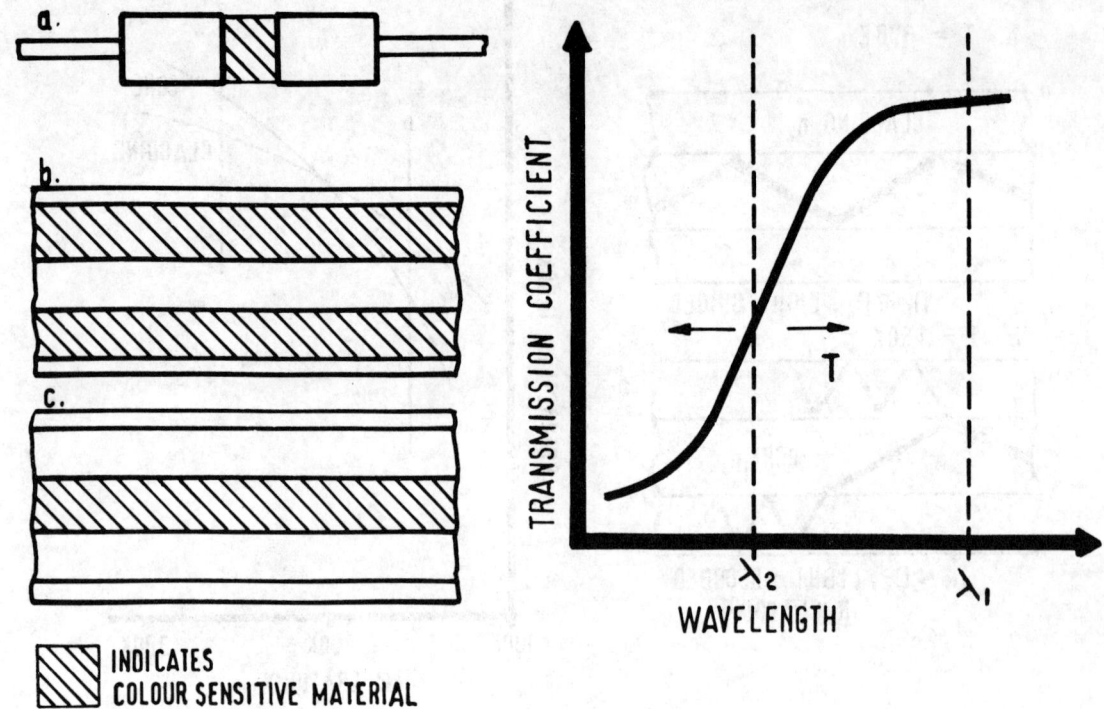

Figure 4 BAND EDGE TEMPERATURE SENSOR SHOWING SPECTRAL OUTPUT AND TWO SUITABLE MONITORING WAVELENGTHS

a. Colour sensitive material sandwiched between two lenses.
b. Colour sensitive fibre cladding.
c. Colour sensitive fibre core.

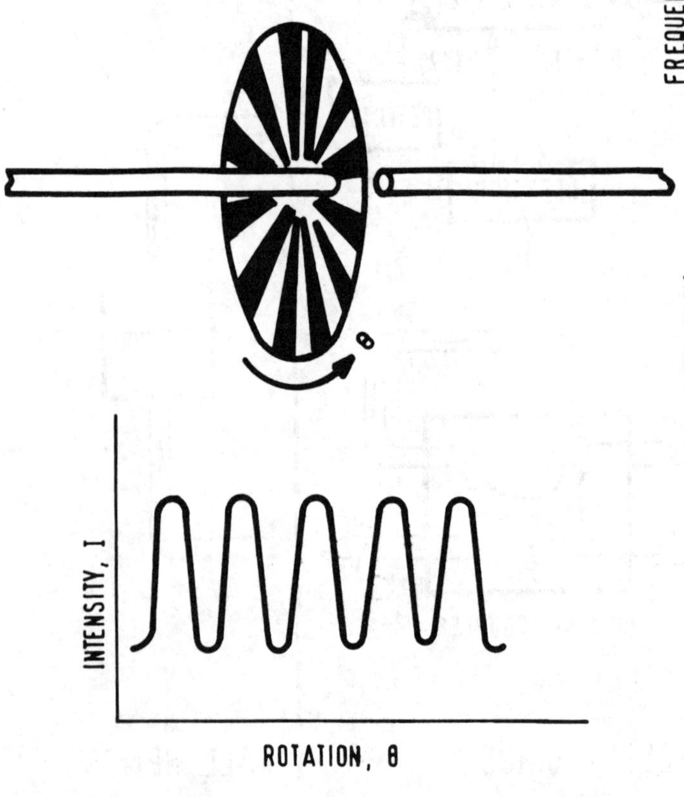

Figure 5 CODING PLATE ROTATION SENSOR

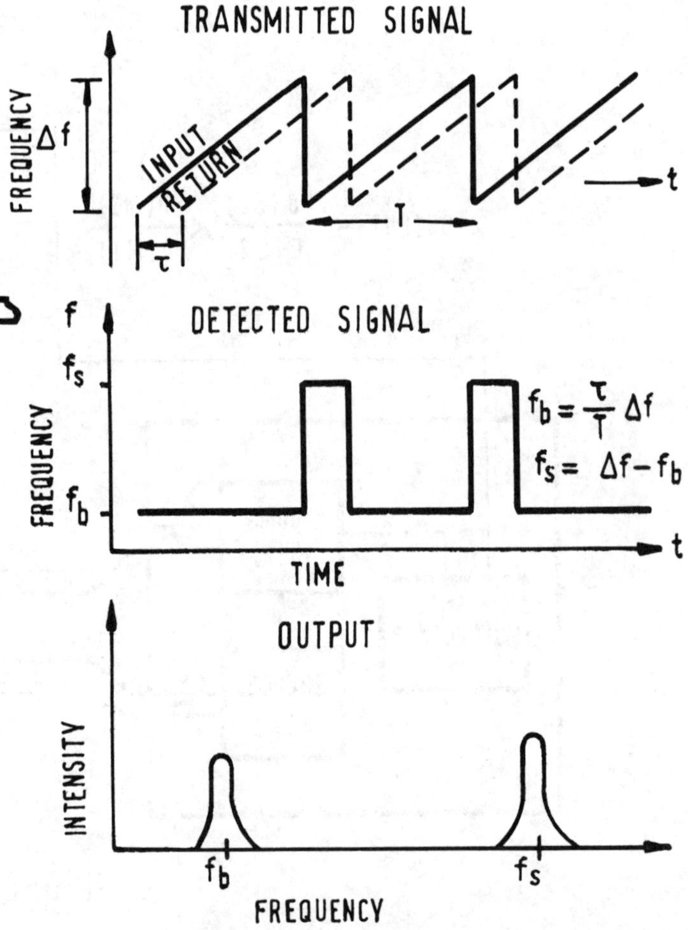

Figure 6 FREQUENCY MODULATED LASER RANGE SENSOR (RADAR)

$$f_b = \frac{\tau}{T} \Delta f$$
$$f_s = \Delta f - f_b$$

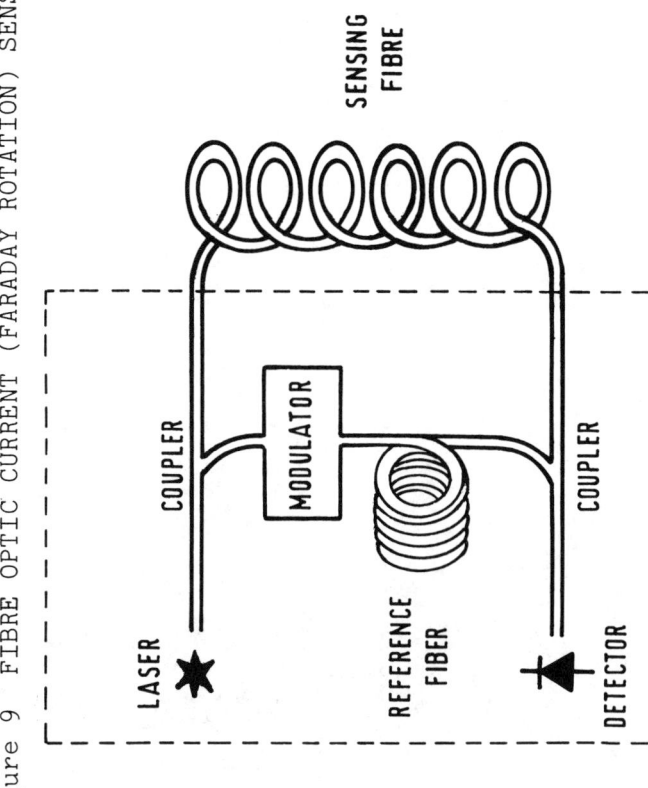

Figure 9 FIBRE OPTIC CURRENT (FARADAY ROTATION) SENSOR

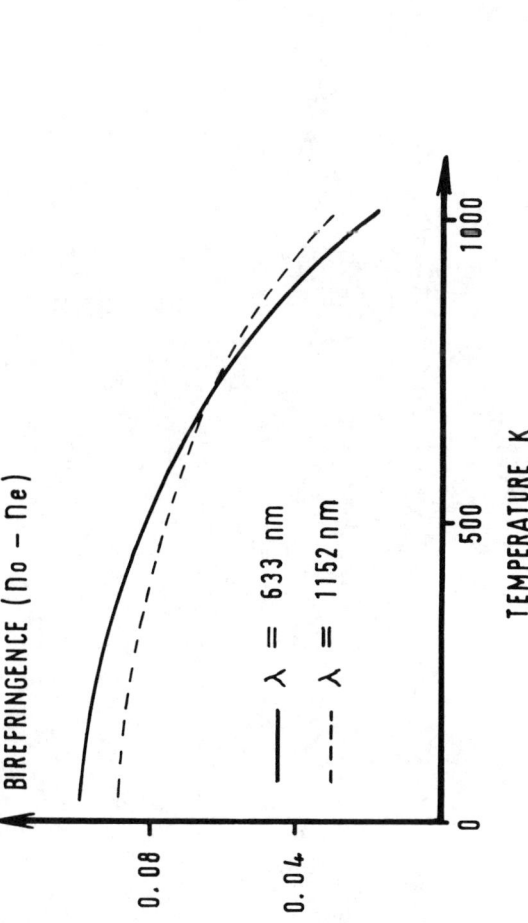

Figure 10 SINGLE MODE FIBRE INTERFEROMETRIC SENSOR

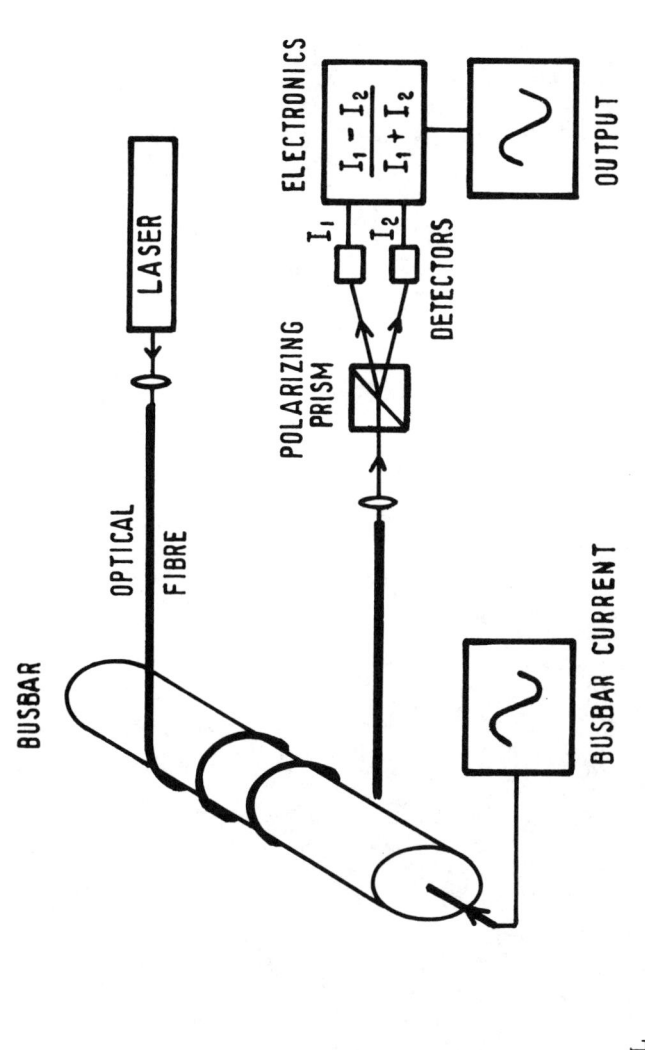

Figure 7 MONITORING BIREFRINGENCE OF AN LiNbO$_3$ CRYSTAL

Figure 8 BIREFRINGENCE AS A FUNCTION OF TEMPERATURE AND WAVELENGTH

Laser diagnostics of combustion devices and chemical reactors using coherent Anti-Stokes Raman spectroscopy

D. A. Greenhalgh
Harwell Laboratories, UK

ABSTRACT

Coherent Anti-Stokes Raman Spectroscopy has been used to measure temperature and species concentrations in research and production devices. In this paper the principals of the technique are briefly introduced. Examples of recent measurements on an oil-spray furnace, a catatlytic chemical reactor and an incandescent lamp are given. The application of CARS to other areas is discussed, these include internal combustion engines, heat transfer and micro-electronics.

INTRODUCTION

Coherent Anti-Stokes Raman Spectroscopy (CARS) is a laser spectroscopic technique for probing gaseous systems. CARS has been widely used for "in-situ" probing of temperature and species concentration in a wide variety of research [1-5] and development [6-10] devices. Current application areas include combustion, chemical engineering, nuclear and micro-electronic. To date its greatest usage has been for probing hostile combustion systems, normally for thermometery but increasingly for species concentrations as well. For diagnostics in systems such as I.C. engines, jet combustors and oil or coal furnaces it offers certain significant advantages over conventional mechanical probes.

(i)   it is non invasive (apart from any windowing when required)

(ii) it is spatially precise

(iii) measurements are made in a single 10 nano-second laser pulse - thus flow fluctuation (turbulence) is frozen

(iv) it is insensitive to background luminosity or fluorescence emissions

(v)   it is durable - laser beams are not irreparably damaged by violent flow transients or particles.

More recently CARS has found use in a wider range of applications and in particular chemical [3,11] and micro-electronic processes [12]. For measurements in these systems the technique has been adapted to probe specific species concentrations. Apart from the advantages noted for combustion systems, CARS offers the potential for probing in certain types of restricted spaces [11]

METHOD

The theory and application of CARS spectroscopy have received wide attention in a number of reviews [13-15]. These reviews provide an in depth treatment to which the interested reader is referred. Briefly the essentials of the CARS process are illustrated in Figure 1. Two lasers provide beams of frequencies ω_1 and ω_s. To generate the signal efficiently the beams are combined in the medium in a suitable geometry to achieve phase or momentum matching*. The geometry illustrated in Figure 1 is commonly known as BOXCARS [16,17]. The result is a coherent generation of a "laser-like" signal at a frequency $\omega_{as} = 2\omega_1 - \omega_s$. Commonly the laser at ω_1 is termed the pump laser, and the laser at ω_s is termed the Stokes laser because it is frequency shifted to the "red" of the pump laser. The signal beam at ω_{as} is termed anti-Stokes because of its "blue" shift from the pump laser. The properties of the anti-Stokes signal are determined both by the properties of the input lasers and the probed medium. Most practical applications of CARS utilise the "broadband" technique where the Stokes laser, normally a dye laser, is designed to operate with a 150 cm^{-1} or 3-4 nanometer bandwidth. Suitable choice of the difference frequency between ω_1 and ω_s allows a particular molecular Raman resonance to be selected. CARS signals are efficiently generated when a Raman resonance is driven by the difference frequency of the input lasers. Because a broadband spectra source is used for the Stokes laser a complete Raman spectrum of a single species may be simultaneously generated; this is illustrated in the lower part of Figure 1. The pump laser source is a Q-switched Nd:YAG laser with a 10-20 Hz repetition rate and frequency doubled to 532 nm. One third of typically 300 millijoules at 532 nm would be used to optically pump a broadband dye laser to provide the Stokes source. The spectral properties of the generated anti-Stokes beam are analysed and recorded using a multi-channel detector. Figure 2 shows a photograph of a commercial CARS spectrometer system (Epsilon Research Limited, Model RLA5). Modern computerised multichannel spectrometers allow individual CARS spectra to be captured at repetition rates up to 100 Hz; however, suitable low divergence high pulse energy laser systems are not yet commercially available.

Referring to Figure 2, the desired test piece or experimental assembly is located between the field lenses which focus and recollimate the laser beams. Optical access to the desired measurement zone would normally be provided by two windows suitably located in the experimental assembly. Optical access has to be "line of sight" since the signal is generated in a forward sense, with respect to the pump and Stokes lasers. On exit the now unwanted lasers are trapped and the generated signal is directed to the spectrograph/detector assembly either via mirrors or, if the experimental assembly is excessively noisy or otherwise hostile, then via a fibre optic line [7].

Temperature measurements

Temperature information is derived from the shape of a CARS spectrum. The principle for such an analysis is illustrated in Figure 3. As is clearly seen the CARS spectrum of, in this instance, nitrogen varies greatly with temperature.

*If the pump and Stokes lasers are properly vectored then the generated CARS signal will be in-phase from all points within the intersection volume of the lasers; thus the CARS signal will grow coherently in a specific direction.

Strong CARS signals arise from transitions between rotational-vibrational modes of
the molecules. Since the population of these quantised energy levels vary with
temperature, according to the Boltzmann function, the shape of the spectrum changes
in a predictable manner.

To obtain accurate and reliable temperature information, from CARS spectra, a
computer model of the appropriate molecular spectrum (usually nitrogen for combustion
thermometry) is least squares fitted to the experimental spectrum, with temperature
as the principal variable. The often quoted accuracy of such measurements is 1-2% of
the measured value. This applies to time averaged spectra when say the spectra from
100 or more individual laser pulses have been summed together. Recently this figure
has been experimentally verified by a series of comparisons between CARS and
calibrated thermocouple temperatures simultaneously recorded in an isothermal
furnace.[18,19] Currently the principal limitations in accuracy are almost certainly
the overall stability of the Stokes laser frequency profile and, to a lesser extent,
overall drift and gain stability of the detection electronics.

A particular attraction of the CARS technique is its ability to perform a single
measurement in 10 nano-seconds. In a turbulent flow the fluctuating components are
almost always found at frequencies of less than 10 kHz. Consequently this represents
essentially instantaneous measurement. In single-pulse CARS thermometry the
temperature accuracy has been determined to be [18,19] of the order of 5% of the mean
value for a well designed experiment. If many single-pulse spectra are recorded, and
each spectrum analysed to give a temperature, then the average temperature of the
ensemble is found to be within 1-2% of the true mean[18]. The principal source of the
additional uncertainty arises from significant spectral noise on the broadband Stokes
dye laser. The noise on broadband dye lasers has been studied theoretically and
experimentally [20]. Although some measures may be employed to reduce the noise [19,20]
due to its fundamental origin in spontaneous emission [20] it is unlikely that it will
ever be completely eliminated. In real, non-isothermal systems it has been shown
that shot noise in the detector is also important [18]. However, neither of these
noise terms present significant limitations for measurements in practical combustors
such as oil spray furnaces[18] or jet engines[7]. This is because the temperature
turbulence in a practical device is typically 10 times greater than the single shot
measurement uncertainty. Also in a real measurement usually 400-1,000 spectra are
collected from a single spatial point and analysed to produce either or any of i) a
mean temperature, ii) a r.m.s. of temperature (i.e. measure of the intensity of
fluctuations) and iii) a probability distribution function (p.d.f.) or histogram.
The uncertainty of 5% on a single measurement thus translates into a 1-2% uncertainty
on the mean or simply a resolution limitation on the p.d.f. Since the r.m.s. is
typically greater than 5% its error is associated principally with the statistical
size of the single shot data set. If the device under test is cyclic, such as an
internal combustion engine, then this limitation can be readily overcome by averaging
together spectra for specific crank angles of the engine together with other
conditions of interest (e.g. pressure, ignition timing or flame position).

For measurements in devices operating substantially above ambient pressure, the
effects of pressure must be fully accounted for. These effects are complex and will
not be discussed here, they have been well reviewed in references [15,21]. However, if
they are not included errors of up to 50% in the CARS temperatures are readily
possible. The key to accurate analysis is a sophisticated computer model which will
automatically account for these effects. Harwell has developed such a model [22] for
use from 0.1-100 atmospheres for nitrogen, other diatomic species and also water
vapour [23]. The model uses a fast computational method first reported by
Gordon et al[24].

Concentration Measurements

CARS signals may be related to the number density of the probed medium. CARS
signal intensity (I_{as}) is given by

$$I_{as} = \left(\frac{4\pi \omega_{as}}{c^2 n_{as}}\right)^2 I_1^2 I_s \left(3\chi^{(3)}\right)^2 L^2 \qquad (1)$$

where perfect phase matching is assumed, c is the speed of light, ω_{as} is the CARS signal frequency, n refractive index, L laser beam interaction length and I_1 and I_s the intensity of pump and Stokes lasers respectively. $\chi^{(3)}$ is the third order bulk susceptibility of the medium and is given by

$$\chi^{(3)} = \chi_{NR} + \frac{4\pi N c^4}{h \omega_s^4} \sum_j \frac{d\sigma}{d\Omega}\bigg|_j \Delta\rho_j \left[2(\omega_j - \omega_1 + \omega_s) - i\Gamma_j\right]^{-1} \qquad (2)$$

where $(d\sigma/d\Omega)_j$ is the so called Raman cross section of the jth line of the probed species, h is Planck's constant, N is number density, $\Delta\rho_j$ the population differences between upper and lower states of the jth resonance of the probed species. Note the line shape term $[\ldots]^{-1}$ is a complex Lorentzian. χ_{NR} is a spectrally flat contribution from two photon electronic resonances from all the constituents of the medium. For strong resonances χ_{NR} produces only a slight modification to the spectrum. For a single resonance $(\chi^{(3)})^2$ becomes

$$\chi_j^{(3)\,2} = (\mathrm{Imag}\chi_j)^2 + (\mathrm{Real}\chi_j)^2 + 2\,\mathrm{Real}\chi_j\,\chi_{NR} + \chi_{NR}^2 \qquad (3)$$

As the concentration decreases the contribution to the CARS signal from the species resonance χ_j will approach χ_{NR}. Thus the last two terms in equation (3) become important. $\mathrm{Imag}\chi_j$ has a simple Lorentzian shape but $\mathrm{Real}\chi_j$ has a dispersive shape. Therefore as species concentration changes, the spectrum shape changes; this is illustrated in Figure 4.

Species concentrations can, in principle, be derived directly from the signal intensity; however, uncertainties in laser intensity and the complex form of equations 1-3 can lead to serious error. Most practical analysis is based on equation 3 and uses the shape of the spectrum[15]. This has the advantage that both temperature and species concentration can be simultaneously determined. Unfortunately, this type of analysis is limited currently to diatomic and simple triatomic species for which accurate spectroscopic data is available. For more complex systems, and for systems which are nearly isothermal, a more convenient method has been developed by England et al [3,11]. This is based on direct measurement of the second and third terms of equation 3; the method is fast and accurate but requires an initial calibration for absolute measurements.

RESULTS AND APPLICATONS

In order to illustrate the use of CARS three applications taken from recent work performed at Harwell are presented.

Incandescent Lamp

During recent collaborative work with Sheffield University and Thorn EMI [25], CARS thermometry was performed on an incandescent lamp. Figures 5 and 6 show a photograph of the modified lamp and a CARS temperature contour plot. The pressure in the lamp was 0.2 atmospheres and the filament temperature 2260K. In such a system mechanical probes would have markedly disturbed the sensitive flow field. CARS thermometry has provided key data and enabled a three dimensional flow field/heat transfer model for the system to be validated. This type of work should enable the final performance of this and similar devices to be critically evaluated at the design stage, and reduce the need for holistic development work.

Oil Spray Combustion

Furnace combustion of fossil fuels forms the major part of industrial energy usage. Careful design and control are required to manage pollutant formation and efficiency. Figure 7 shows a temperature map taken over the centre part of a burning oil spray in an oil furnace. This data, together with equivalent maps of the fluctuating component of temperature are currently being used to design complex combustion and fluid mechanic models of these systems. CARS can also be used to measure concentration in these systems. Figure 8 shows spectra of water vapour and nitrogen. Note the excellent agreement in temperatures. This data can be used to measure the completeness of the combustion chemistry.

Chemical Reactors

Although the CARS technique has been extensively developed for heat-transfer and combustion applications it is also well suited to the study of chemical reactors. Recently England et al [3,11] have completed a study of the methanation reaction ($CO + H_2 \rightarrow CH_4 + H_2O$) in a catalytic tube wall reactor. An illustration of the reactor is shown in Figure 9. Typical profiles of the CARS measurements of CH_4 concentrations are shown in Figure 10. Also shown in Figure 10 are theoretical plots of a full chemical kinetic, fluid dynamic model. As with the lamp study, mechanical probes would have been too intrusive to allow realistic measurements to be made in this 6 mm diameter tube.

Other Areas

At Harwell other current studies using CARS include temperature measurements during combustion studies in petrol engines and temperature measurements in high pressure, high temperature steam [23]. In the future, the study of neutral species in devices such as chemical vapour deposition and plasma vapour deposition reactors is likely to be an important area. In these systems CARS will permit "in-situ" probing of neutral species and provide spatial information on species distribution with respect to the substrate or device.

SUMMARY

The CARS technique has been described for gas phase thermometry and species concentration measurements. The use of CARS in practical devices and development systems is illustrated with reference to recent work on incandescent lamps, oil spray combustion and catalytic chemical reactors.

ACKNOWLEDGEMENTS

The chemical reactor studies were supported by the materials, chemicals and vehicles requirement board of the Department of Trade and Industry. The oil spray combustion studies were supported by the Department of Energy as part of an international collaboration via the International Energy Agency.

REFERENCES

1. Greenhalgh, D.A., England, W.A. and Porter, F.M. "The Application of CARS to Turbulent Combustion Thermometry". Combustion and Flame, Vol.49, pp.171 (1983).

2. Klick, D., Marko, K.A. and Rimai, L. "Broadband Single-Pulse CARS Spectra in a Fired Internal Combustion Engine". Appl.Opt. Vol.20, pp.1178 (1981).

3. England, W.A., Milne, J.M., Jenny, S.N., and Greenhalgh, D.A. "Application of CARS to an Operating Chemical Reactor". Appl.Spectrosc., Vol.38, pp.867 (1984).

4. Pealat, M., Taran, J.P.E., and Moya, F. "CARS Spectrometer for Gases and Flames". Opt. Laser Technol. Vol.12, pp.21 (1980).

5. Rahn, L.A., Johnston, S.C., Farrow, R.L., and Mattern, P.L. Temperature, Vol.5, Part 1, pp.609, (American Institute of Physics), New York (1982).

6. Majiyama, K., Sajiki, K., Kataoka, H., Maeda, S. and Hirose, C. "N_2CARS Thermometry in Diesel Engine". SAE Technical Paper, 821036 (1982).

7. Eckbreth, A.C., Dobbs, G.M., Stufflebean, J.H., and Tellex, P.A. "CARS Temperature and Species Measurements in Augmented Jet Engine Exhausts". Appl.Opt., Vol.23, pp.1328-1339 (1984).

8. Ferrario, A., Gabi, M., and Malvicini, C.
Paper WD2, Technical Digest, Conference on Lasers and Electro-optical Systems (Optical Society of America), Washington, D.C. (1983).

9. Hartford, A., Cremes, D.A., Loree, T.R., Quigles, G.P., Radzienski, L.J., and Taylor, D.J.
Proc.Soc. Photo-Opt. Instrum. Eng. Vol.411, pp.23 (1983).

10. Murphree, D.L., Cook, R.L., Bauman, L.E., Beiting, E.J., Stickel, R.E., Daubach, R.O., and Ali, M.F.
AIAA Paper 82-0377 (1982).

11. England, W.A., Glass, D.H.W., Brennan, G., and Greenhalgh, D.A. "Study of a Tube Wall Methanation Reactor using CARS Spectroscopy". Submitted for Publication, 1985.

12. Hata, N., Matsuda, A., Tanaka, K., Kajiyama, K., Moro, N., and Sajiki, K. "Detection of Neutral Species in Silane Plasma Using Coherent Anti-Stokes Raman Spectroscopy". Japanese J. Appl. Phys., Vol.22, pp.L1-L3, (1983).

13. Druet, S., and Taran, J-P.E., "Coherent Anti-Stokes Raman Spectroscopy". In Lasers (Moore, C.B., Ed) Academic, New York (1979).

14. Nibler, J.W., and Knighten, G.V. "Coherent Anti-Stokes Raman Spectroscopy, in Raman Spectroscopy of Gases and Liquids", (Weber, A., Ed) Springer, Berlin (1979).

15. Hall, R.J., and Eckbreth, A.C. "Coherent Anti-Stokes Raman Spectroscopy (CARS): Applications to Combustion Diagnostics", in Laser Applications, Vol.5, (Erf, R.K. Ed) Academic, New York (1984).

16. Eckbreth, A.C., "BOXCARS : Crossed Beam Phase-Matched CARS Generation in Gases". Appl.Phys.Lett., Vol.32, 421-423 (1978).

17. Greenhalgh, D.A. "Comments on the Use of BOXCARS for Gas-Phase CARS Spectroscopy", J. Raman. Spec., Vol.14, pp.150-153 (1983).

18. Porter, F.M., and Greenhalgh, D.A. "Analysis and Application of CARS Spectroscopy to Oil Spray Combustion". In preparation (1985).

19. Pelat, M., Bouchardy, P., Letebrvre, M., and Taran, J-P.E. to be published.

20. Greenhalgh, D.A., and Whittley, S.T. "Mode Noise in Broadband CARS Spectroscopy". Appl.Optics., to be published (March 1985).

21. Hall, R.J. "Coherent Anti-Stokes Raman Spectroscopic Modelling for Combustion Diagnostics". Opt.Eng., Vol.22, pp.322-329 (1983).

22. CARP is a computer code for analysis of CARS spectra developed at Harwell. Licences for the code are available through Epsilon Research Limited, Paynes Lane, Rugby.

23. Greenhalgh, D.A., Hall, R.J., Porter, F.M. and England, W.A. "Application of the Rotational Diffusion Model to the CARS Spectra of High Temperature, High-Pressure Water Vapour". J.Raman Spectros. Vol.15, pp.71-79, (1984).

24. Gordon, R.G. and McGinnis, R.P. "Line Shapes in Molecular Spectra". J.Chem.Phys. Vol.49, pp.2455-2456 (1971).

25. Devonshire, R., Dring, I.S., Hoey, G., Greenhalgh, D.A., Porter, F.M. and Williams, D.R. "A Comparison of CARS Measurements of the Gas Temperature Profile Around an Incandescent Filament with Predictions of a Fluid Flow Model", to be published.

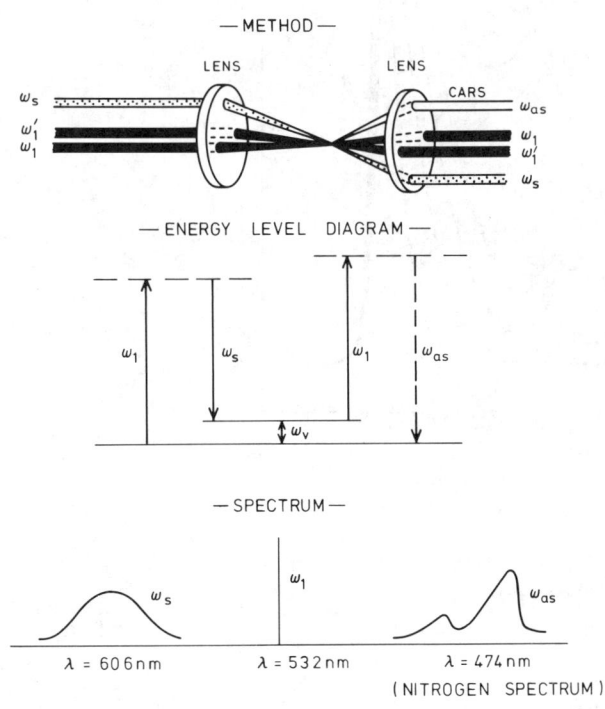

Fig. 1.

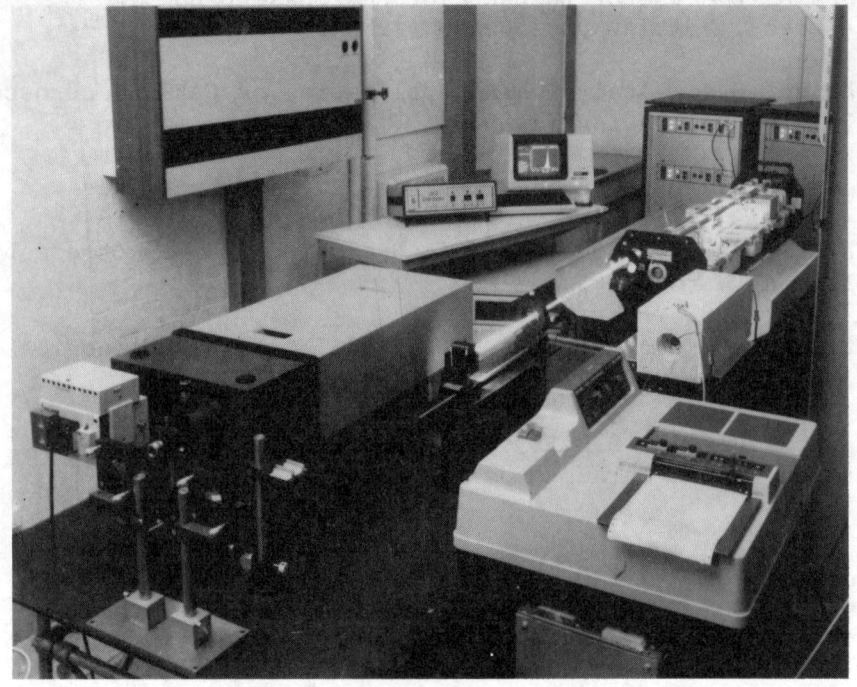

Fig.2. Photograph of Epsilon RLA5 commercial CARS system.

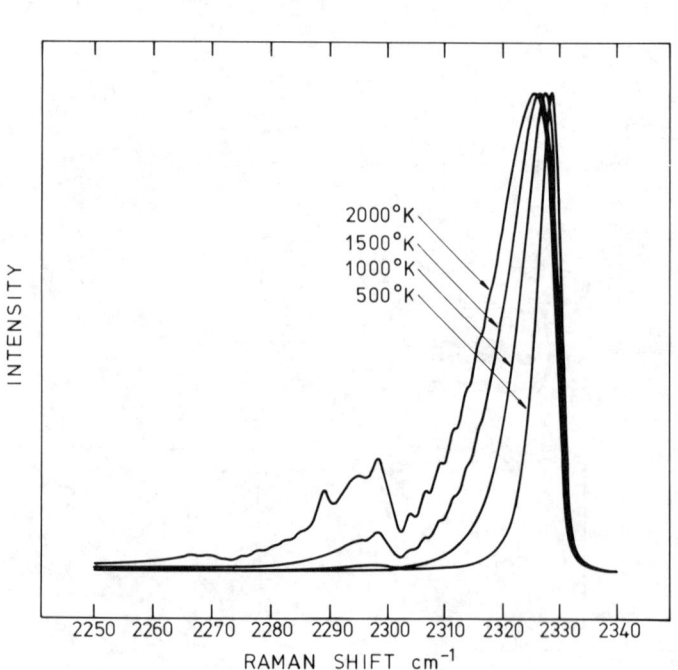

Fig.3. CARS spectra of nitrogen at various temperatures.

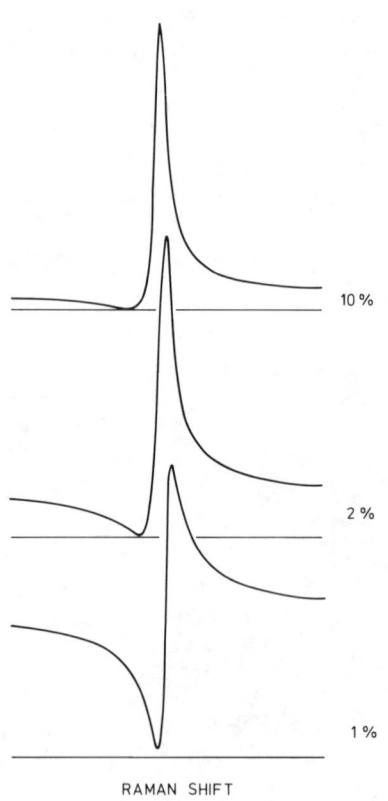

Fig.4. Hypothetical example of a CARS spectrum as a function of fractional species concentration.

Fig.5. Photograph of a CARS experiment on an incandescent lamp.

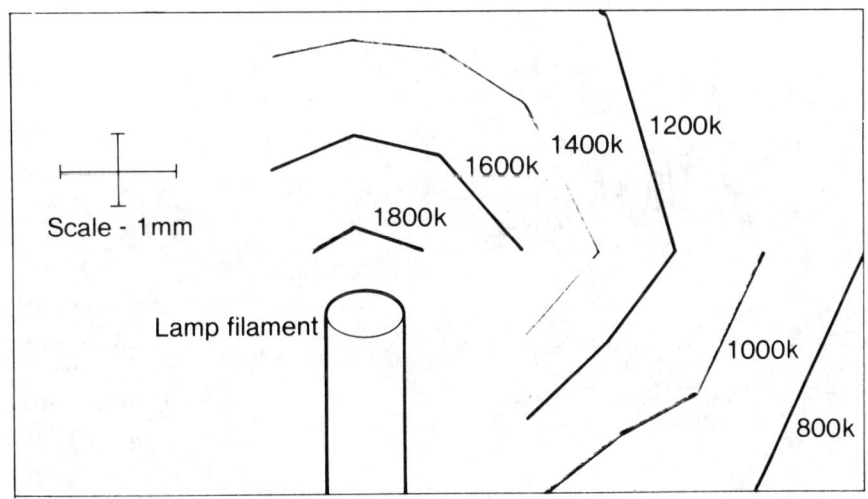

Fig.6. Contour map of temperature in radial plane around lamp filament in Figure 5; filament temperature 2260K.

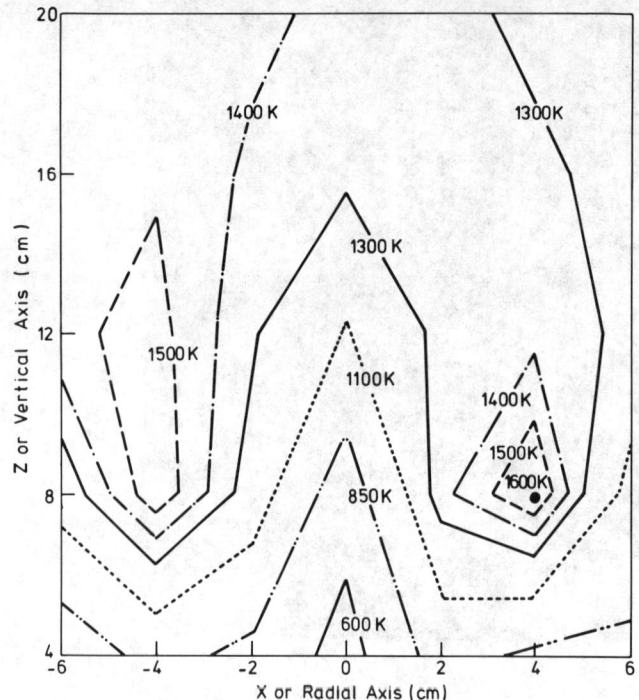

Fig.7. Mean temperature contour map of radial section through a 0.3 metre dia. oil spray furnace

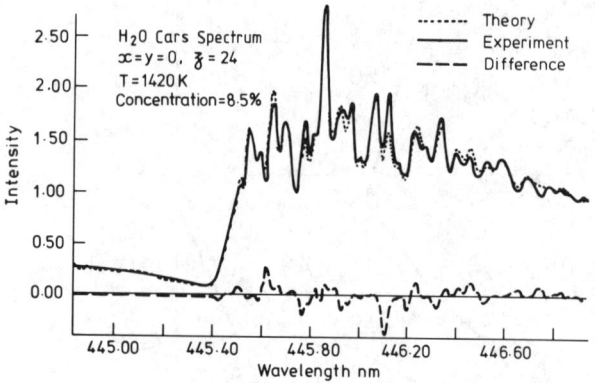

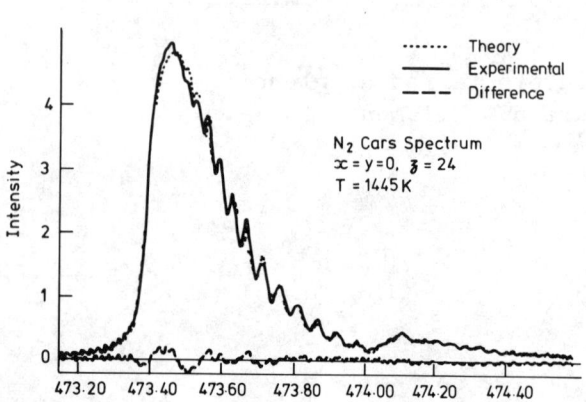

Fig.8. Examples of CARS spectra of H_2O and N_2 taken in the same oil spray furnace as Figure 7.

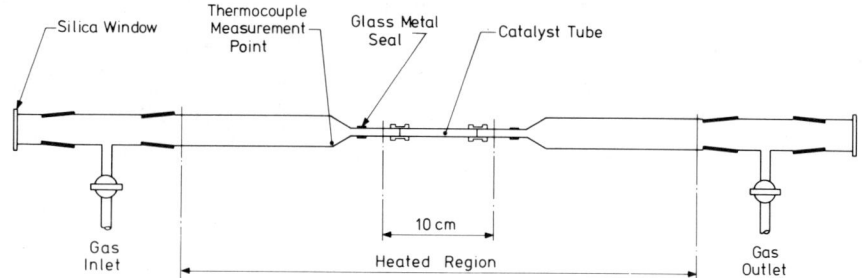

Fig.9. Schematic of laboratory tubular catalytic reactor for methanation studies; tube diameter = 6.3 mm.

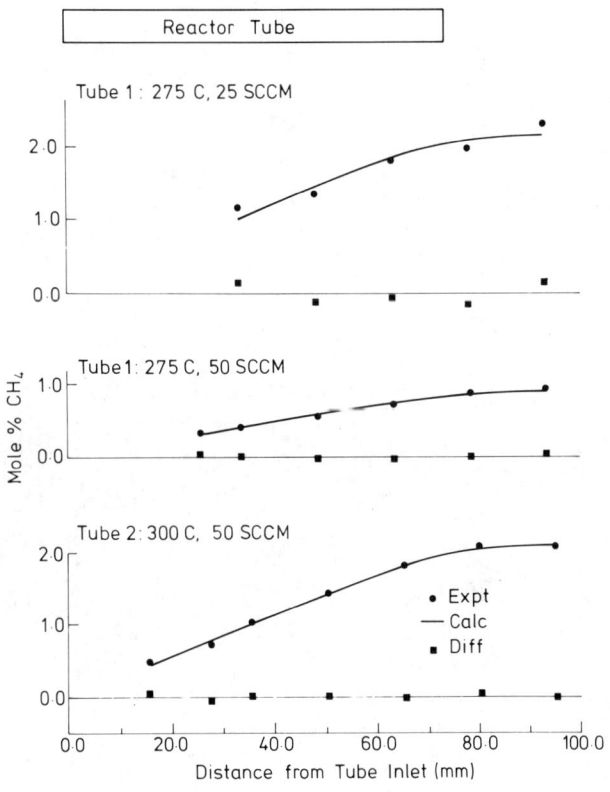

Fig.10. Comparison of CARS methane concentration measurements against fluid/chemical kinetic model predictions for catalyst tube in Figure 9.

EQUIPMENT

New CO_2-lasers of 1-4 kilowatt power with fast axial gas flow
L. Bakowsky
Messer Griesheim GmbH, West Germany

Messer Griesheim has developed a new series of high-performance CO_2-gas-lasers for materials processing applications (EUROLAS). This series comprises a modular system which includes four performance classes at 1 kW intervals from 1 kW up to 4 kW. All four models feature the following advantages:
- simple, robust design
- high mode and power stability
- high specific laser power (1 kW/m).

Because of the high specific power it is possible to get 1 kW laser from only two discharge tubes of 500 mm length each. The laser can be operated both continuously (cw) and pulsed.

Introduction

Applications of high power CO_2-lasers to industrial material processing show that high processing velocity and quality are only guaranteed in case of a stable intensity profile on the workpiece. This is only achieved by laser systems without any power and mode fluctuations and by handling systems of high accuracy. Three types of CO_2-lasers are used in industry slow flow systems, fast axial flow systems and transverse flow systems.

The systems with best beam quality are the slow flow systems. They usually emit in the fundamental mode (TEM_{oo}). Specific laser power however is limited to approximately 50 W/m caused by the restricted cooling of the laser gas. The total maximum laser power is limited to $P \simeq 1000$ W by mechanic stability reasons, which is too low for nearly all welding and surface treatment applications.

Transverse flow systems are typical multikilowatt systems with powers up to 20 kW. The main disadvantage is the fact, that transverse systems cannot be pulsed. Pulsing is necessary for cutting sharp edges and small geometries. Messer Griesheim has concentrated on the fast axial flow systems. The new series (EUROLAS) comprices a modular system which includes four performance classes at 1 kW intervals up to 4 kW. The aim was to get high mode and power stability with a simple, robust design (minimum number of mirrors and wearing parts).

Construction of the Lasersystems

The basic laser unit (EUROLAS 1000) consists of two discharge tubes with a common cathode head and two anode heads (Fig. 1). The resonator is formed by mirrors arranged at both ends
- one fully reflective curved end mirror (copper) and
- one partly translucent output mirror (GaAs).

To get an high thermal stability the mirrors are directly water-cooled. The mechanic stability of the resonator is given by a granite foundation. A fast flowing mixture of helium, nitrogen and carbon dioxide serves as the laser medium. The gas is pumped by a Roots blower. The heat generated during discharge is conducted away by heat exchangers and by the water-cooled walls of the discharge tubes. The main point of the new systems is the special turbulent gas flow. Design of gas inlet and anode geometry leads to a very homogeneous discharge at the high pressure side of the tubes. To maintain the turbulence flow over the whole discharge length so-called "Nipple-Diffusor" tubes are used, with two Carnot-diffusors and some blades inside. By these means it is possible to increase the electric input power without any hot filaments or arcing in the discharge, resulting in high specific laser power of 1000 W/m and good beam quality. Figure 2 gives an idea of the size of 1 kW laser system EUROLAS 1000. This system is usually used for cutting applications. Typical values for cutting speeds and edge roughness in case of St 37 plates shows figure 3.

The higher power versions EUROLAS 2000 - 4000 are built up by several units of the EUROLAS 1000 type (Fig. 4).

Because of the high specific power Messer Griesheim succeeded in building a 3 kW axial flow laser system with only two mirrors. The 4 kW version is folded, caused by the limited available sizes of the granite foundations.

All systems are running in a stable low order mode (TEM_{10} or TEM_{20}). Divergence is better than 1 mrad, power stability better than ± 2 %.

Each discharge tube is supplied by a separate mains unit. The lasers can be operated both continuously and pulsed with pulse peak powers twice the maximum cw-powers (superpulsing).

Combination with Handling systems

To get a complete laser-processing tool the laser systems have to be combined with workpiece and beam manipulators of high accuracy. Messer Griesheim has developed a system of coordinate tables, positioners and multi-axes workpiece or beam manipulators especially for the use with the EUROLAS laser systems.

Figure 5 shows the five-axes manipulator LASCONTUR QUINTA for cutting and welding spatial contours with the aid of five CNC-controlled axes of motion. This system is used in the automotive industry to cut carriage bodies.

Three main points are the basis for cutting quality and velocity
- stable beam parameters (mode, divergence, power and pointing stability)
- high accuracy of the guiding system
- CNC control able to control 5 mechanical axes and the laser parameters (power, pulsing) with high updaterate and short record-circle-time.

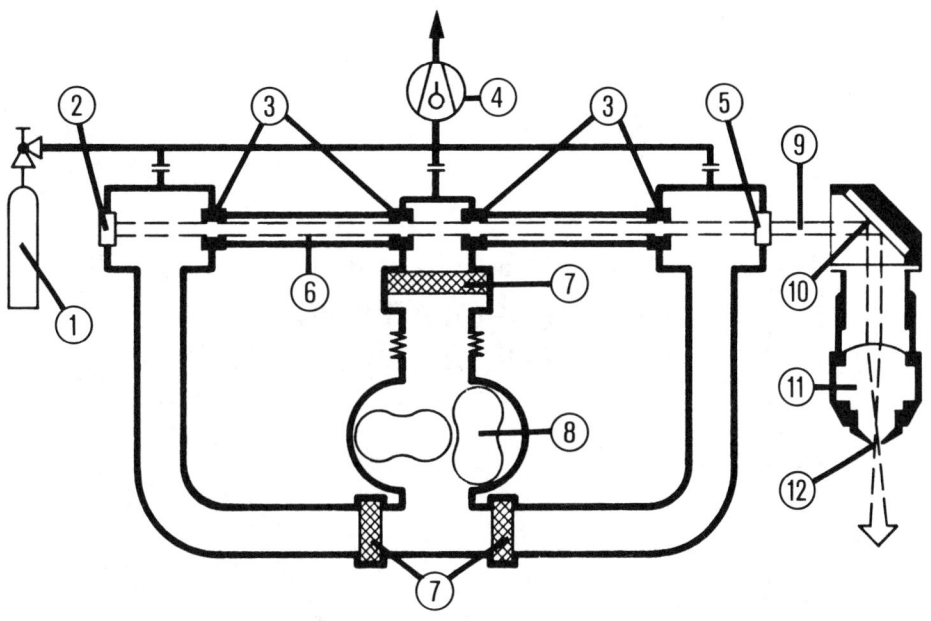

Fig. 1: Functional diagram of a fast axial flow CO_2-laser

Fig. 2: Resonator of the EUROLAS 1000 laser system

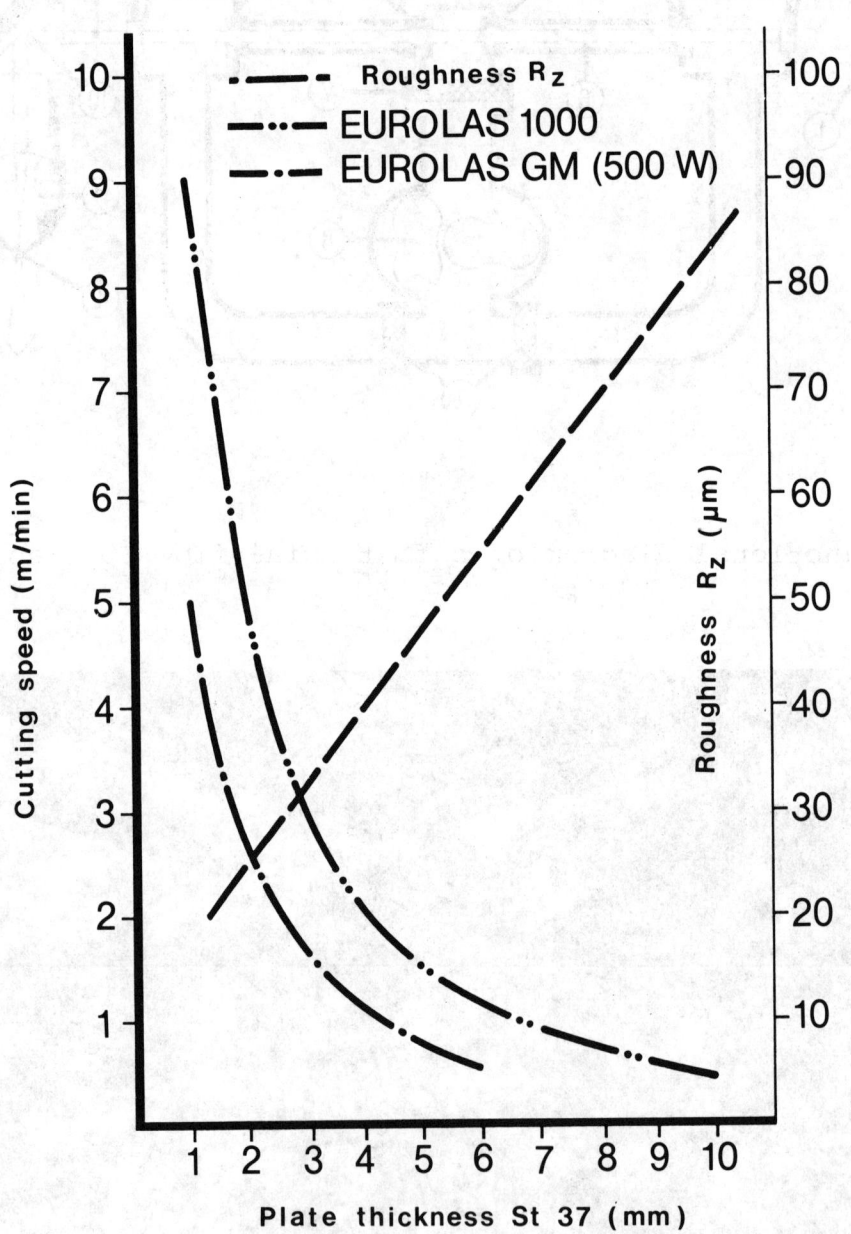

Fig. 3: Cutting diagram

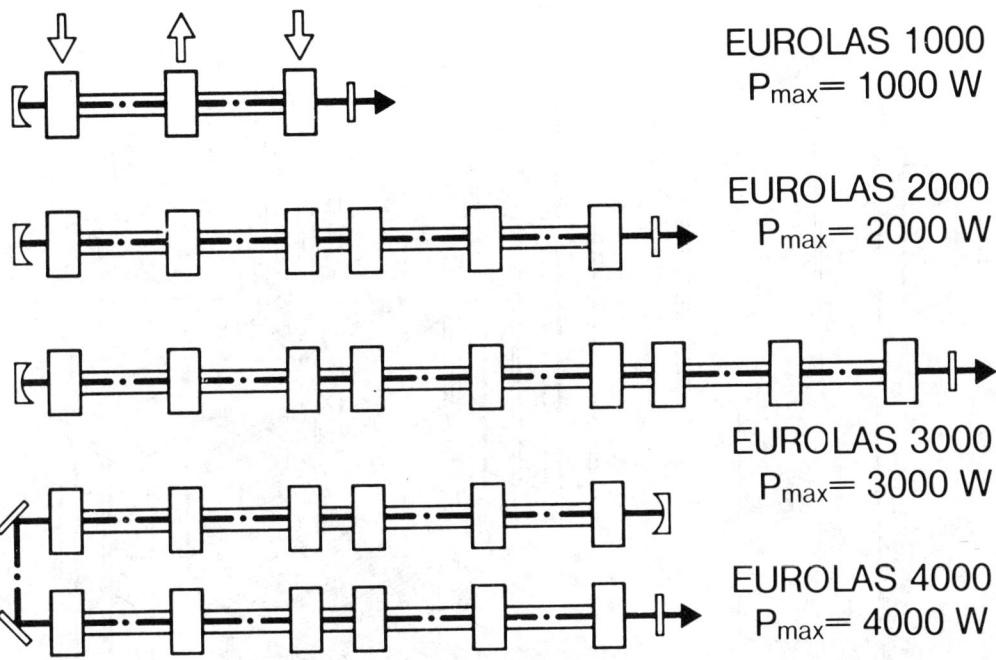

Fig. 4: Scheme of the EUROLAS modular laser series

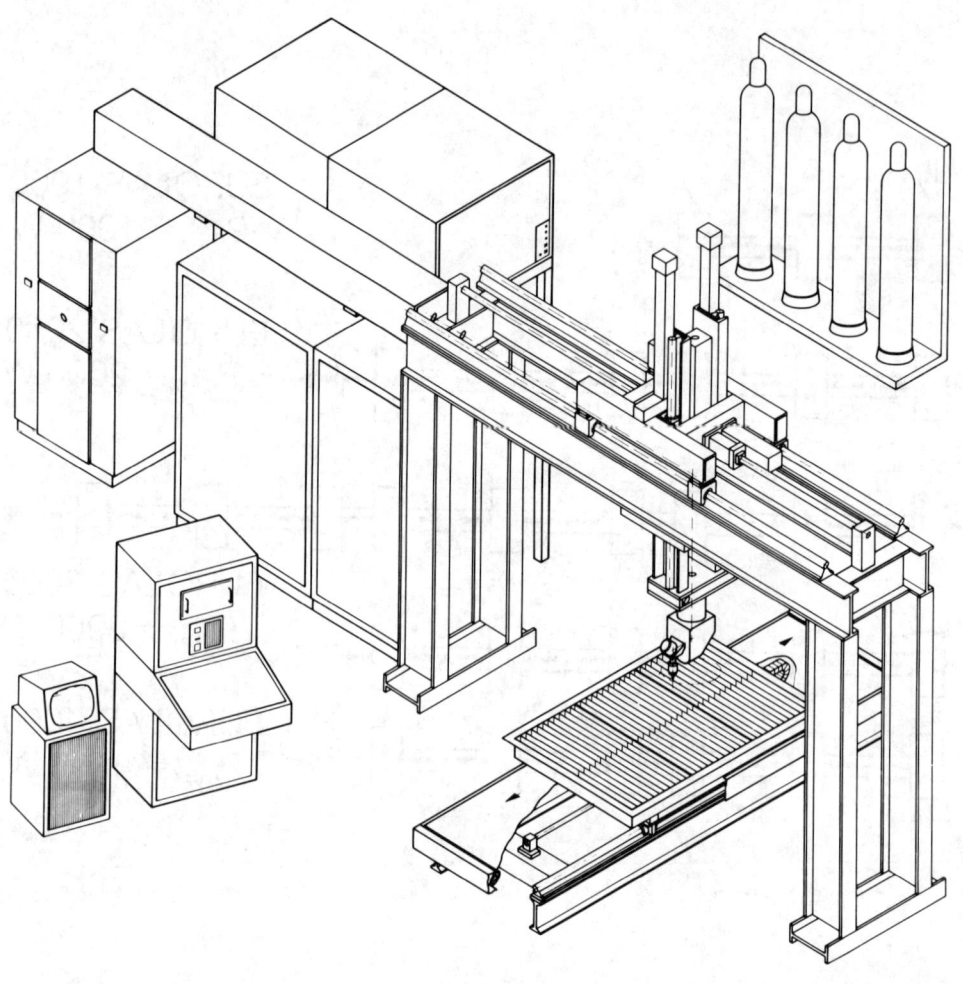

Fig. 5: 5-axes manipulator LASCONTUR QUINTA

The start of a new generation of CO2-lasers for industry
P. Hoffmann
Laser Innovation GmbH & Co KG, West Germany

The present, first generation of industrial CO_2-lasers originated in the period 1966 - 1970 and is differentiated into axial flow-axial discharge and transverse flow-transverse discharge types, all using DC glow discharges to generate the required population inversions. We describe here one of the first of the new, second, generation of devices, utilizing an RF-excited transverse flow-transverse glow discharge, 3-axis configuration which provides greatly increased power density in the discharge plasma. This approach permits significant reductions in both head and power supply sizes and weights, simplicity of construction, and reduced manufacturing costs and maintenance requirements.

INTRODUCTION

The first CO_2-laser was built by Patel in 1965, and by 1966 versions with several kilowatts output power had been developed in the laboratory. The possibilities for materials working using this then-new type of laser were clear, and several groups set out to engineer "industrialized" units for the foreseen broad range of applications. The first of these appeared in 1967, and by 1970 the range of design had been fixed in its present form. Three basic types of configurations for high-power CO_2-lasers emerged: slow axial flow-axial discharge, fast axial flow-axial discharge, and transverse flow-transverse discharge. All of these use DC-excited glow discharges in optimized $He-N_2-CO_2$ gas mixtures. The slow axial flow design, which requires meters of discharge length to reach powers of 1 kW, is presently fading away in favor of the fast-flow device, which offers greater compactness and eliminates the need for folded optical paths. The transverse flow configuration has many advantages and is capable of reaching output powers up to 20 kW, but has thus far been too expensive and bulky to be commercially viable at output powers below about 1.5 kW.

These first-generation devices are generally characterized by high bulk and weight, complexity, and a reliability record which falls short of the needs and demands of industry. Over the past 10 - 12 years, work has been underway on the basic features of a new generation of CO_2-lasers for industry which will be much more compact, lighter, more simply constructed, and should offer the reliability and maintainability which industry expects from production equipment. This work has been centered on increasing power densities in transverse flow configurations to permit a compact and versatile unit for the output power range 300 W - 3 kW, which meets the needs of at least 80 % of all industrial applications.

This work evolved from early efforts to develop supersonic-flow, transversely excited, RF- or microwave discharge gasdynamic lasers.[1,2,3] The results of this work were then applied to the case of subsonic flows utilizing transverse, low-frequency RF-excitation in a more conventional configuration better suited to industrial use. The primary goal is the achievement of highly uniform, high energy density discharges in useful volumes in relatively simple flow channels. Realizing this goal took some years of exploratory work.

DESIGN CONSIDERATIONS

General

The following discussion shall concentrate on the above mentioned questions about an optimum design of a CO_2-laser in the 1 kW power range. It will therefore deal with three topics:

1. RF versus DC-excitation. Selection of an appropriate frequency.

2. Fast axial flow versus fast transverse flow. Flow channel design and mode control.

3. Other design considerations concerning compactness, low weight, simplicity, reliability and reduced costs of laser head and periphery.

RF Discharge Concept

The advantages of RF-excitation over the DC-excitation used in the first generation of industrial CO_2-lasers lie in their stability at higher discharge power densities. DC glow discharges are restricted to relatively low gas pressures and become unstable at relatively low discharge power densities. Further, the fixed polarity of the electrodes creates a basis for thermal instabilities in the cathode region and places restrictions on electrode design. The alternating RF-fields, however generate temporally alternating discharge phases while maintaining a relatively high, and substantially constant, free electron density in the discharge plasma. The free electron density is recombination-dominated, which shifts the onset of large-scale instabilities to greater discharge energy densities. The alternation in polarity of the electrodes which take place each cycle periodically interrupts the cathodicity of the electrodes at time intervals which are short in comparison to the typical appearance times of thermal instabilities in cathode regions. This inhibits their appearance, and greatly simplifies electrode design.

RF glow discharges may be ballasted by pure reactances without any ohmic resistance to prevent a negative-slope voltage-current characteristic, thus eliminating the need for the series ballast resistors required for the DC glow discharge. This means that the ohmic power dissipation across the ballast resistor bank, which typically amounts to about 30 per cent of the total electrical power consumption, is not present, and the overall electrical efficiency of the device should be better than for DC-excitation.

The ballast reactance, partly consisting of capacitances, is simultaneously used to prevent too high current densities at the electrode surface which otherwise tend to cause arcing. This limitation in current density may be done in two ways: either by splitting the electrodes into a manifold and decoupling them by capacitors. The electrode segmentation and their decoupling by resistors is well known from DC discharges. This first way could be considered the RF equivalent. The second way would be to use so-called dielectric electrodes, i. e. metal electrodes covered with a suitable dielectric.

This way is the technically more elegant because it avoids the effort for segmenting the electrodes. On the other hand, the problem lies in the suitability of the dielectric. Without going further into details, I would maintain that to date there is no distinct advantage for one of both.

To further justify this statement, we must now consider the frequency used for excitation. From a physical point of view, the optimum frequency should lie in the upper megacycles range because of the short time scale for field strength variation and polarity change compared to typical recombination times and the time needed for the development of instabilities. But taking into account technical aspects, the upper kilocycles range will be preferred for the following reasons:

1. Commercial solid state RF-generators working at frequencies above 50 kHz are available since lately. They are based on the long history of RF-induction furnaces which led to very reliable devices with high efficiencies (> 95 %) compared to tube generators. Furthermore, they can be built very compact and lightweight.

2. At the high power used (typically tens of kilowatts), there arise severe shielding problems with RF in the megacycles range. Lowering the frequency by, let me say, a factor of hundred drastically reduces shielding problems.

3. Solid state devices can easily be switched and pulsed.

4. Solid state devices are mechanically much more rugged than tube generators.

Flow Channel Design

Disciples of the axial-flow laser concept often argue that the transverse-flow concept

be inefficient because the excited volume inherently is of a large cross-section, so that overlapping with the mode volume of a stable resonator is bad; folding of the optical path is complicated, and the use of an unstable resonator to improve overlapping results in a worse far-field distribution;

secondly, be of bad mode quality because of the uneven gain distribution in flow direction. There is indeed a strong asymmetry with respect to small signal gain and saturation intensity.

On the other hand, one should argue that the transverse-flow concept has the advantages

that power density is higher because of better convective cooling; and if flow velocity is raised in axial-flow systems in order to meet also this cooling condition, the requirements for high differential pressure to drive the gas flow lead to costly blowers with a question mark concerning size and/or reliability.

furthermore, that the applied voltages are inherently much lower so that the well-known safety problems are substantially reduced. (By the way: RF is much less dangerous than DC!)

So, if we successfully come up to reduce or even overcome the above mentioned disadvantages of the transverse-flow system, then we should consider the transverse-flow system the concept of choice.

Other Design Considerations

With respect to the economy of a system, one should choose a construction which is able to work as far as possible with proven subsystems which are - for different applications - already produced in large series at low cost and with high reliability. Therefore we use for instance only slightly modified blowers where we do not expect any trouble because of the long-term know-how of the manufacturer. This is also applicable for our integrated cooling system or the RF-generator as a whole.

THE LASER INNOVATION CONCEPT

We started on our project in 1982 having a 10-year background in the preliminary work needed. The primary design goal was a 1 kW (customer specification) output power unit, with a head weight under 100 kg, a head volume of no more than 0.3 m³, a simple outer structure for easy mounting, built up from a small number of simple, reliable, easy to replace, parts. Commercially available parts were to be used so far as possible.

We came close to this goal already in July 1983 with a downscaled version of more than 300 watts within a volume of only 0.15 m³, but at a weight only slightly below 100 kg.

Since that time, the work has concentrated on refining the initial successful configuration to obtain higher output power without significantly increasing dimensions, to reduce weight, to simplify manufacture and to increase reliability of the whole system.

We achieved a modular design, which will later be scaled in output power upward to 3 kW and more in order to provide a full range of devices which will meet the needs of virtually all industrial users.

The schematic design of the laser head is shown in fig. 1. Clearly visible is the basic principle of the design: the two active channels with opposite flow directions. These channels, together with their integrated heat exchangers for cooling the gas, the blower and the gas turn-back form the closed gas cycle system. The laser-active discharge regions of both channels are optically connected in series by a u-type resonator. This design has the following advantages:

- o Because of the gas flow in opposite direction in the two channels, the total integrated gain distribution in flow direction becomes symmetrical, thus resulting in an excellent mode compared with other transverse-flow systems. This theoretical prediction was already verified by the first experiments.

- o The electrodes form a kind of subsonic nozzle, combining gas dynamic requirements for increasing the gas velocity with the conditions necessary for a well-shaped discharge region, taking into account parameters such as gas velocity, overlapping degree of discharge cross-section and mode volume, control of the field distribution for a homogeneous discharge, development of boundary layers and gas heating, etc.

- o By the shape of the resonator a stable linear polarized beam is provided, which can easily be handled by a phase-shifter to convert it into a circularly polarized beam.

- o The design as a whole allows a very compact embodiment of the laser head.

Fig. 2 shows the prototype of the Laser Innovation laser head without a case around the central unit. The two channels can be seen easily behind the resonator structure with the coupling mirror in the upper part. At the right hand side, a side channel type blower for gas circulation is mounted and at the left, the gas flow is simply turned back. Other types of blowers may be suitable, too.

The hand adjusting the coupling mirror gives a good impression of the small size of the device.

The laser head is connected to two 19" equipment racks, containing the power and gas supply, control electronics, and system interlocks by an umbilical cord incorporating a flexible pipe to the vacuum pump (located inside one rack), laser gas and cooling water supply, a coaxial RF line, auxiliary lines for powering the blower, etc. To reduce operation costs, the gas is reprocessed by a catalyst, also located in one of the two 19" racks.

The further design goal is to reach a final head size and weight which would permit mounting the head on the arm of a typical heavy-duty industrial robot. This sets an upper weight limit of about 50 - 60 kg.

DISCHARGE STABILITY

Since the fundamental work of Nighan[4] on instabilities in laser discharges, there has been a lack of deeper insight into the basic problems although many papers have been published on this subject. There exists a broad empirical knowledge but little is really understood, for instance about the role of the boundary layer at the electrodes, with respect to the development of instabilities especially in RF discharges. Here, we only want to draw attention on these problems, not to offer solutions. We only want to show two empirical results.

Fig. 3 shows laser output power versus the reactance of the impedance matching circuit, ballasting the discharge purely reactive. The electrodes are segmented and decoupled from each other. Decreasing the reactance of the system means increasing power deposition into the discharge at fixed E/n, so that laser power increases, too, until gas heating caused by the high current densities leads to the irreversible occurence of arcs. This behaviour is completely equivalent to the situation of resistor ballasted DC discharges.

Utilizing an appropriate circuit reactance for stable laser operation, one can now vary the partial pressures of the gases CO_2, N_2 and He up to levels where instability again occurs. This instability is of the same thermal nature, but its macroscopic behaviour is different. There is no irreversible breakdown into an arc but low frequency (some kHz) fluctuations, leading to fast moving filaments finally ceasing laser operation. In accordance to Nighan's theory, the onset of these instabilities is mainly dependent on the partial pressure of the molecular gases only so that a large pressure range may be spun by varying the partial pressure of Helium. So one should keep in mind that the total pressure often used in literature to describe the stability of a system is an irrelevant measure, unless the gas mixture ratios are given.

The typical dependance of laser power on partial pressure of the molecular gases at constant loading is shown in fig. 4. p^2-dependance should be achieved according to Fowler[5].

SUMMARY

Work on high power density RF-discharges in CO_2-laser gas mixtures in subsonic flow channels has led to the design of a commercial CO_2-laser for industrial applications having extremely compact head dimensions with a total head volume below 0.3 m³, a head weight of less than 100 kg, for a specified output power of 1 kW. The simple construction and use of well-proven industrial hardware for many critical components indicates that the high reliability expected by industry will be achieved in actual use. The great advances in reducing size, weight, and complexity, and the expected much higher reliability at somewhat lower capital costs can rightfully be designated as the start of a new, second generation of CO_2 materials processing lasers for industry.

REFERENCES

1. Schock, W., Schall, W., Hügel, H. and Hoffmann, P. "CW Carbon Monoxide Laser with RF Excitation in the Supersonic Flow". Appl. Phys. Lett., Vol. 36, No. 10, pp. 793 - 794 (1980)

2. Hoffmann, P. "Microwave Discharge in a Supersonic Flow for the Excitation of a Gas-Dynamic CO-Laser". European Space Agency, ESA-TT-833, (July 1984)

3. Schock, W., Hügel, H., and Hoffmann, P. "Ein neues Laserkonzept: Quergeströmter CO_2-Laser mit Hochfrequenzanregung." Laser + Electro-Optik, Vol. 2, pp. 76 - 77 (1981)

4. Nighan, W. L. "Stability of High Power Molecular Laser Discharges" in Gases Electronics of Lasers. MIT-Press, Cambridge, Mass. (1975)

5. Fowler, M. C. "Influence of Plasma Kinetic Processes in Electrically Excited CO_2 Laser Performance". J. Appl. Phys., Vol. 43, pp. 3480 - 3487 (1972)

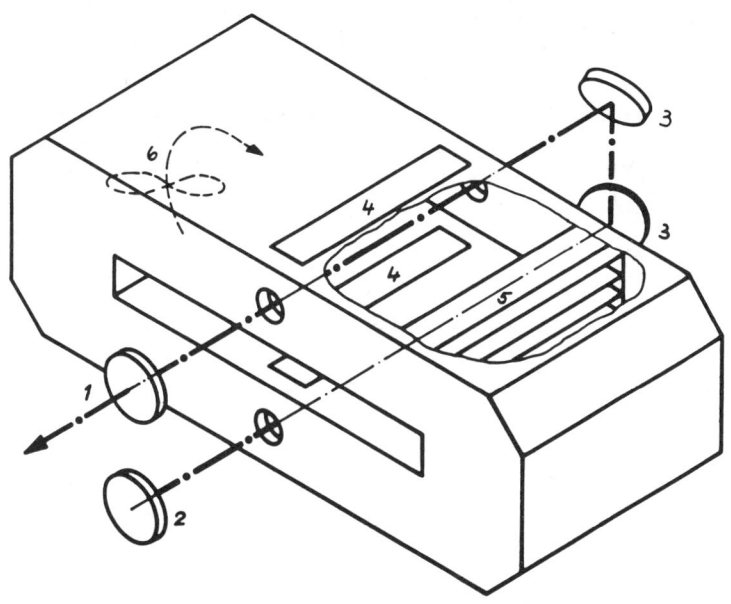

Fig. 1 Schematic representation of the laser head.
1: coupling mirror, 2: total reflector, 3: bending mirrors,
4: electrodes, 5: heat exchanger, 6: blower
Gas flow is recirculated through a parallel flow channel which
is also excited. A u-type folded resonator combines the output
powers of the two channels while providing for a linearly polarized output beam.

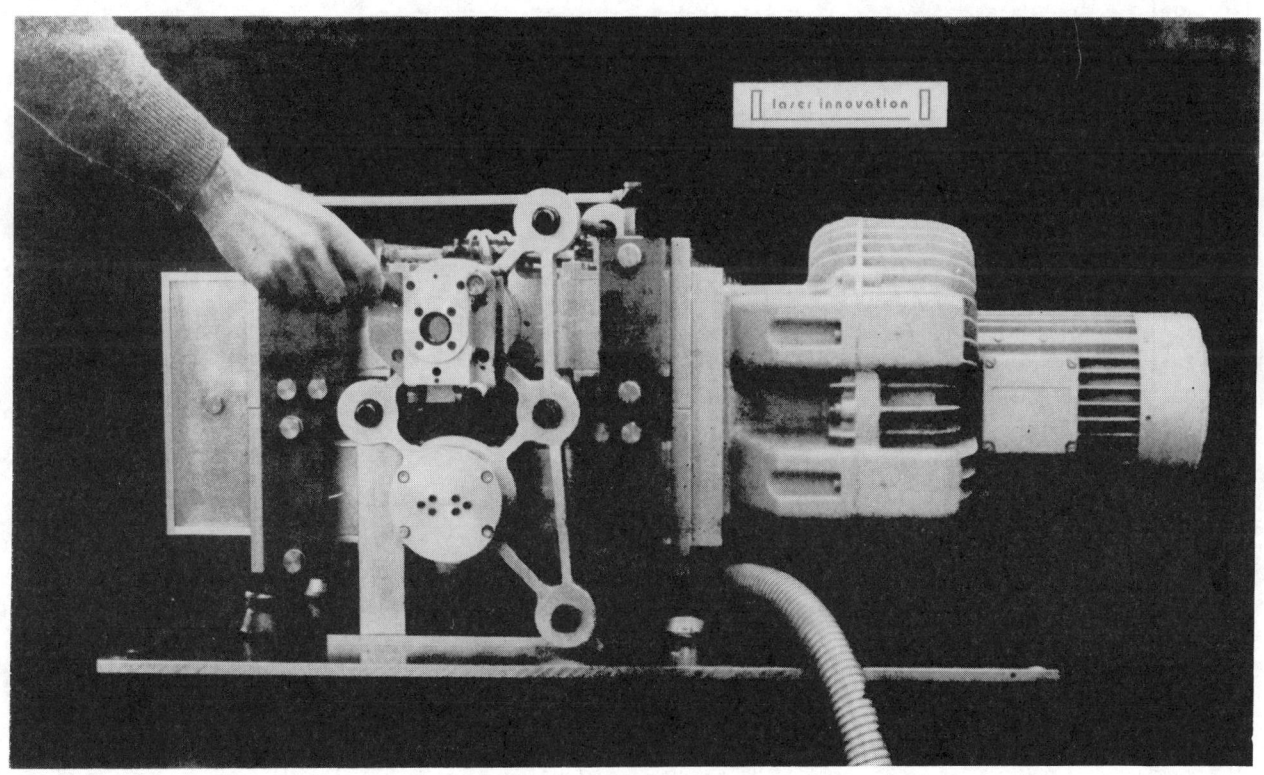

Fig. 2 The prototype laser head

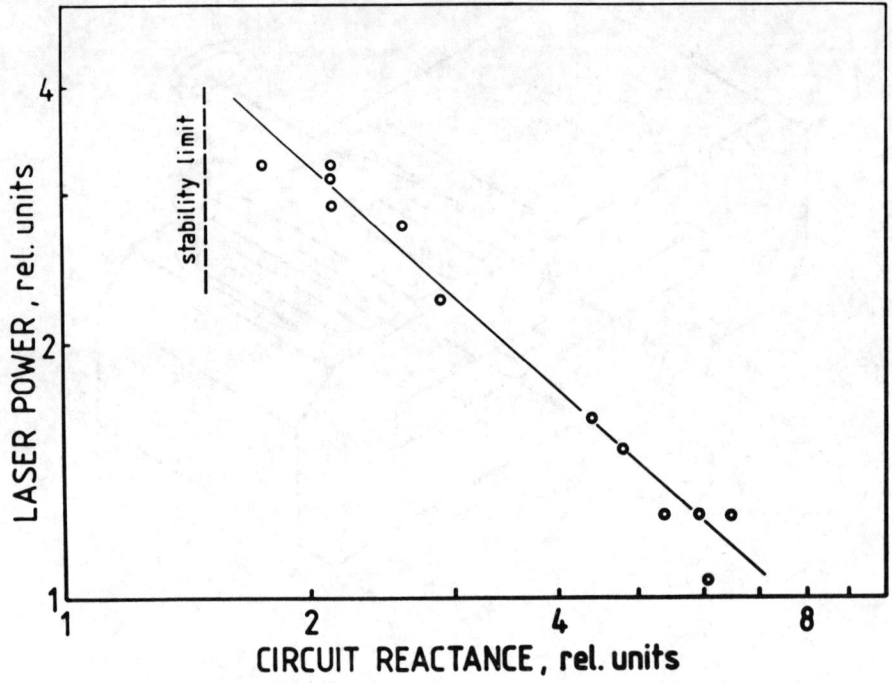

Fig. 3 Laser power versus stabilizing circuit reactance. Onset of arcing beyond the stability limit.

Fig. 4 Laser power versus partial pressure of molecular laser gases.

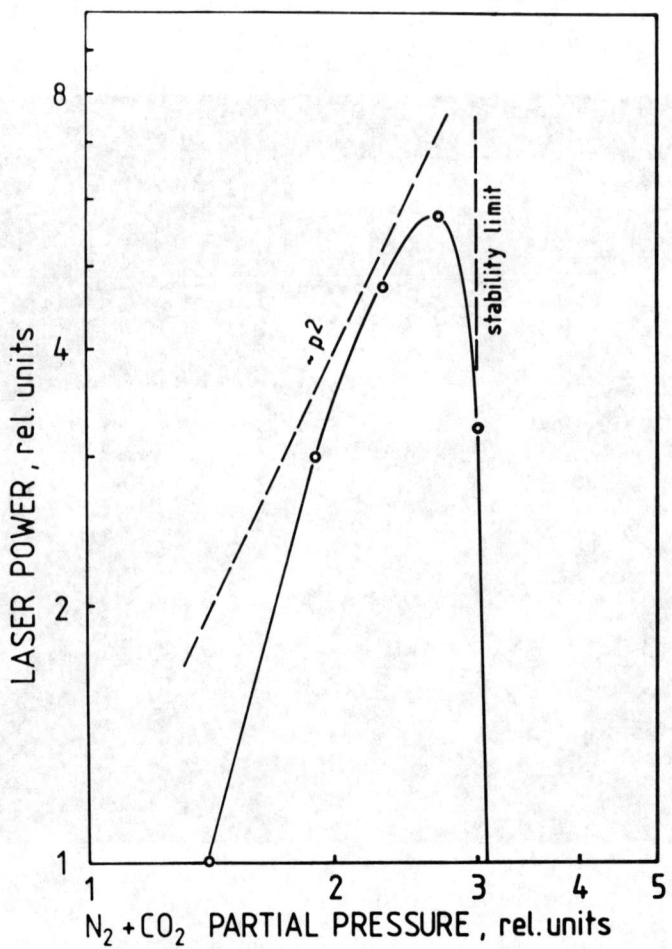

The CO2 waveguide laser (a laser with designs on industry)
I. E. Ross
Ferranti plc, UK

The design of lasers from 2 to 50 watts has previously been totally unsuited to the industrial environment. Glass tubes, flowing gases, unstable alignment, high voltages and temperamental controls have prevented their use in a myriad of applications (from micro-soldering, through plastic cutting and engraving to optical fibre splicing, not to mention the life saving uses in medical surgery.)

The CO_2 waveguide laser has developed from military laser technology which has solved all the above problems and has recently proven itself as a tool for manufacturing industry. The laser is based on rugged ceramic technology, has a sealed long life gas fill, giving excellent MTBF, and is extremely controllable with on/off switching and pulsing on command.

The development of the technology, the specific advantages in performance and a taste of the possible applications will be described.

THE DESIGN OF A LASER FOR INDUSTRY

The design of lasers from 2 to 50 watts has previously been totally usuited to the industrial environment. Although there are numerous applications for CO_2 lasers in this power range, it has not proved possible to exploit them due to the previous nature of the devices themselves. Figure 1 shows a conventional CO_2 laser. The design is based on a glass tube which encloses the laser gas mix (CO_2 plus additives). This glass tube is in turn enclosed in a water cooling jacket. The end of the tube is normally sealed by a Brewster angle plate of an infra-red transmitting material which is epoxied in place. Mirror mounts are supported external to this tube assembly to complete the laser resonator. The laser gas mix would be excited by a high voltage discharge between metal electrodes inserted into the glass envelope. This design suffers from many disadvantages: it is glass and fragile; it is large; the external cavity makes it sensitive to misalignment; it requires water cooling; it has short shelf life due to epoxy seals; it has short active life due to chemical effects at the metal electrodes.

The waveguide laser uses a totally different design philosophy which circumvents all of these problems. The design is based on a rugged ceramic tube which contains the laser gas and also contains the laser beam within the cavity by guiding the infra-red light between the mirrors. Because of the high thermal conductivity of the ceramic the device can be conduction cooled, giving the option of either air or water cooling. The ceramic structure can utilise all metal sealing technology (as has been known for many years in the manufacture of high reliability power valves or 'tubes') and in consequence exhibits both very long shelf and active lives. The laser optics can be mounted directly on the ceramic tube without an external support structure which reduces the size of the device. This assembly is shown in Figure 2: a. shows the metal seals at the end of the tube; b. shows the laser bore (containing the gas) drilled in the rugged ceramic; c. shows the laser mirrors mounted directly on the end of the tube.

In addition by using a Radio Frequency excited discharge in the laser gas it is possible to remove the requirement for metal electrodes in the gas. The ceramic from which the tube is manufactured has a high dielectric constant which allows RF power to be capacitively coupled through the wall of the tube. This also allows the use of a transverse rather than a longitudinal discharge which reduces the required voltages by an order of magnitude. The removal of metal from the discharge region has allowed very long active lives to be achieved in tests. Devices have been operated continuously (24 hours per day) for periods in excess of a year.

The CO_2 Waveguide laser has the following advantageous features:
* Rugged
* Small (devices can be housed within 10% of the volume of the conventional counterparts)
* Long shelf life
* Long active life
* Can be air or water cooled
* Easily controlled by CNC machines

In consequence this new type of CO_2 laser which was initially designed and developed for stringent military requirements has inherently meets most of the requirements of the industrial user.

PERFORMANCE OF A LASER FOR INDUSTRY

The performance of CO_2 waveguide lasers has been studied in depth as part of military and industrial development programmes. Devices have been produced from 2 to 30 watts of infra-red power (at 10.6 microns). Each of the important performance parameters are detailed below:

Power Output Performance

The CO_2 waveguide laser has proven capable of producing output powers in excess of 30 watts. These powers have been produced from devices of a size considered impracticable from conventional devices. Three comparisons can be made in terms of power output per unit active discharge length, power output per unit total length, and power output per unit volume.

Power output per unit active discharge length:
Waveguide=   40 to 70 watts per metre.
Conventional= 20 to 30 watts per metre.

Power output per unit total length
Waveguide= 20 to 30 watts per metre.
Conventional= 5 to 20 watts per metre.

Power output per unit volume
Waveguide= 10 watts per litre
Conventional= 1 watt per litre

In every case the advantage of the new technology is remarkable. In addition the waveguide laser has proved to be more efficient and improvements of up to 25% have been achieved in overall wall plug efficiency.

Active Life Performance

Due to the proven construction technology transferred from power tube manufacture and the use of RF excited discharges, the lives of the devices have been excellent. The active life is defined as the _operating_ hours before the output power falls to 60% of the initial performance. At this time the laser can be returned to the factory to be refilled to recover the initial performance. This refilling is a relatively simple and inexpensive process. Figure 3 shows a life test chosen from one of the many devices on test. As can be seen this 4 watt laser has operated successfully for almost a year of continuous operation (24 hours per day). Of course it has been shown that the devices can exhibit just as long lives with intermittent operation. This operation does not require the use of vacuum pumps, gas bottles or any other external supply to the device.

The laser power supply consists of a simple RF amplifier, similar to those used for communications transmitters or even Citizens Band radio. In consequence they are small, inexpensive and reliable. Thus, overall, the laser devices have a very long M.T.B.F. which is effectively only limited by the life of the laser gas itself.

Beam quality

For many applications the laser output beam has to be focused to a small diameter spot, to obtain the maximum power density on the work piece. To allow this to be achieved successfully it is essential that the laser emits a beam of sufficient optical quality to be focussed by a simple lens configuration. The optimum beam 'mode' is the Gaussian profiled TEM_{00} mode. Waveguide lasers of good design emit a beam which almost perfectly matches this requirement as can be seen from Figure 4, where the output has been scanned with a detector to portray the intensity profile in its cross section. As can be seen the beam quality is free from any perturbations from the required Gaussian profile.

Rugged

The lasers are very rugged due to their construction from ceramic. In fact examples of this type of laser have been able to meet the requirements of a full military environmental test programme (shock, bump, vibration) without any degradation. In consequence the lasers can be expected to perform in most industrial environments and to survive a high degree of malevolent handling.

The CO_2 waveguide laser has been designed to meet the requirements of industry. This design has allowed performance characteristics to be achieved with which conventional devices (in this power range) could not compete.

THE USES OF A LASER DESIGNED FOR INDUSTRY

The CO_2 waveguide laser emits a beam of infra-red light at around 10.6 microns. The beam is of sufficient quality to be focused into a small high power density spot. Powers from 2 to over 30 watts are available. The devices are light, rugged, easy to operate, require few services (normally mains electricity or batteries only), require little maintenance (there are no moving parts, and the only lifed item is the laser tube), and are relatively inexpensive. The CO_2 laser has an inherent advantage over lasers such as YAG operating near the visible region of the spectrum and which can damage the retina of the eye. Protection is obviously required against the danger of burns from the CO_2 devices but at power levels at which heating does not occur, the laser is effectively eye safe and requires no further protection. (Note: Class IV Laser)

In consequence there is a mushrooming range of applications in manufacturing industry, in addition to several possible military uses and several life saving medical applications (as a non contact, highly accurate, and self cauterising scalpel).

A selection of CO_2 waveguide lasers from 2 to 20 watts is shown in Figure 5. Figure 6 shows a 20 watt waveguide laser. The first set of applications all depend on the high power densities that can be achieved in the focused spot of the laser. (4 Giga-watts per metre^2). This involves passing the beam through a focussing system as shown in Figure 7. The focused beam passes through a nozzle which can be pressurised with air (or an inert gas) to protect the lens from vapours or smoke produced at the workpiece. There is no mechanical contact with the workpiece, the system can be extremely accurate (when coupled with CNC moving table, moving laser or moving optic systems), and total control of the power of the beam can be readily achieved. The following sections give a taste of the many applications where this intense beam can be used.

Cutting Plastics

The CO_2 light beam interacts strongly with most organic materials and is capable of cutting most plastics. With powers up to 20 watts a range of thickness up to a few millimetres can be cut at speeds up to several metres per minute (not necessarily at the same time) The multi-directional cutting ability of the laser beam and its accuracy allow complex shapes to be easily cut on a variety of work handling stations. Most of the applications described have used an optical line following system which only requires a draughted line drawing of the shape to give an accurately cut product.

* Perspex   1.2mm thick cut at 0.4 metres per minute at 10 watts.
* Plastic filters   0.5mm thick cut at 2 metres per minute at 10 watts.
* Reinforced plastic 0.2mm thick cut at 2 metres per minute at 10 watts.
* Plastic sail cloth 0.1mm thick cut at 3 metres per minute at 10 watts.
* Self adhesive vinyl cut at 11 metres per minute at 10 watts WITHOUT CUTTING BACKING PAPER.
* Transparent vinyl 0.15mm thick cut at 9 metres per minute at 10 watts.
* Rubber clad aluminium cut at 1 metres per minute at 15 watts without damaging the aluminium.

The above gives an indication of the wide range of plastics that can be cut with the laser. This allows applications in the motor industry (cutting interior trim), in the sign industry (cutting perspex for signs, or self adhesive vinyls), in the aerospace industry (trimming protective cladding on air frame components), in the plastics industries (cutting plastics, removing flash from moulded components)

Cutting Other Materials

The CO_2 laser beam will also interact with other non-metallic materials. It can cause burning or vapourisation and hence cutting of natural fibres, wood and paper. Cloth cutting has been demonstrated at powers above 10 watts, with sealing of the edges where small amounts of man-made fibre are included in the yarn. Fine wood veneers can be accurately cut into complex shapes with these devices and paper is readily cut at speeds up to 2 metres per minute (0.25mm black dust paper).

There are numerous applications in cutting where utilising CO_2 waveguide lasers might be a significant advantage. The cutting parameters of any material can easily be assessed by trials performed with the CO_2 waveguide lasers.

Marking and Engraving

Akin to the cutting applications are marking and engraving. In most instances if the laser will cut a material, the beam can also be used to mark the surface with a company logo, a serial number, or an anti-forgery code. The laser beam can be steered under numerical control by using galvanometric mirrors, for high speed writing, or the work piece can be moved by CNC techniques. This allows the appropriate patterns or letters to be generated. Two processes of marking can be considered. Some materials react with the laser beam to form a different colour at the surface. Materials such as plastics, rubber and paper are especially suitable for this technique. For example white polyvinyl chloride (PVC) turns brown when irradiated. Chemical optimisation of the material (for example by the use of additives) is however required to ensure the mark is sufficiently prominent and durable. The second technique involves melting or "damaging" the surface to leave an indentation. Dependent on the laser power and the material used the process may involve melting, vapourisation, or explosive removal of material. In each case a deep grove can be formed which gives a strong visible mark. In extreme cases it is possible to completely remove a surface layer to reveal an alternative colour or material beneath the surface which further enchances the visual effect. Painted items, and mutilayer plastics can be engraved in this manner.

CO_2 lasers will never replace YAG solid state laser for marking metal, however the performance on plastics is generally superior and on glass and quartz is without competition from the YAG alternative. When associated with a simple micro-processor controlled system the CO_2 waveguide laser will have an important role to play in marking products for many industrial and commercial applications.

Micro-Soldering

The CO_2 laser can be used to melt tin-lead solders. Despite the problem of the reflectivity of the metal the laser can accurately apply a contolled level of power to manufacture a reliable joint between miniature electronic components, without the recourse to having to contact the area to be processed. The problem of reflectivity can be reduced by applying flux to the surface of the solder to enhance its absorption thereby reducing the power levels of laser required. A suitable alternative is to use solder paste which already includes the flux. Generally lasers between 20 and 80 watts are ideal for soldering applications and the CO_2 waveguide laser is the optimum choice of device for the lower part of this range.

The laser power level and the exposure time can be accurately contolled at the joint and, with the focused beam, pad areas down to 0.2 mm can be accessed. With this non contact method the handling and soldering of fine wires becomes more reliable. In consequence the technique of laser soldering is particularily applicable to both thick and thin film hybrid circuits and to the rapidly increasing technology of surface mounting components, where high accuracy, high reliability and speed are of the utmost concern.

Pollution Detection

The CO_2 laser has properties other than its inherent power. The wavelength of the light lies in a region of the infra-red spectrum where many chemicals exhibit strong absorption characteristics. The laser can operate on several discrete wavelengths around 10.6 microns and using the appropriate selection techniques a specific wavelength can be emitted. By either switching the wavelength of the laser or by using two lasers it is possible to detect the presence of small quantities of highly absorbing gases. The technique is based on passing the two laser beams across or through the area of interest (for example through the air above a chemical plant) to detectors at the other side. In the absence of the gas to be detected, both detectors will measure the same power. When the polluting gas is present however one of the beams will be absorbed and the power reaching the detector reduced. This signal will then be less than the beam which is not absorbed and an alarm can be signalled.

There is an obvious requirement for such systems as environmentalists become increasingly concerned about the safety of chemical plants.

CONCLUSION

The CO_2 Waveguide laser is a laser with designs on industry. The laser has been developed to have the specific properties required by most industrial users. The device is small, reliable, rugged, and requires the minimum of external services. The laser is there, "on command" whenever required. Already a wealth of applications for these devices is being unearthed. Users have gained a more cost effective production technique, simplified on-line production technology, or simply the only method of achieving their manufacturing goals. Imagination only limits the other uses for this technology and in the years to come this expanding technology could find its way into many more interesting applications areas.

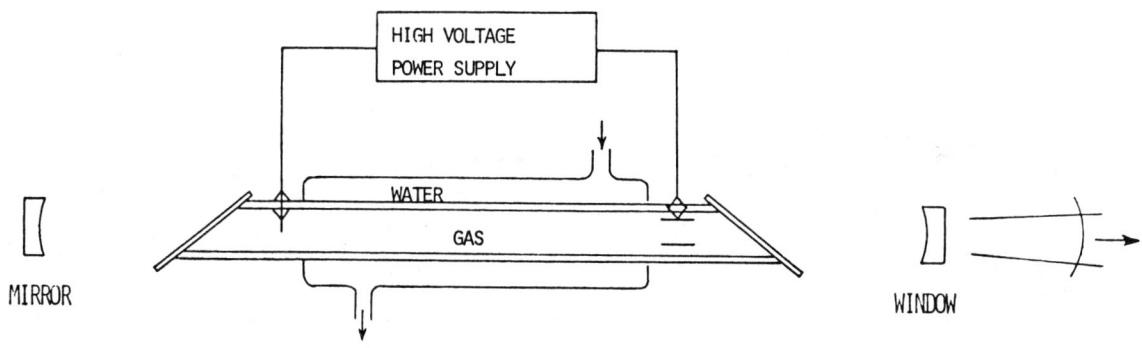

FIGURE 1   CONVENTIONAL CO$_2$ LASER (10 WATTS)

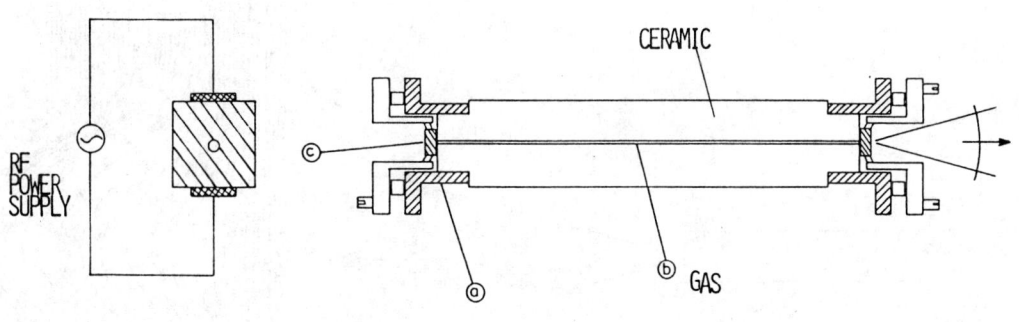

FIGURE 2   WAVEGUIDE CO$_2$ LASER (10 WATTS)

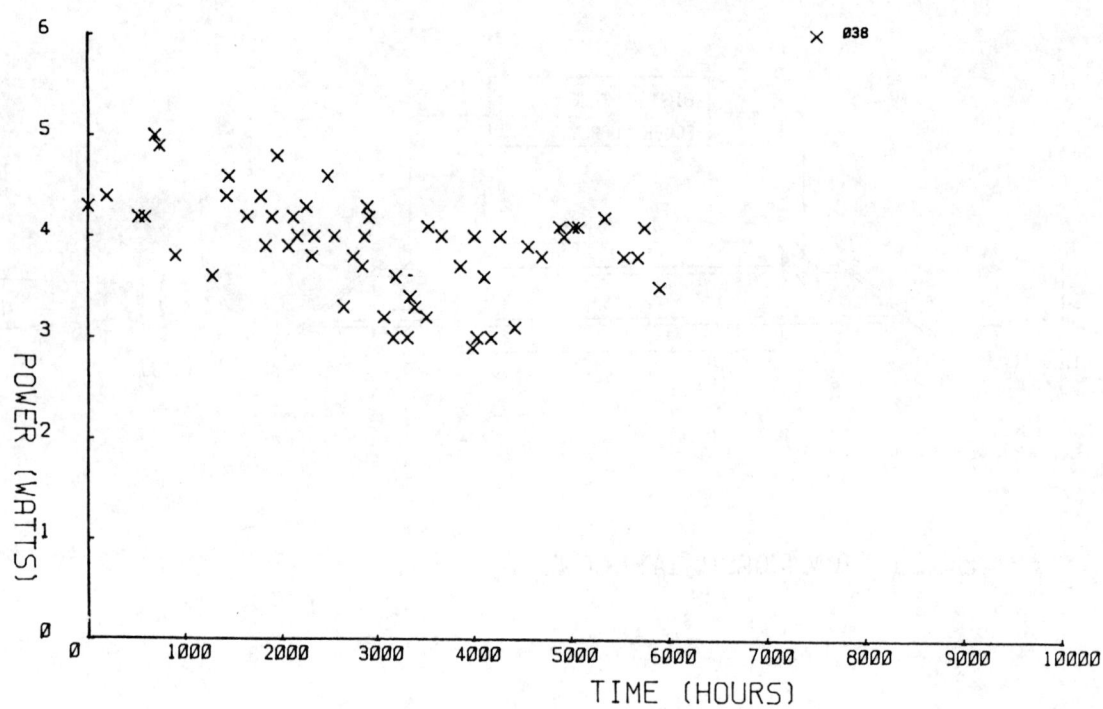

FIGURE 3   LASER ACTIVE LIFE TEST

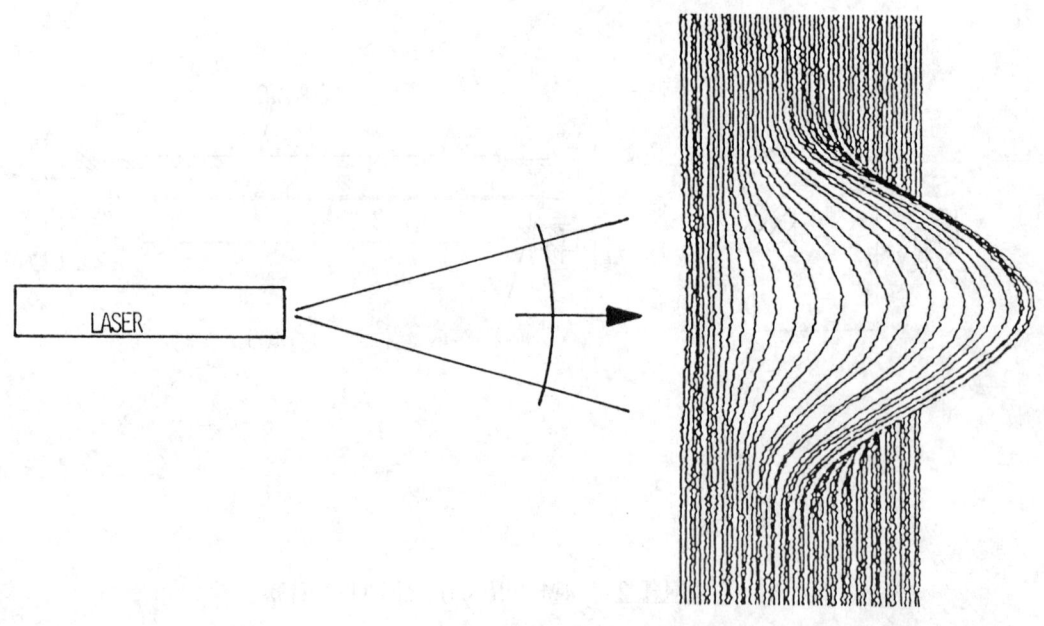

FIGURE 4   THE OUTPUT BEAM OF A CO_2 WAVEGUIDE LASER

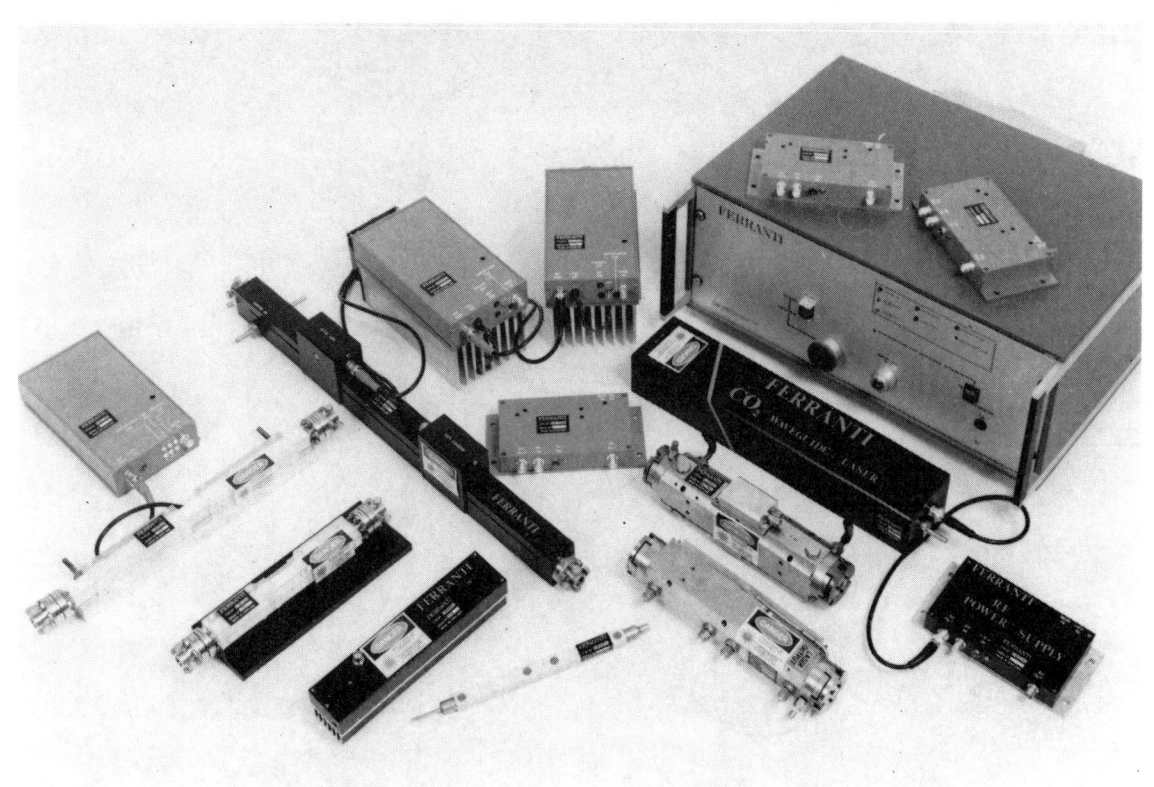

FIGURE 5. A SELECTION OF CO_2 WAVEGUIDE LASERS FROM 2 TO 20 WATTS

FIGURE 6: A 20 WATT CO_2 WAVEGUIDE LASER

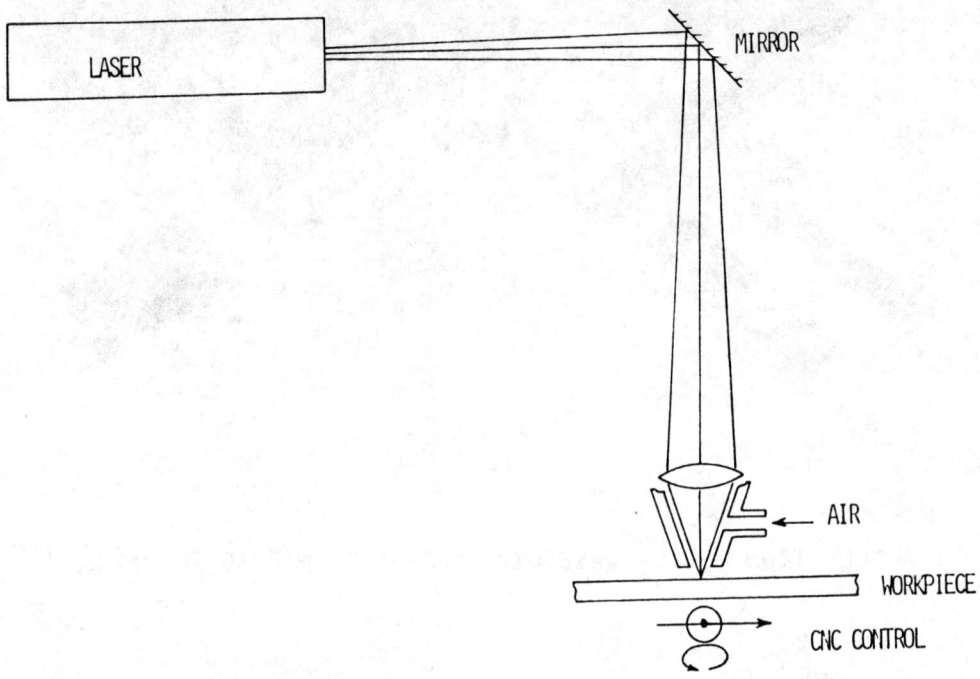

FIGURE 7   A FOCUSING NOZZLE SYSTEM

A laser robot for cutting and trimming deeply stamped metal sheets
A. Delle Piane
Prima Progetti SpA, Italy

0. ABSTRACT

The present paper describes a robotized system, called ZAC, designed for cutting and trimming pre-formed metal sheets and deeply stamped plastic parts by means of a focused CO_2 power laser beam as a non contacting tool.

The system allows the cutting of really 3 - D pieces, moving and orienting the focused beam with 5 NC controlled movements (3 rectangular translations and 2 angular rotations).

Furthermore the ZAC performs the adaptive behaviour of the focusing point on the actual position of the metal to be cut.

The robot is programmed by self teaching with a portable gun.

The loading/unloading of the workpiece is automatically performed outside the operating volume.

The system is conceived in such a way to be easily integrated in a production line.

In the motor car industry the robotized laser system ZAC finds the following fields of application:
- the making of prototypes
- the small batches of experimental cars
- the limited quantities of high class cars
- the spare parts production
- the support at the start-up of production lines

and in the general, when the quantities are limited, production changes are frequent, and the part shapes are subject to modifications.

1. GENESIS OF THE MACHINE

At the beginning, the ZAC robot was designed, at the end of 1977, to solve the following problem: the trimming of 3 - D thin plastic pieces, covered with carpets, for internal fittings of motorcars.

The parts under consideration are obtained by vacuum thermoforming. When the moulded part is obtained, the problem is to trim it clean from the scrap: this operation can not be performed with traditional blancking dies (apart from economic considerations), it is difficult and complex to be obtained with other kinds of machines, it is hard and unpleasant to be done manually.

To face the problem we preferred to think a new robotic system of sequential cutting. We have chosen this solution aware of the fact that the cutting time, though inevitably longer compared with a parallel cutting, would have been largely rewarded by the flexibility that only a sequential cutting performed by a robot could have given to the system. We chose a non contacting tool because the particular kind of material wears out very quickly traditional tools such as knives, small saws or cutting mills; moreover the semi-flexible nature of these materials does not allow the use of a tool which may cause on the workpiece shearing stress, because this would imply the use of expensive supporting and locking masks. Hence, between laser and water jet we chose the former firstly because the CO_2 laser had, at that time, a longer background as industrial product and secondly because the quality of the cut edge was better.

Finally we decided that five axes were necessary: three rectangular in order to operate in a large working volume and two rotations in order to oriente the laser tool with respect of a deeply preformed workpiece.

A certain number of considerations were made in order to decide how many movements had to entrust the laser beam and how many the workpiece.

Moving and orienting suitably the workpiece to be cut under a fixed focusing head, could have appeared a solution without technical surprises but too expensive because of the large dimensions and semi-softness of the workpiece. On the other hand, leaving the workpiece fixed and moving the tool upon it, following the cutting path would have required moving the deflection mirrors and the focusing lenses along an optical path of a too changeable lenght from point to point of the working volume.

Moving the laser generator itself as sometimes it is done, would have limited the choice of the generator, that on the contrary we wanted as independent as possible from the robot structure.

The appreciation of all these circumstances gave birth to the ZAC robot. In it two movements are entrusted to the workpiece which is set on a pallet moving on a cross table X-Y. The third cartesian axis, Z, is represented by a vertical column on which a focusing head, with two hortogonal rotations, is fixed. This architecture reduces at less than one meter the maximum variation lenght of the optical path and enables, within certain limits, an easy tailoring of the working volume, especially for X and Y axes, without being obliged to provide expensive devices for the workpiece to be cut. Built and successfully installed since the end of 1979, the first machine, in a production line for 24 hours work per day, we realized that new and different uses of such a system were possible.

ZAC robot is ideal in cutting thin and deeply preformed plastic materials because its architecture allows working on large pieces and the tool may be oriented in any diretion. Typical operations are cutting and trimming plastic materials such as PVC, ABS, methacrylate, polypropilene, polyethilene, polyuretane, elastomers, glass and carbon fibers, rubber, felt, cork, impregated textile fibers, cardboard, fibers and wood-pulp (woodstock).

Of course we began to think to metal sheets in order to face the following considerations: 1) the shape of the pieces to be cut requires a true 3-D machine. Traditionally the cut is made up by blancking dies, but, when the shape is a complex one, with multiple shearing planes the dies become very expensive; furthermore they are tied to a particular piece with a consequent tying up of big investments. For the large series the cost of the blancking dies can be divided on such a number of pieces to become acceptable; but the small series emphasize the problem and the inadequacy of the traditional solution. In particular the prototypes, the start-up of new productions, the personalisations of large series, the spare parts manufacturing are all field in which an alternative cutting system is useful.

2) the focused power laser beam gives, used as a tool, a lot of advantages among which it has to be remembered the quality of the cut edge (clear and accurate).

3) The large variety of pieces to be made needs a solution the most flexible to the trimming problem. Only a robotized solution can be used, especially in the prototype production, in which the manual cut seems to be the unique alternative (at least from the economic point of view).

4) In addition to the drastic reduction of the manpower, the robotized cut ensures an absolute uniformity to the final product that allows to face in a different way the problem of the quality control on the finished pieces. With the exception of few aluminium or copper alloys, for which surface treatments are sometimes required in order to reduce reflection, metallic materials of common industrial use can be cut with laser. The ZAC system has therefore excellent opportunities of applications to trim preformed metal sheets and particularly deeply preformed steel parts of bodies for motorcar industry.

In this area we have the following applications:

- prototypes - all the big car manufacturers spend, for marking and cutting only prototype body-shell parts many thousands hours per year. The number of parts to be produced goes from a single piece to a few tens for experimental pre-serial: they are always 3-D parts, sometimes of large size. With a laser cutting robot the marking is avoided because replaced by the machine-programming una tantum; the trimming is executed in a very short time and the finishing is eliminated because of the excellent quality of the laser cutting.

- production of small series - it is the typical case of lorries and coaches: production of 50 to 100 specimen a day for pieces of large dimensions. If we consider that very often the average life of these models, for buses and considering coaches in particular, is no more than 3 or 4 years, this explains how difficult could be to pay these expenses, for an investment totally non-reconvertible, when the limited series is ended. A further case for small series is given by the customer tailoring of large production models: we cite for instance the sunshine roof version of any motorcar, or the right-hand drive cars that have simmetrical holes for clutch and brakes pedals as to those of large production series.

- spare parts - motorcar manufacturers for many years (8, 10, sometimes 12) must ensure the market of spare parts availability, also for body parts.

The problem is solved either by establishing expensive spare parts stocks, or by retaining the forming and the cutting dies. The latter solution is the one commonly pursued, but while the costly forming mouldings steel-made can be replaced by the far more economical moulding plastic-made owing to the limited number of pieces to be marketed, the cutting mouldings in steel will remain an unreasonable capital immobilized as against the production of only a few hundred pieces per year.

ZAC will completely solve this problem.

2. THE ZAC SYSTEM

2.1. The robot

ZAC is a 5-axes N/C robot. The purpose of the robot is to allow the focused laser beam to follow any possible path in a large working volume with the proper angular attitude. With five coordinated movements (three perpendicular translations and two angular rotations) the axis of the laser beam can reach the part and it is always orthogonal to its tangent plane along the cutting itinerary. Two movements are made by the workpiece: an X-Y table on the floor moves horizontally the part. Three movements are performed by the laser beam: a vertical column Z (the third cartesian axis) leads a focusing head against the workpiece. The focusing head is able to deviate the laser beam with two rotations: A (360° continuous) around the column axis and B ($\pm$ 90° with respect to the vertical position) perpendicular to A.

Three movable mirrors deflect the beam into the column and into the focusing head constituing an optical chain which terminates on a focusing lens.

The metal cut with the laser requires a very good focusing of the beam on the workpiece. Unfortunately the deeply stamped parts have, very often, unpredictable deformations due to non constant behaviour of the material and other typical reasons as for instance the stocking conditions of them.

It is then possible to obtain a good quality of such cuts without giving to the machine some kind of adaptivity to the actual position in which the steel to be cut is. This is obtained with a sensor on the nozzle which is interfaced with the electronic control of the robot: automatically, during the movement of the machine axes along the path, the focusing lens is moved in such a way to guarantee the focus to follow the actual position of the piece. This device can be considered as a sixth axis of the robot: its low inertia and an adequate stroke allow the robot to be adaptive and correct the deformations of the parts to be cut.

Through the conic nozzle a flow of oxygen is injected in the cutting point; this injection is of primary importance for the cutting technology. On the other hand, the gas flow is used to cool the lens, and pressurizes the projection of microparticles on the lens itself. On the upside of the lens there is an injection of pressurized air: in this way all the optical chain, before the lens and the three mirrors, is protected against dust.

To feed the gases till to the nozzle and to bring the electrical connections to the servo-motor of the rotation B there is a rotary pneumatic and electrical collector that allows the head to rotate continuously around the A axis.

2.2. The laser generator

The laser generator stands on the same mechanical structure which supports the Z column and the focusing head. This architecture allows an easy and stable alignment of the optical path. The ZAC system was conceived in such a way to be independent from the laser generator choice.

In fact any CO_2 laser source up to 1.5 KW can equip the ZAC system. Therefore, the choice of the generator is usually made up depending on the application requirements, in terms of speed and quality of the results, or under considerations like the price, the servicing facilities in a particular country and so on.

However, the following recommendations must be followed:
- maximum power: 1.5 KW
- maximum beam diameter: 25 mm
- maximum divergence: 2mrad
- availability of an external input for the analogic power control
- availability of external controls of the mechanical and electronic shutters of the laser cavity
- availability of external outputs for shutter status and diagnosis

Furthermore the beam energy distribution is of particular importance and it is necessary that the distribution approximates the fundamental mode TEM_{00}: this is an essential requirement to obtain a good penetration in the material to be cut with a consequent good quality of the cut itself. Under these conditions the laser generator can be integrated in the system with the proper safety interlocks with the ZAC robot.

2.3. The loading/unloading system

For safety reasons the working volume is usually totally enclosed in a protection cabin

in which the operator does not enter during the cutting operations.
For that reason the loading and unloading of the workpiece must be performed outside the working area. In the standard version of the ZAC system, the workpiece is palletized on a dedicated fixture which refers and block it. The pallet is then introduced automatically in the working area through a moving door operated by the electronic control of the robot on a request made by the operator with a push-button. In order to minimize the time related to the unloading of the cut part and the loading of a new part on the fixture, usually two pallets, with two fixtures, are used in the system. During the cutting operation on a pallet, the operator changes the part on the other pallet: when the cutting cycle is terminated the robot exchanges automatically the two pallets. There is a unique moving door and the two pallets move inside and outside the working area from a single side of the robot. Generally on the two pallets two different parts are mounted: the robot automatically recognizes the pallet and recalls from memory the correct cutting program.

The whole loading/unloading operation, that is the movement of the transfer system, the opening and closing of the moving door, the laser shutter control and the safety interlocks, is under the control of a dedicated PLC structure contained in the control cabinet.

2.4. The electronic control of the robot

The robot is controlled by a control unit specially conceived and designed for industrial robot control, called ROBOPRIMA.

The control is based on a master CPU and slave intelligent boards: the master manages the part-programs, the man-machine communications, the movements of the robot in the absolute axes. A slave unit transforms the absolute movements relevant to the programmed path itself. Each axis is locally controlled by an intelligent board. In the same way, as a 6th axis, is controlled the adaptive axial movement of the lens and the nozzle, following the actual position of the workpiece by the capacitive sensor. Also the power output of the laser beam is handled by the ROBOPRIMA with an intelligent board, as a 7th axis, thus allowing the constance of the ratio between the speed and the power focused on the workpiece. This fact allows the possibility to obtain an high quality standard of the cutting edge, independently from the more or less complicated shape of the path to be cut. The control performs the full space linear and circular interpolation with programmable velocity connection between the trajectory points. The ROBOPRIMA is contained in a IP54 waterproof cabinet with heat exchanger. The power supply units as far as the motor drive units and the PLC structure are contained in the same cabinet.

2.5. The programming of the robot

Usually the ZAC is programmed by self-teaching. When a new program must be obtained, a sample of the new workpiece, on which the cutting path is traced, is loaded on the relevant fixture and is then introduced on the robot table.

The power laser being off, the operator enters into the working area and moves the axes of the robot bringing the nozzle of the focusing head along the cutting path and recording the points of the path itself. For this purpose the programmer uses a portable handbox with jogs for moving each axis, functional keys and a 20 character alphanumerical control, with the display, guides the operator in entering the auxiliary functions when required, asks for the type of interpolation needed, etc. During this phase the machine can be moved, at the operator's choice, in absolute axes, in the axes of the tool (the oriented laser nozzle), or rotating the head axes A and B around the focusing point (TCP mode: tool centre point rotation). Furthermore during this phase the speed of the machine is automatically reduced at the maximum of 2 m/min and it is software controlled and limited for safety reasons. To materialize the point in which the focus of the power laser will be in the following cutting phase, two alternatives are possible. The first consists in the use of the alignment HeNe laser which is coaxial with the power beam. The optical chain allows the visible light of this laser to be focused in a red spot on the path to be programmed. As a second alternative it is possible to unscrew the conic nozzle of the head and to substitute it with a traditional tracer tool, point shaped and spring loaded, which has the same conic shape of the nozzle.

The teaching phase on the part elapsed, the technological data relevant to the particular material to be cut (e.g. the maximum allowed speed, some local variations of the established ratio power/speed, etc.) can be added to the program by means of a portable programming unit. On it, which is equipped with a powerful program editor, it is possible to edit the program, adding, deleting, modifying some blocks of the programmed trajectory; The program so obtained is stored on a magnetic cassette.

The described way of programming is normally used because it is simple, intuitive, easy to learn, and fast and because it allows ZAC to be used standing alone, independently from the environment in which it is situated. On the other hand it is easy to imagine ZAC as an output peripheral of a CAD/CAM system performing its task automatically and following the cutting paths calculated by and sent from a remote computer.

2.6. The ancillary equipment

The ZAC system requires some ancillary equipment relevant to the plant installation, the laser generator and the safety.

The robot needs a foundation; the laser generator requires a He, N, CO_2 supply, and O_2 is required as covering gas for the cutting: usually the gases are contained in compressed bottles equipped with suitable pressure switch reducers, etc.

To pressurize the optical chain dry and clean, air is also required. The laser generator needs a water cooling system. The ZAC system uses CO_2 laser generators of CLASS IV. For this reason and to comply the safety regulations usually applicable (like, for instance, the BRM regulations) when using this type of laser, the system was conceived since the beginning in such a way as to be operated outside the working volume in which the beam can be activated, that can be consequently totally enclosed. For this purpose a protection cabin is always mandatory. The walls are fire proof, and they can have methacrylate windows, this material being transparent to the visible light but not the infrared wavelenght of the CO_2 laser.

The doors of the protection cabin are of course lockable and interfaced with the control unit by means of safety microswitches.

3. CONCLUSIONS

We hope that this brief analysis of our 3-D laser cutting system has given a sufficient idea of its applications. Obviously the potential market for this machine can only be evaluated country by country according to their industrial structure and characteristics. Notwithstanding we believe that this new technology has so many points in its favour to make feel obsolete in everyone mind the scepticism, doubts and mistrust that till now, in a way, were justified: the laser should have finally found, together with the ZAC robot, a right partnership.

Laser safety in perspective
E. A. Cox
Consultant, UK

The growth of lasers into all aspects of industrial application over the last decade or so demonstrates the great value of these devices but, in a society now very conscious of the potential hazards of any new technology, inevitably the question of safety is raised. In the absence of any specific legislation manufacturers of lasers and users alike have to turn to British Standards to assist them in determining and discharging their legal responsibilities under the Health and Safety at Work Act 1974. Providing the control procedures set out in these standards are followed safety should be assured.

INTRODUCTION

Lasers continue to find ever wider fields of application in all branches of industry, research, medicine, communications and even entertainments. Many of the lasers used in such applications produce powerful beams of radiation which are potentially hazardous. This paper will review briefly the range of hazards, the standards and the safety legislation which together establish both the appropriate engineering controls which should be built into the laser product by the manufacturer and the control procedures to be instituted by the laser user to secure its safe use. It is important, however, that these various safety measures are tailored both to the manufacturing performance requirements of the specific laser product or installation and the working practices and constraints at the location where the laser product is to be used.

GENERAL

Lasers are sources of very intense optical radiation and exposure to the main beam or its reflections may cause damage to the eyes or the skin depending on the power or eneregy of the radiation, its wavelength, the duration of the exposure and the tissue site which is irradiated.

The radiation emitted by a laser may, depending on the type of laser, be in the wavelength range which can be percieved by the human eye (light) or in the invisible (ultra violet or infra red) regions of the electromagnetic spectrum (see Fig 1). For these reasons the term "laser radiation" should generally be used instead of the term "laser light" except in the special circumstances of when a laser with an output beam in the specific wavelength range 400-700 nm is being described.

The wavelength of the laser beam is related directly to the chemical composition of the "lasing medium" which is the substance in the heart of the laser into which energy is pumped to bring about electronic excitation of the atoms of which the substance is composed. The output beam may be in the form of a continuous beam whose duration lasts for longer than 0.25 seconds which is called continuous wave or CW. Alternatively the beam may be in the form of a short pulse or train of short pulses. Lasers are generally described in terms of the lasing medium, the beam output power or energy and the pulse length eg a 4 watt CW argon ion laser or a 20 millijoule pulsed ruby laser.

Unlike other sources of optical radiation the laser typically produces radiation at a single specific wavelength (monochromatic) although in some types of laser it is possible to adjust or modify the output wavelength within fairly narrow wavelength limits. The laser beam normally has a very low angular divergence and so is capable of producing a near parallel beam of radiation and, as a consequence, exposure to a laser beam may present a hazard over a very considerable range.

BIOLOGICAL EFFECTS OF EXPOSURE

There are two organs at risk following exposure to the radiation from industrial lasers the first, and most important, being the eye and the second being the skin.

The eye is designed to receive visible radiation and to transmit it through the various structures within the eyeball and bring it to a focus onto the surface of the retina at the back of the eye.

Lasers which emit radiation in the visible or near infra red regions pose a potential threat to the retina, which is very much greater from a laser than from a conventional incandescent light source by virtue of the monochromaticity of the radiation. Radiation from a laser in fact will be brought to a much sharper focus on the retinal surface, whereas, a light source producing radiation by a thermal process emitting a wide range of wavelengths would, as a result of chromatic aberration in the optics of the eye, be brought to a much less sharp focus and hence a significantly lower power or energy density per unit area at the retinal surface. The optical gain of the eye is the ratio of the area of the input aperture (pupil) to the area of the focussed spot at the retina in the case of a visible laser this is calculated to be about 200,000 hence an apparently modest irradiance level at the cornea would result in a very high power or energy density at the focus raising the temperature of the cells and causing their destruction. The nature and the mechanism of damage to the retina is related to the exposure duration. Long exposures (many seconds) gives rise to photochemical damage whereas shorter exposures result in thermal damage and very short exposures (nanoseconds) deposit the energy in such a short space of time that a small explosion may occur causing extensive damage.

The particular tissue of the eye at risk is related to the wavelength of the radiation, in the above case only visible (400-700 nm) and near infra red radiation (700-1400 nm) can reach the retina. Due to the transmission characteristics of the optical media in the eye far infra red (1400- 1000,000 nm) and ultra violet (200-400 nm) radiations can not pass beyond the anterior optical components. This means that not only are different tissues at risk from different types of laser but, because there is no focussing effect, the risk from exposure changes by a large factor and consequently the maximum permissible exposure limits for the invisible radiations are significantly relaxed over those for visible and near infra red radiation.

The biological effects of exposure to the skin are much less complex than those for the eye, there are no focussing elements although the depth of penetration of a laser beam into the skin will vary with the wavelength of the radiation. The most normal reaction is one of surface burning which may occur following acute exposure to laser beam powers in excess of 0.5 watts. Unlike the eye, skin burns may be treated and will heal, albeit slowly, in the case of deep burns.

VIEWING CONDITIONS

Depending on the power or energy of the laser, a hazard may be presented by either direct intra beam viewing or by viewing reflections of the beam. Such reflections may come from mirror like surfaces (specular reflections) that do little to reduce the irradiance and simply redirect the beam or they may come from rough surfaces (difuse reflections).

Most circumstances that involve direct viewing of either the beam (along its axis) or its reflections result in the beam being seen as a point source (intra beam viewing). If, however, a laser diode array or difuse reflections from an illuminated screen are veiwed at fairly close quarters a point image would not be formed (extended source viewing) and in these circumstances different criteria are used in assessing the safety of any viewing situation.

LEGAL PROVISIONS

The Health and Safety at Work etc Act 1974 by virtue of Sections 2, 3, 6, and 7 places various duties on employers, designers, installers, manufacturers, and persons at work respectively. These duties are aimed at securing, as far as is reasonably practicable, the health and safety of both persons at work and others who may be affected by the work activity. Additionally, in those premises which are subject to the Factories Act 1961 there are the Protection of Eyes Regulations 1974 made under this Act, Regulation 28 calls for the provision of "Approved" eye protectors where there is a reasonably forseeable risk of damage to the eyes from laser radiation. Certificate of Approval No 2 (General) made under these Regulations gives the design criteria which laser protective eyewear must meet in Factories Act premises.

Apart from this last Regulatory provision there is currently no specific legislation relating directly to the safe use of lasers and so all control measures have to stem from the general provisions of the Health and Safety at Work Act. Clearly some more guidance is required to assist the industry and users to adopt a unified approach to safety which leads to the recently introduced British Standard.

BRITISH STANDARD 4803:83

The British Standard has been published in three parts and replaces completely the previous standard (BS 4803:72) it is closely in line with the International Electrotechnical Commission standard (Committee TC76) which is scheduled to be published later this year (1985). The BS is intended to provide information and recommendations to assist manufacturers, suppliers, and users of lasers to comply with their legal responsibilities imposed by the HSW Act. Although British Standards do not have the force of law they may be cited as a national recommendation which has been subject to widespread consultation.

Part 1 of the standard gives general definitions and information on biological effects of laser radiation.

Part 2 of the Standard gives detailed information to laser manufacturers on the classification of laser products into one of five classes depending on the level of laser radiation to which access is possible and the potential hazard that such exposure could incur. As the beam power or energy and wavelength range will vary widely between different types of laser it is clear that lasers cannot be regarded as a single group to which common standards can, or should, be applied. Aspects of laser application that must be taken into account and influence overall hazard evaluation and thereby any control measures to be applied include; the laser's potential for causing injury, the potential exposure or viewing situation, the environment where the laser is to be used, the degree of beam enclosure that is reasonably practicable, and the training and knowledge of the personnel who will operate, maintain or service the laser.

While some of these aspects will fall within the province of the manufacturer others will be outside his knowledge or control. But because the responsibility to classify a laser product is placed on the laser's manufacturer some standardised procedure was required so that he could be allowed to make reasonable assumptions as to the laser user's potential exposure times etc. Hence classification of laser products is not based directly on maximum permissible exposure levels (MPEs) but on derived accessible emission levels (AELs). In this part of the standard a series of tables are given based on wavelength and emission duration which establish the boundary conditions for each class of laser product. The measured or calculated AELs are compared to these tables.

The time basis for classification for all CW and repetitively pulsed laser products is 30,000 seconds except for those emitting a wavelength in the visible or near infra red regions and the design and use of the product is such that there is no need to look into the beam or to be exposed to laser radiation, in such cases the time basis is reduced to 300 seconds. In the special case of class 2 laser products (see below) the time basis is reduced to 0.25 seconds.

The definition of the various laser classes may be summarised as follows:-

Class 1 Laser Products

These are safe under all viewing conditions so that the MPEs given in part 3 of the standard cannot be exceeded under any viewing situation because either, the output beam from the laser is of a very low power or the laser system is totally enclosed such that no laser radiation in excess of the AEL leaves the enclosure. This is the ideal situation for most industrial laser systems but there are a number of applications where such enclosure is not practicable.

Class 2 Laser Products

These are low power devices emitting radiation in the visible region. The maximum output power of this class is limited to 1 milliwatt (collected via a 7mm diameter aperture ie the maximum possible pupil diameter). Safety with such lasers is normally afforded by the eye's aversion response including the blink reflex, which limits the exposure of the eye to less than 0.25 seconds. Deliberately looking into the beam for periods in excess of 0.25 seconds is hazardous.

Class 3A Laser Products

These are again restricted to the visible region only and rely on the eye's aversion response for protection in accidental viewing situations. the output power of this class is restricted to 5 mW (collected via an 80 mm daimeter aperture) and a maximum irradiance of 25 Wm^{-2}. this restricts the amount of radiation which could pass through a 7 mm daimeter aperture to 1 mW and so it will be seen that class 3A is a special case of class 2. The difference being that viewing the beam with optical aids, unless they have been specially designed or fitted with suitable filters, is liable to expose the eye to a hazardous level of radiation.

Class 3B Laser Products

These may emit in any part of the electromagnetic spectrum and are hazardous when viewed intra beam. They may be viewed safety as extended sources under specified viewing criteria. Lasers in this class are restricted to a maximum output power of 0.5 W for CW lasers and up to 10^5 $J m^{-2}$ for very short pulsed lasers.

Class 4 Laser Products

These are high power devices whose AELs exceed those of class 3B. They may emit in any part of the electromagnetic spectrum. Intra beam exposure of the eye or skin to the beam and specular reflections are hazardous. Difuse reflections may also be hazardous. The beam may pose a fire risk. Use of lasers in this class requires extreme caution.

Once the manufacturer has classified the laser product the standard then goes on to specify a range of engineering control features such as protective housings, key switches, beam stops, warning signals interlocks to be built into the product (Fig 2) and outlines the range of information to be supplied to the user. Additionally the standard details the type of labelling to be affixed to the laser product This was thought to be in line with the EEC Directive on labelling and the Safety Signs Regulations but, nevertheless, the complexity of colour schemes and the multiplicity of labels which this introduces is clearly most inappropriate for lasers. As a result of very recent discussions it has now been agreed that an amendment will be issued which will in effect mean that the labelling requirements will revert the the simple yellow and black single labels which were shown in the draft standard which was published for public consultation several years ago.

Part three of the standard is written for the end user of the laser product it gives details of the MPEs and recommends a range of engineering and administrative controls for the user of the laser product to follow to ensure its safe use (Fig 3). It also gives a number worked laser calculations

Although the standard is written in three parts it is essential that all three parts are read together and to this end a special discount price is offered by the British Standards Institution for the complete package.

Since publication of the standard the Committee have been engaged in updating and correcting the standard and have written a new and separate standard for diode lasers and light emitting diodes. It is hoped that this new standard will be available for public comment in 1986.

It is likely also that the definition for class 3B laser products will be changed shortly to bring it into line with the forthcoming IEC standard this mean that this class will be extended to include lasers emitting invisible radiation.

PROBLEMS REVEALED AT INSPECTION

Routine inspection of a large number of industrial laser systems throughout the UK has revealed several aspects of design and operation that frequently require improvement. These are detailed individually below.

Failsafe Interlocks

The most common problem encountered is the choice and operation of electrical interlock switches. All interlocks which are fitted to enclosures should be of such design and fitted in such a way that they fail to safety. In other words, a simple plunger type microswitch fitted in a negative mode relying on the operation of a simple spring to move the contacts into the safe position is not acceptable. Simple magnetic reed switch systems are also generally unacceptable, however, a properly constructed magnetically operated limit switch fitted in a positive mode would be acceptable. Other alternatives for interlocking work station enclosures etc include the use of trapped key exchange control systems, captive key switches with a coded male key captive to the removable panel interfacing with a captive female component fixed to the structure of the enclosure. If "plug and socket" type interlocks are used a diode link and an associated slugged relay should be used in the circuit to ensure that the interlock cannot be defeated by inserting a simple jumper across the "socket", it should go without saying that mains voltage should not be available at this socket. On the question of interlock defeat the standard recognises that for certain setting up and beam alignment operations it is necessary to be able to defeat enclosure interlocks so that direct access to the beam may be obtained. In such situations it is required that when the interlock is defeated an audible or visible warning signal is given automatically. Additionally the defeat "key" should be captive when in the "defeat" position and should be arranged in such a way that it is not possible to re-establish the enclosure without first removing the "defeat" key. Such keys should be available only to authorised persons and made available only when necessary.

Enclosure Fastners

The fact that the requirement for any parts of the laser or workstation enclosure which are not interlocked, to be secured by fastners requiring the use of special tools for their removal, is not given very prominantly in the British Standard perhaps explains why so many such enclosures are to be seen secured by simple slot headed screws. The intention of the requirement in clause 6.4 of Part 2 of the standard was that any persons wishing to partially dismantle an enclosure could not do so easily without first getting a tool kit to release the fastners, as would be the case for any fixed guard. Clearly any guard or panel has to be capable of being fitted and removed as necessary. At the same time it is not reasonably practicable to fit interlocks to every such panel hence requirement for special tools. In practice this will normally be taken to mean spanners and cross head or hexagon socket screw drivers.

Beam Enclosure

As far as possible manufacturers and users should should aim to enclose any beam paths and arrive at the class 1 situation. While such total enclosure is not always practicable an attempt must be made to enclose all hazardous beams. Local beam enclosures may often be simply and inexpensively constructed from lengths of plastic or metal pipe. In the case of laser products like profile cutters a simple metal or polycarbonate shield can be fitted to the cutting nozzle head containing and reflections of laser radiation from the workpiece. If beam enclosure is not at all possible then some other form of physical constraints should be considered. Simply relying on protective eyewear alone will not generally be accepted. This is particularly so in maintenance and servicing situations when the manufacturers covers have perhaps been removed. In such cases it will be necessary to give thought to the design of some special covers or beam enclosures that may be fitted during these operations which, while not fully enclosing the beam will, give limited access to part of the beam for the task to be performed safely.

Electrical Safety

All the fatal accidents known to have occurred with lasers haved been due to electrical problems. This is not of course surprising in view of the high voltages and currents used to drive the majority of industrial lasers. The most frequent problem encountered is one of failure to provide adequate secondary covers shielding high voltage terminals within the laser or the power cabinet. Opening a locked access door should not give immediate access to hazardous voltages, the engineer should have to remove a secondary cover gain access only to those terminals necessary for the task in hand all others should be shielded against accidental contact. Where regular test probe measurement are made the designer should arrange to either hard wire to a special bank of test points or place small holes in the secondary covers so that only limited access to a test probe is available.

Mechanical Safety

Some lasers have within them belt driven pumps or blowers and, as would be the case for any other similar industrial equipment, such belt drives, rotating key way or splined shafts, or rotating spoked wheels should be enclosed within close fitting secondard guards. This is the same philosophy as for electrical safety, simply releasing locked or secured covers should not give immediate access to all the hazardous equipment contained within the enclosure.

Laser Safety Officer

The appointment of a Laser Safety Officer is not spelt out very clearly in the British Standard. He is defined in Part 1 and he is mentioned in the sections dealing with hazard control. While the majority of laser products supplied for use in industry will be manufactured to conform to class 1 specification which would not require the services of a Laser Safty Officer, most laser manufacturers and people developing laser systems will be working with less well enclosed laser systems and it is certain in such cases that the appointment of somebody who is knowledgeable in the hazards and ways of control them is a necessity. Any person so appointed from within the ranks of the company should be given adequate training. Suitable training courses are available in the UK being arranged through Universities such as Loughborough and Essex.

CONCLUSION

The safety record for lasers in the field of industrial application is to date quite good and providing that standards are kept under review as the range of laser applications develops and adequate attention is given to the relatively simple control measures it should remain so.

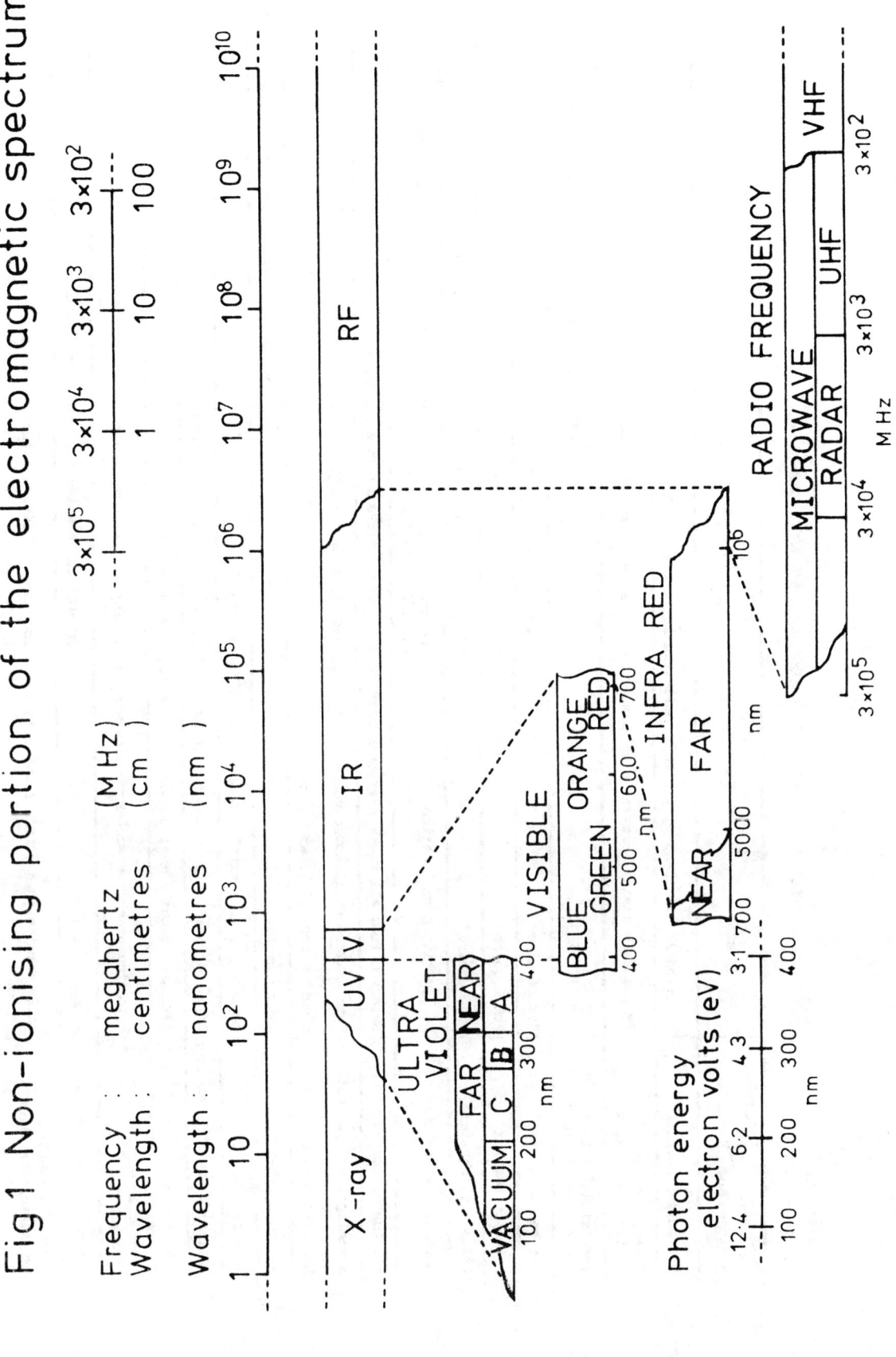

Fig 1 Non-ionising portion of the electromagnetic spectrum

FIG. 1. LASER PRODUCTS — SUMMARY OF MANUFACTURER'S REQUIREMENTS

Class / Requirements	Class 1	Class 2	Class 3A	Class 3B	Class 4
Protective housing	Required in all classes; prevents radiation in excess of AEL				
Safety interlock in protective housing	By failsafe design, prevent displacement of cover until output below Class 1 or Class 2 AEL				
Provision for interlock connection at laser	Not required			To allow connection of external interlock.	
Key switch control	Not required		Laser inoperative when key is removed		
Emission warning	Not required		To give warning when laser is on		
Attenuator/shutter	Not required		To give means beside ON/OFF switch to temporarily block beam		
Location of controls	No limitation on location		Controls to be located so adjustment does not require access to AEL above Class 1 or 2.		
Viewing optics	Emission from all viewing systems must be below Class 1: AELs as applicable				
Scanning laser products	Scan failure shall not cause product to exceed its classification				
Label to show class	Required for all laser products				
Aperture label	Not required			Required to show where beam will emerge	
Service entry label	Required as appropriate to the class of laser exposed				
Interlock override warning	Required under certain conditions as appropriate to the class of laser exposed				
User information	Operation manuals must contain instructions for safe use				
Purchasing and service information	Promotion brochures must reproduce warning labels; service manuals must contain safety information.				
Medical laser products	No requirements			Special calibration and beam power meter requirements	
Laser fibre optic transmission systems	Cable service connections require tool to disconnect if disconnection breaks protective housing.				

FIG. 3   LASER PRODUCTS   -   SUMMARY OF USER REQUIREMENT

Class Requirements	Class 1	Class 2	Class 3A	Class 3B	Class 4
Key switch control	No requirement		Key to be removed from laser and held by authorised person while laser not in use.		
Beam shutter or attenuator	No requirement		When in use stops beam		
Emission indicator	No requirement		Indicates when laser is energised		
Remote interlock connector	No requirement			Connect to enclosure, room or door circuits where appropriate	
Warning signs	Not required	Specific precautions shown on warning signs to be followed			
Beam stop	Not required	To terminate beam at end of its useful path.			
Control of specular reflections	No requirements		Care to be taken	Prevent unintentional reflections	
Protective eyewear	Not required			Required if foreseeable risk of ocular exposure	
Protective clothing	Not required			Sometimes required	Specific requirements
Training	No special requirements	Required for all operator and maintenance personnel *			

* <u>NOTE</u>: Many laser products will have a different classification when protective covers, etc, are removed for service, maintenance or repair. Maintenance personnel should be properly equipped with correct tools, protective eyewear, local beam shields, attenuators, neutral density filters, etc, as appropriate to reduce their risk of exposure as far as practicable.

LASER SYSTEMS

The use of lasers in manufacturing – relevant research at the Welding Institute

P. J. Oakley, M. N. Watson, C. J. Dawes
and
N. R. Stockham
The Welding Institute, UK

The Welding Institute has been involved with research and development in the field of high power lasers for materials processing since the late 1960's. This work has been guided by the requirements of manufacturing industry as the available laser equipment and the potential applications have evolved. Initially experimental prototype carbon dioxide laser equipment was developed and the ensuing designs have been licenced to several companies for manufacture. More recently, however, research has mainly been devoted to the development of a better understanding of materials processing with lasers. The research has generated practical data to aid the production engineer in making a realistic assessment of the applicability of laser processes. It has also led to the development of techniques to extend the versatility of laser processes.

This paper will briefly review the development of carbon dioxide laser equipments, and then describe typical results from our process research work with Nd-YAG and carbon dioxide lasers relating to microjoining, welding of sheet and plate, and to surfacing.

AN INTRODUCTION TO THE WELDING INSTITUTE LASER SECTION

The Welding Institute is an international research and development organisation and its services are available worldwide to companies and organisations in Research Membership. The laser laboratory is one of the specialist sections of the Institute's Research Centre at Abington Hall, near Cambridge UK. As an integral part of the largest welding research and development organisation in the Western World the Laser Section has, in addition to its own specialist staff and equipment, backup from the other departments of The Welding Institute covering the range of welding processes, metallurgy, engineering, manufacturing, computing and process control.

The Laser Section was formed at The Welding Institute in 1968 to develop and exploit the then embryonic laser materials processing technology. A continuous programme of basic and applied research has since been supported by the UK Government and by industrial Research Member companies. The work of the section has therefore been guided by the requirements of industry. Initially the main accent of the work was concerned with the development of carbon dioxide laser equipment, although some process development work was also carried out. More recently, as a range of commercial laser equipment has become available, research has been mainly involved with the development and characterisation of laser processes.

CARBON DIOXIDE LASER DEVELOPMENT

The four basic types of carbon dioxide laser design are shown schematically in Fig. 1.

The sealed tube carbon dioxide laser is limited in both its power output and its operating lifetime and this type of laser was not considered suitable for development for materials processing.

The first two lasers used at The Welding Institute were of the slow flow type (Fig. 1b) where the excess heat from the laser gas mixture is lost through the laser tube walls by conduction, usually into a jacket of cooling fluid. Decomposition products generated by the effects of the high voltage discharges on the gas mixture are removed by flowing the gas along the laser tube and then exhausting them. A maximum power per unit length of lasing tube of approximately 100W/M can be achieved. Lasers of 400W and 800W based on this configuration were used at The Welding Institute, having originally been built by what was then the UK Government Services Electronic Research Laboratory.

Two design philosophies can be adopted to increase the power per unit length of a carbon dioxide laser over that produced by the slow flow designs, and thus to develop a higher maximum output power from a realistically sized machine. These are the fast axial flow (Fig. 1c) and crossflow (Fig. 1d) types. Both of these rely on recirculating the laser gases through heat exchangers. A series of five lasers were built based on the fast axial flow configuration, which was chosen in order to maintain the good mode structure previously found with the slow flow lasers. These lasers also incorporated turbulence generators upstream of the discharges to maintain a glow discharge in the lasing gas mixture. Output powers ranged from 2kW to 5kW. Details of the 2kW laser are given in references 1 and 2.

CURRENT WELDING INSTITUTE LASER EQUIPMENTS

The 2kW and 5kW fast axial flow carbon dioxide lasers built at The Welding Institute are still available for process development. The 2kW has been recently refurbished and upgraded to a maximum rated power of 2.5kW (see Fig. 2). In order to offer as wide a range of laser equipment as possible, and also to gain experience with the different types of laser representative of those used by industry, these lasers have been supplemented by commercially available equipments, e.g. one commercial carbon dioxide laser and two neodymium/YAG lasers. Details of all the lasers are given in Table 1. A full range of workpiece handling facilities are available to complement the laser equipment including precision rotary and X-Y tables, a large CNC table and robots.

TABLE 1   CURRENT WELDING INSTITUTE LASER EQUIPMENTS

Machine Type	Maximum Rated Power	Maximum Pulse Energy (if appropriate)	Capability
Pulsed Nd-YAG	100W	10J	Welding up to 0.5mm thick steel
Pulsed Nd-YAG	300W	45J	Welding up to 2mm thick steel
Pulsed or continuous fast axial flow CO_2	500W	-	Welding up to 2mm thick steel Cutting up to 8mm thick steel
Continuous fast axial flow CO_2	2500W	-	Welding up to 6mm thick steel Cutting up to 12mm thick steel
Continuous fast axial flow CO_2	5000W	-	Welding up to 10mm thick steel

RECENT LASER MATERIALS PROCESSING RESEARCH

Some laser process development was conducted in parallel to the laser equipment development, and the late 1960s and early 1970s saw the development of coaxial gas jet laser cutting[3,4] and the Institute's first deep penetration laser welds. More recently, research has been carried out in four main areas. These areas, which have been dictated by the most pressing interests of industry, are microwelding, welding of sheet materials, welding of thicker section steel, and surfacing. Concurrently with these projects, specific application studies have also been undertaken for individual companies.

Microwelding with a Nd-YAG Laser

Welding of small components (<2mm penetration) with pulsed Nd-YAG lasers (<500W) is one of the largest application areas for materials processing lasers (Fig. 3 illustrates a typical production application). Yet there is almost no information generally available to indicate the weld quality that is obtainable and the effects of welding parameters on weld quality and properties. The results of recent work which has investigated process parameters and gas shielding when spot welding 0.5mm thick stainless steel with the 300W Nd-YAG laser illustrate the capabilities of the equipment and the weld quality that can be achieved.

Spot welding trials were conducted on BS 1449 type 316 stainless steel, 0.5mm thick in both butt and lap joint configurations. Three gas shields were evaluated: a simple 7.5mm diameter copper pipe directed at the top of the weld; a copper manifold over the weld zone with a central hole for transmission of the laser beam; and a tube completely enclosing the laser beam which allowed the beam and shielding gas to impinge coaxially onto the workpiece through a small hole. Argon and helium were evaluated as shielding gases.

A typical result of tensile testing the welds (using a Hounsfield machine) is shown in Fig. 4. Weld strengths increase with increasing pulse energy, because larger weld nuggets are formed by higher energies. Lap welds are weaker for the same pulse energy because the fusion zone must penetrate through the thickness of the top sheet before reaching the joint. The laser beam energy is therefore utilized less efficiently when making lap rather than butt joints.

The best gas shield in terms of weld strength was the coaxial tube, but this gave only slightly better results than welds made in air. The other two shields gave reduced penetrations and strengths, because, it is believed, a turbulent interface existed between the air and shielding gas above the weld which scattered the laser beam passing through it. For all three shields there was a certain minimum gas flow needed to prevent oxidation of the surface: in this respect the manifold was the most efficient (needing a flow of at least 0.25 l/min of argon), followed by the copper pipe (0.75 l/min) and the coaxial tube (2.0 l/min). However, experience on real components shows that the best gas shield to use can depend on the component design. Argon provided a much better shield than helium, and is cheaper. These observations are in contrast to the experience with carbon dioxide lasers, which indicates that helium is the better shielding gas, and that welding without a gas shield gives poor results.

The work described is continuing by investigating other materials of interest, by examining seam welds, and by comparing the performance of the Nd-YAG laser to a carbon dioxide laser of similar output power and capital cost.

Carbon dioxide Laser Welding of Low Carbon Steel Sheet

The high welding speeds achieved and ease of automation make laser welding very suitable for the mass production industries. Recent work has examined the techniques suitable for welding sheet steel (up to 4mm thick) with lasers of up to 5kW output power. The basic sheet metal joints that can be laser welded are shown in Fig. 5. All these joints can be welded by the laser with much less distortion than is encountered with conventional fusion welding techniques.

Laser power and welding speed are the most important welding parameters as they control the weld heat input and hence fused zone width and penetration depth. Focused spot size, which depends on the f number of the focusing system, is also important as it controls the maximum power density at the workpiece surface. Figure 6 shows for T butt joints in 2mm thick steel how these factors are interrelated. Note that the f3.8 focusing system gives much higher welding speeds for the same output power because it produces a higher power density. A systematic study has not yet been undertaken, but it appears that there may be an optimum f number for each material thickness and joint configuration.

The f number of the focusing system governs the depth of focus, which is greater for larger f numbers. The typical tolerance to variations in position of the focus with respect to the workpiece surface was found to be ±2mm for the f7.5 system, ±1mm for the f3.8 system. Thus the use of low f numbers to obtain high welding speeds in sheet materials demands careful control of the focus position.

High welding speeds produce narrow fused zones and so the distance by which the laser beam can be displaced sideways from the joint line is small. Figure 7 shows macrosections of butt joints in 4mm thick steel. In one case the beam was displaced by 0.25mm and the section shows that the fused zone encompassed the joint. When the beam was displaced by 0.5mm, however, although the top and bottom beads appeared sound, the fused zone had not spread across the joint at mid-thickness.

A further problem with laser welds can be joint fitup. Gaps at the joint present the biggest problem, since if they are larger than the focused spot size the beam will pass straight through. Trials on butt joints in 2mm steel have shown that joints with gaps ≦0.12mm can be welded consistently, and for 4mm material gaps ≦0.18mm can be tolerated. To improve the tolerance to gap, and to beam/joint misalignment, a wider fused zone is needed. This can be achieved by defocusing slightly, but trials conducted recently indicate that spinning the laser beam is a better approach as it maintains the maximum power density. Using a spin radius of 0.5mm the tolerable joint gaps increased to 0.25mm on 2mm material and 0.3mm on 4mm material. Beam spinning also gave good quality welds in butt joints on guillotined edges.

Current work is investigating the role of gas shielding when laser welding sheet materials. Helium is generally accepted as being the best gas as it has a high ionisation potential and is therefore less likely to form a plasma above the weld, which reduces the laser power density reaching the workpiece. Argon is particularly prone to this effect at high focused beam power densities, although it is usually adequate for laser powers of less than 2kW. The quality of gas shield strongly influences weld quality, but the gas is an expensive consumable and developments to reduce gas usage are very important.

Laser Welds in Thick Sections

As the power available from commercial high power lasers becomes greater, the number of potential applications of laser welding increases, particularly in the fields of structural and general engineering in which heavy section steels need to be welded. Single pass laser welds in steels of 25mm thickness can now be achieved. Potential applications in section thicknesses greater than 6mm cover a wide range, including truck axles, earth-moving equipment, ships and pipelines. However there are two areas in which new work is required before laser welding is widely adopted for large structures. These are, firstly, that in the section thicknesses under consideration it is important to achieve good impact toughness, which can be difficult with laser welds, and secondly, that with large structures close fitting joints are difficult to achieve, so techniques for welding joints with gaps are required. The research described below covers both these points.

The Charpy impact properties of laser welds in 12.5mm thick C-Mn-Si-Nb-Al steels to BS 4360 grade 50D have been evaluated. Figure 8 shows a typical weld. Charpy properties were found to vary with the welding speed. At slow speeds (0.35 m/min) poor results, with 27J transition temperatures of 20°C were obtained. At high speeds (0.9 m/min and above, using laser powers of 8kW) good results, with 27J transition temperatures of below -40°C, were obtained. This change in properties was a result of a change in weld metal microstructure from mixed martensite and bainite to fully martensitic, accompanied by an increase in hardness from c.300 HV to c.450 HV. Steel composition variations in the range 0.11-0.19%C had very little effect on these results, but S contents below 0.02% were necessary to ensure the good impact values at the higher speeds.

The high welding speeds necessary to achieve the good impact values mentioned above have the disadvantage that the weld beads produced are narrow, which exacerbates the joint fitup problem mentioned above. In addition, the high hardnesses are unacceptable to many fabrication codes. An alternative technique has recently been developed to overcome these difficulties for toughness critical applications. Low welding speeds (e.g. 0.35-0.5 m/min with a 5kW laser) are used, so ensuring acceptable hardness values, and a filler wire addition is used to fill the joint gap. The filler composition is chosen to give the required impact properties. Trials have shown that this is a simple and straightforward technique, and that joints with gaps of at least 2mm can be welded with high reliability. The technique has the added advantage that multipass procedures can be used, see Fig. 9. The overall welding speed of the four pass weld illustrated in Fig. 9 is 5.45 m/hr (not allowing for setting up). This speed is not as great as can be achieved by high deposition rate arc processes, such as multiple wire tandem submerged-arc welding, but the multipass laser weld has the advantages of low overall heat input and minimal distortion that are not characteristic of the high heat input, high deposition rate arc processes. Additionally, the use of multipass procedures allows low power lasers to cover a wider range of thicknesses of materials, so making the process more attractive in situations where a range of thicknesses must be welded.

Laser Surfacing

The same properties of the laser that make it attractive as a method of welding also make it suitable for a range of surface treatments such as transformation hardening, alloying, and cladding. The laser seems to be ideal for treating discrete areas on components.

Although laser transformation hardening experimentation and a limited number of industrial applications have been reported, there is again a need for systematic data on the process to assist in its exploitation. A programme of work was therefore conducted on the transformation hardening of a medium carbon steel using the 2kW carbon dioxide laser. This steel was typical of that used for components requiring hardening. Hardening trials with a defocused beam examined the effect of laser power, beam diameter, and traverse speed on the morphology and microstructure of the hardened zones. The reproducibility of hardening was also investigated.

The hardened zones obtained with all tests were the expected lenticular shape that is found when laser surfacing with a defocused low order mode beam. The trends in the dimensions of the hardened zones for a laser power of 1.0kW can be seen in Fig. 10. The maximum hardened depth that can be achieved without surface melting increases as the beam diameter is increased. However, at any given traverse speed and power, the hardened depth increases as the beam diameter decreases. An increase in the width of the hardened zone, as well as the depth, is observed as the traverse speed is decreased or power increased for a given beam diameter. These results follow a logical sequence in terms of the depth and width of hardening, with variations that can be attributed to the reproducibility in the setting of the nominal beam diameter, or possibly differences in the efficiency of beam absorption.

A much more significant finding, when considering production use of the process, is the variation in hardness and microstructure of hardened tracks made at nominally identical conditions. For each of three sets of conditions eight hardened tracks were made. In each case, four of the hardened zones had a light etching microstructure and four a dark etching microstructure (Fig. 11a and 11b). In spite of this marked difference in etching behaviour, the dimensions of the four dark etching tracks were mostly only slightly smaller than those of the four light etching tracks. Hardness measurements made to complement the microstructural examination gave values of 880-1000 HV 2.5 for the light etching tracks but much lower values (420-640 HV 2.5) for the dark etching tracks.

Two possible reasons can be postulated for this difference in hardening occurring at nominally identical conditions. These are variation in the parent material, or variation in the time/temperature exposure of the material. An examination of the parent material microstructure showed no discernible difference between the regions where light etching and dark etching hardened zones occurred. Also parent material variations are thought to be unlikely as hardened tracks with one type of etching behaviour occurred on only one side of the 12.5mm steel disc test specimens.

A shorter time at temperature, or exposure to a lower temperature would reduce carbon diffusion and therefore could account for these differences in hardness and etching behaviour. However, these differences occurred in trials where the laser parameters were kept constant. Suspicion must therefore fall on variations in the efficiency of absorption of the laser beam by the coated specimen surface. This is supported by the circumstantial evidence that for the reproducibility trials the three discs used for test specimens were sprayed with matt black paint, as a beam absorption coating, at the same time and on each, one side showed light etching and the other dark etching tracks. As the coating material was not changed for these tests, then the thickness of the coating could be responsible for the differences encountered. It can be argued that too thin a coating will not reduce the surface reflectivity sufficiently, or conversely, a coating that is too thick will provide a barrier to heat conduction into the disc.

The practical implications of this unpredictable difference in hardening behaviour are significant when considering the process for production use. If, as suspected, beam absorption coating variations are the cause of this difference in hardening then, having established optimum coating thickness, the control of this parameter will be an important part of the laser transformation hardening process.

CONCLUDING REMARKS

The research of the Laser Section of The Welding Institute has been conducted with the objectives of providing systematic data and a detailed understanding of laser processes, and also developing innovations to increase the applicability of the laser in manufacturing. Industry requires the support of research organisations to produce this information, on which potential applications can be realistically assessed, but research organisations also require feedback from industry to guide their research work into the most worthwhile areas.

ACKNOWLEDGEMENTS

The majority of the work discussed in this paper was funded jointly by the Materials, Chemicals and Vehicles Requirements Board of the United Kingdom Department of Industry and by Research Members of The Welding Institute.

REFERENCES

1. CRAFER R C "A 2kW CO_2 laser system for welding sheet material". Welding Institute Conference 'Advances in Welding Processes', Harrogate, England, 7-9 May 1974, Paper 26, pp 178-184.

2. CRAFER R C "Improved welding performance from a 2kW axial flow CO_2 laser welding machine". Welding Institute Conference 'Advance in Welding Processes, Fourth International Conference', Harrogate, England, 9-11 May 1978, Paper 46, pp 267-278.

3. UK Patent Specification No. 1.215.713 "Improvements relating to thermal cutting apparatus". A B J Sullivan, P T Houldcroft and K W Brown. Complete specification published 16 Dec 1970.

4. UK Patent Specification No. 1.215.714 "Improvements relating to thermal cutting apparatus'. P T Houldcroft, K W Brown and A B J Sullivan. Complete specification published 16 Dec 1970.

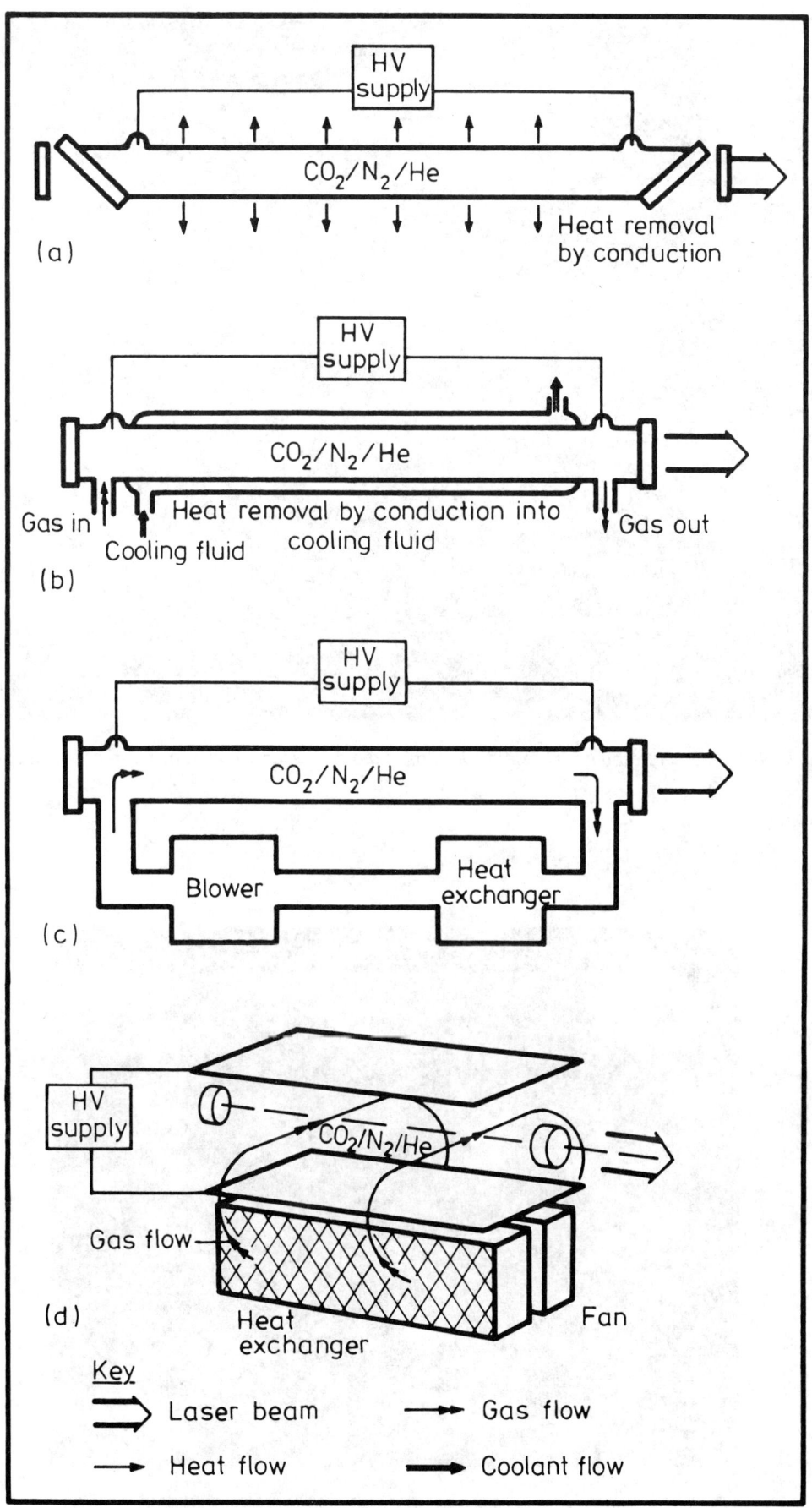

Fig.1. Schematics of CO_2 laser types:

a) sealed tube
b) slow flow
c) fast axial flow
d) cross flow
(NB Diagrams are not to absolute or relative scale).

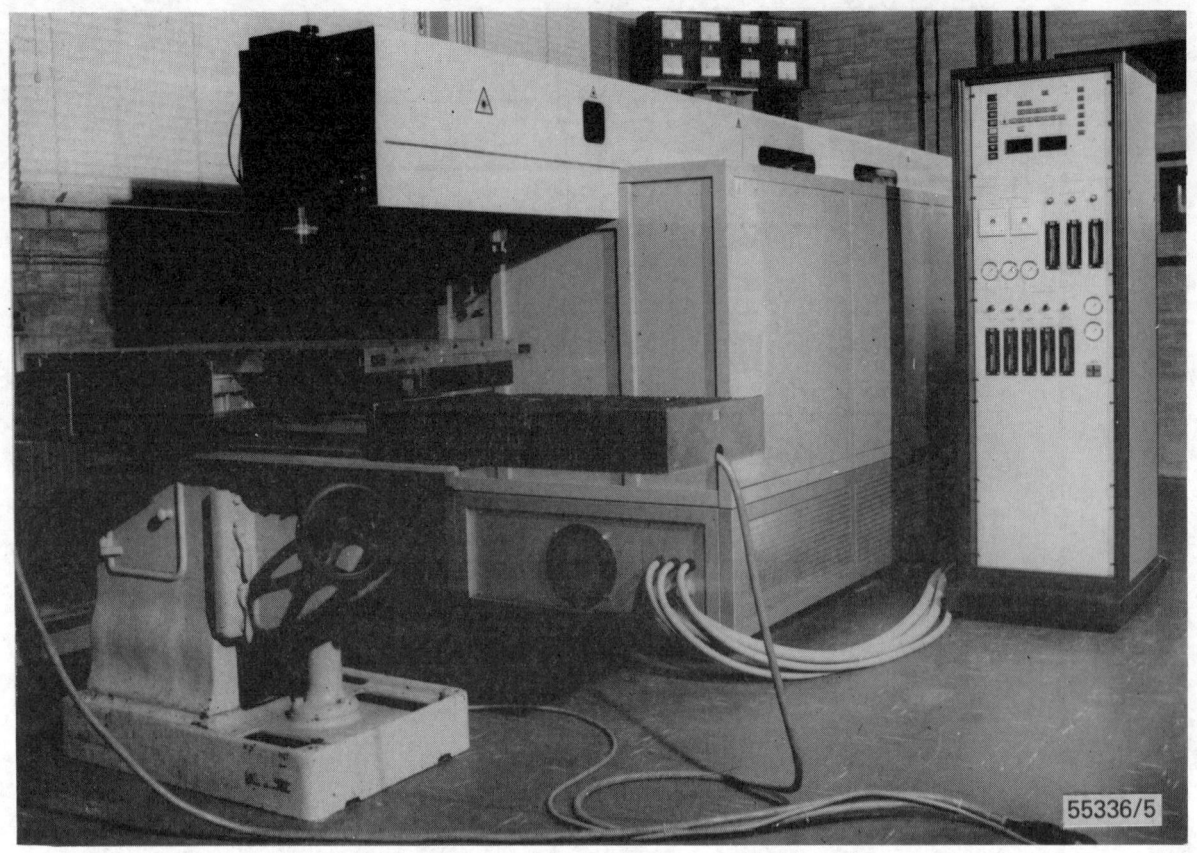

Fig.2. Welding Institute built fast axial flow carbon dioxide laser. Maximum rated power 2.5kW.

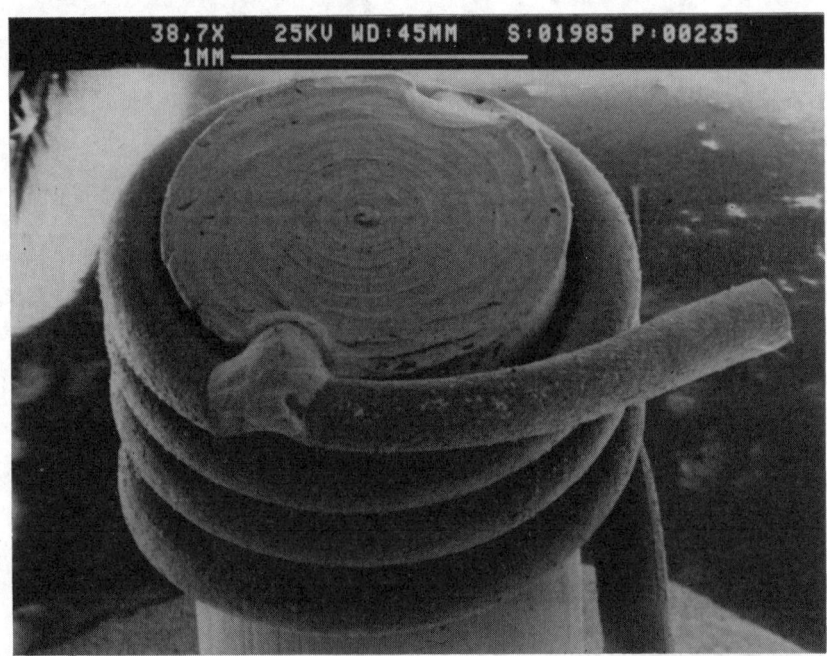

Fig.3. Scanning electron micrograph of laser spot welds joining gold alloy wiring to gold terminal pin for high reliability electronic circuitry in surgical implant.

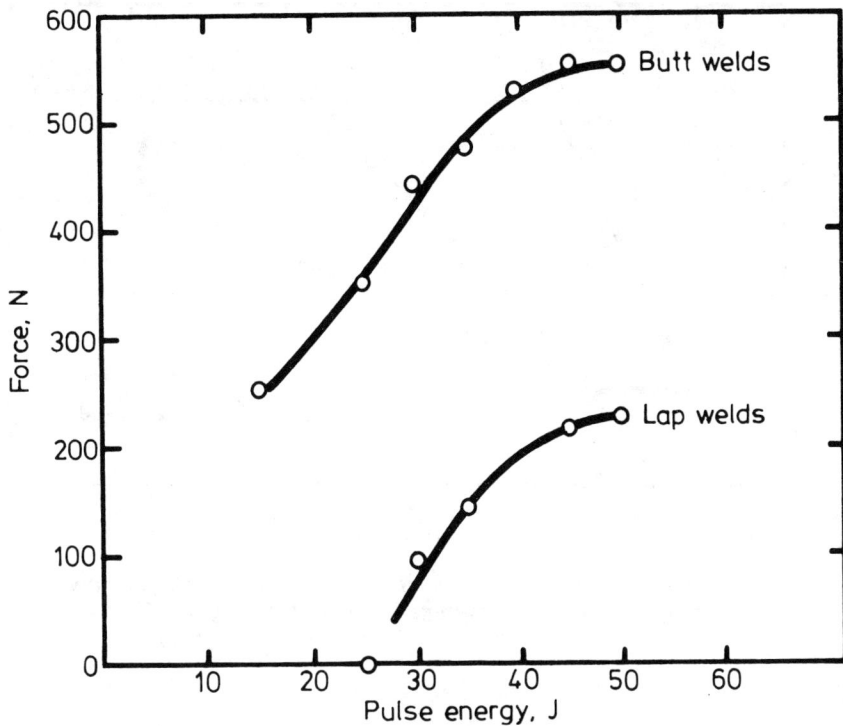

Fig.4. Weld strength v. pulse energy for solid state laser spot welds made in 0.5mm thick stainless steel with the copper manifold gas shield and argon gas (0.5 l/min), 5Hz repetition rate and 20msec pulse duration.

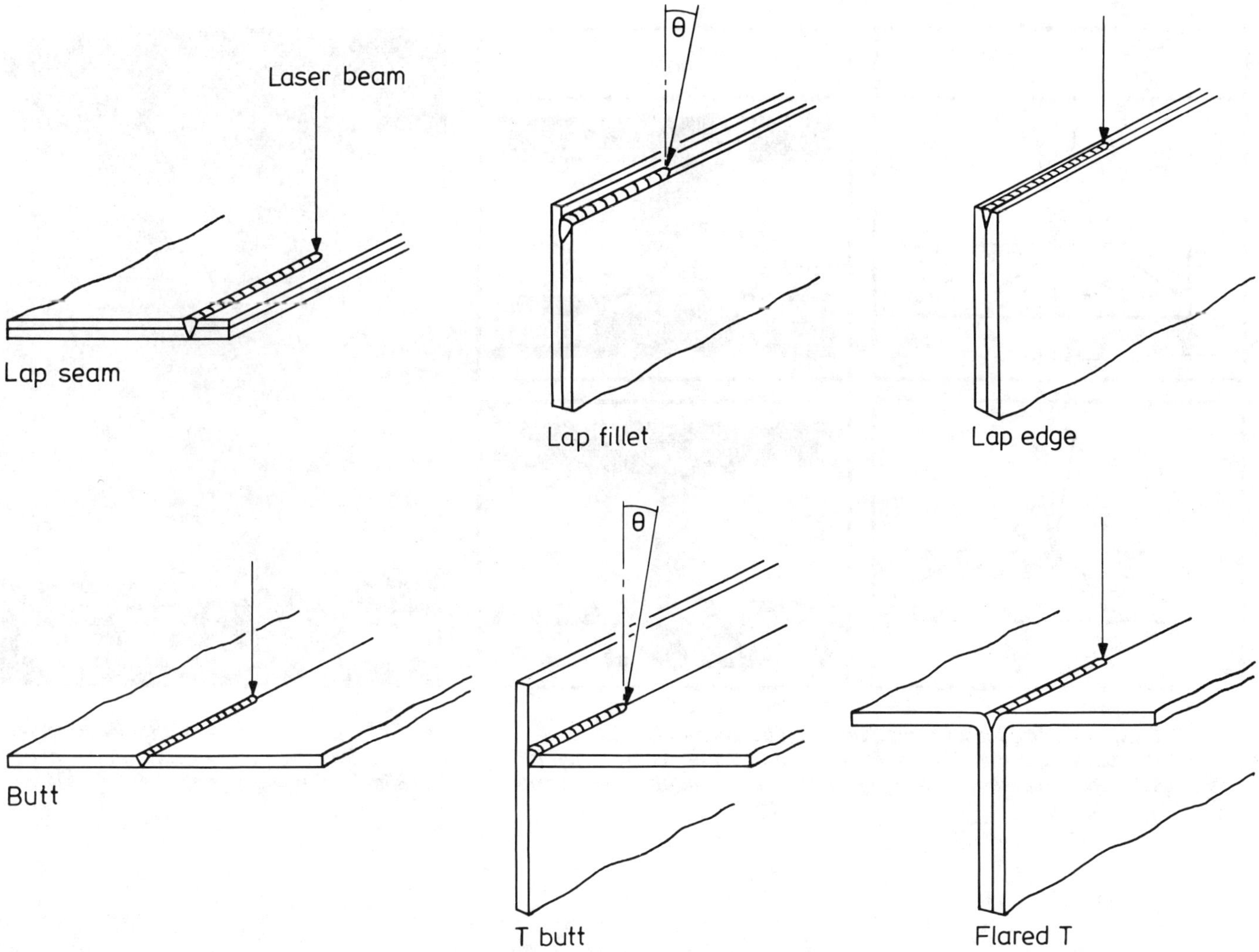

Fig.5. Basic sheet metal joints that can be laser welded.

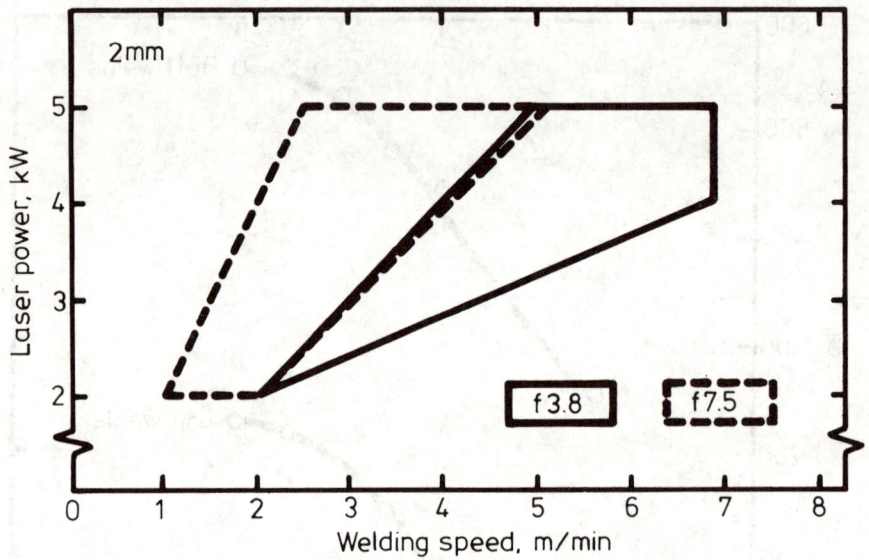

Fig.6. Weldability lobes for T butt joints in 2mm thick fully killed steel, showing the effect of using different f number focusing conditions.

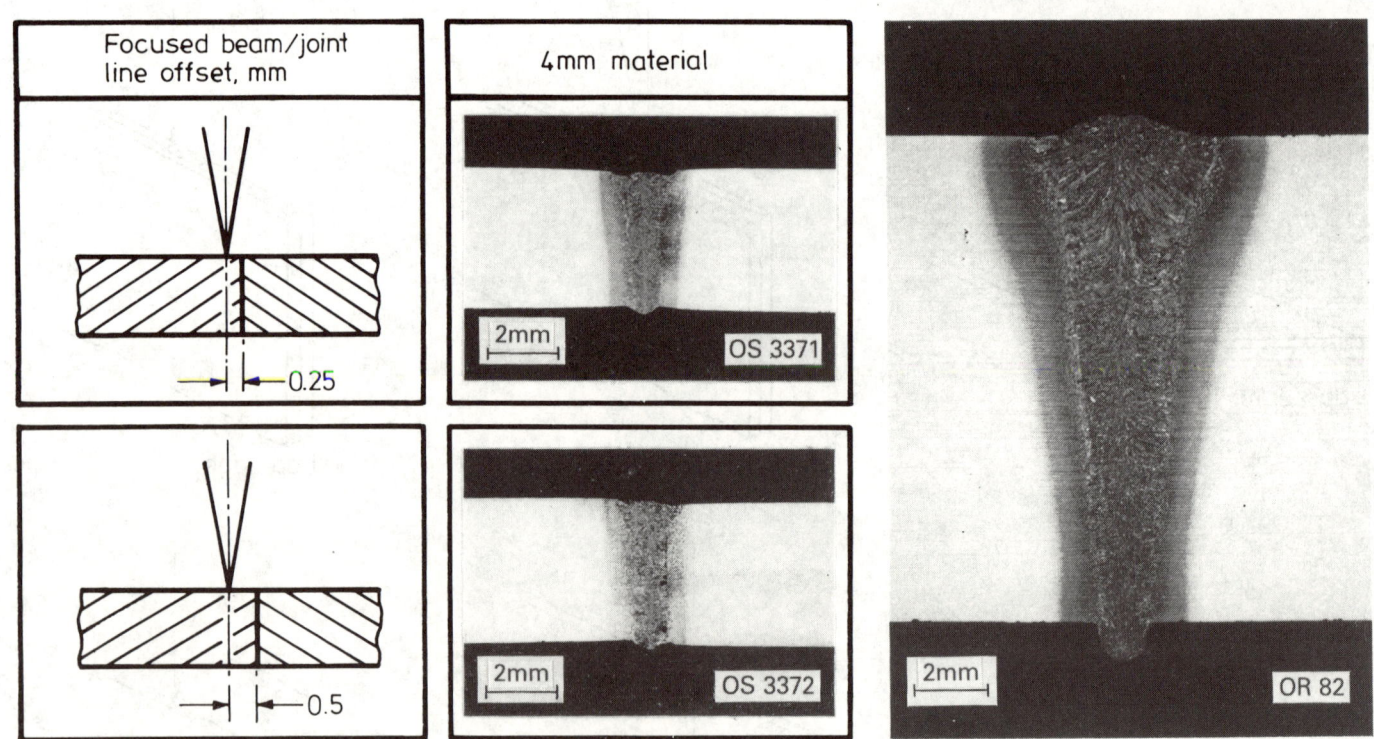

Fig.7. Effect of focused laser beam/joint line offset. 4mm low C steel welded at 5kW, 1.5m/min. Note than an offset of 0.5mm is too much for the fused zone to spread across the joint line at the mid-thickness position.

Fig.8. Laser weld in 12.5mm thick BS 4360 Grade 50D structural steel. Welding speed 0.35m/min, power 6.2kW.

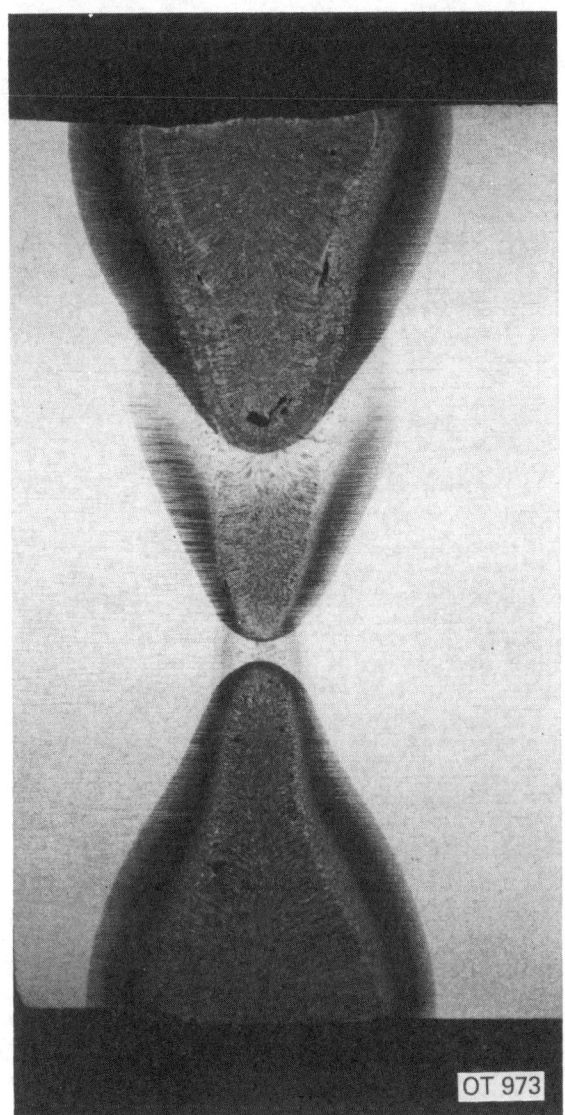

Fig.9. Four pass laser weld with wire feed in 25mm thick steel, ×5.

1st pass: 5kW, 0.5m/min
2nd and 3rd passes: 5kW, 0.35m/min, with wire feed
4th pass: 5kW, 0.30m/min, with wire feed

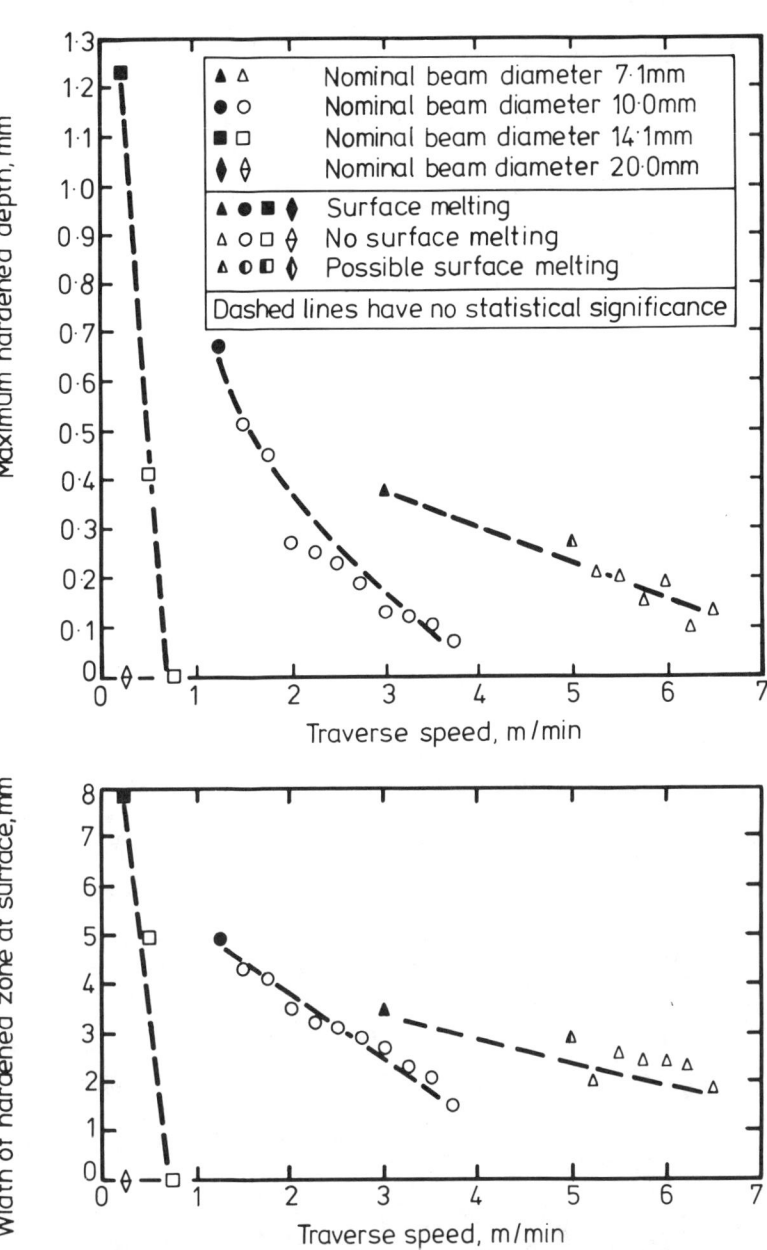

Fig.10. Depth and width of laser transformation hardened zones obtained when using a defocused laser beam at 1.0kW power.

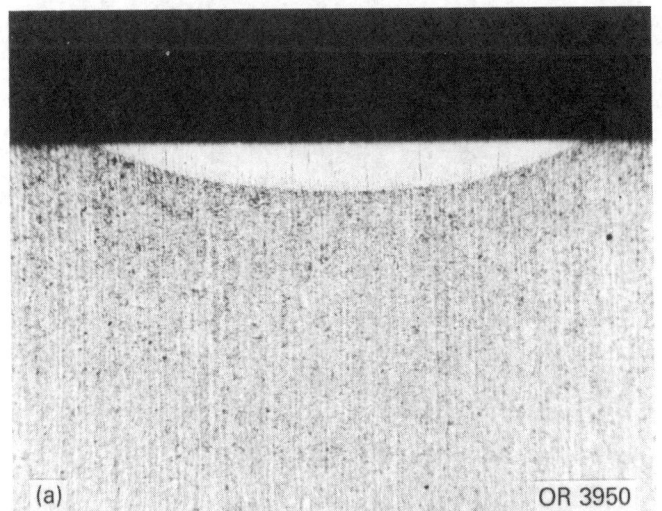

Fig.11. Light and dark etching laser transformation hardened tracks from consistency trial with defocused beam. Condition 1 (1.0kW, 1.5m/min, 10mm nominal beam diameter), ×15:

a) light etching hardened track   b) dark etching hardened track

A flexible laser manufacturing system based on the composition of several laser beams

V. Fantini, L. Garifo, G. Incerti
CISE SpA
and
I. Franchetti
and
L. Grisoni
Alfa Romeo Auto SpA, Italy

An original configuration of a flexible laser manufacturing system is described. This system is based on the concept of the sum of several high power CO_2 laser beams in order to obtain a larger amount of laser power in the workstations starting from relatively low power industrial laser sources.

In the system there are several sources whose beams can be delivered to the workstations either independently or suitably composed. At any moment the total available laser power can be sent to different workstations which can operate simultaneously.

Moreover in each workstation it is possible to change the technological process performed simply by automatically substituting the beam manipulating optics. Finally the system configuration allows optimised performances even in the case of malfunctioning.

1. INTRODUCTION

The industrial world is now facing a deep transformation aiming at substituting, in many cases, the rigid transfer concept with a more flexible way of manufacturing, in order to meet the continuously changing requests of the market.

In this situation the Flexible Manufacturing Systems (FMS's) are growing more and more popular in many industries and, in particular, in the automotive industry.

The laser tecnology is very appealing for integration in FMS's because it exhibits a twofold flexibility: a) toward the product, in the sense that the same laser unit can process pieces with different geometry, dimensions, materials and b) toward the process, in the sense that the same laser unit can make different technological operations, such as cutting, welding, heat treating and so on.

Furthermore a laser unit is easily matched with a numerical control and suitable to be combined with a manipulating robot.

So in many applications the introduction of the laser in a FMS can optimize the performances of the system as a whole.

2. LASER SYSTEM CONFIGURATIONS

There are several possible configurations for introducing a laser system in a production process. In the simplest (and most widespread) configuration the laser is used as a dedicated system and often substitutes another machine-tool. Obviously in this case the laser potentialities in term of flexibility are underexploited.

A more flexible way to use lasers is to consider a laser multistation system, in which the same laser beam is directed, on a time-sharing basis, into several workstations, where the same technological operation is performed.

A further step in the flexibility direction is achieved by considering a configuration in which a single laser unit is used to feed several workstations performing different technological operations (for example a station for welding applications, one for cutting, and so on). In this case every vorkstation must contain an optical system to munipulate the laser beam accordingly to the particular process involved. Also in this case the laser beam operates on a time sharing basis.

In this configuration the nominal maximum power of the single laser unit must be chosen according to the maximum laser power required in a single process, even if this process is not performed many times.

In other words, the single laser unit may results overdimensioned with respect to the power requirements of the majority of the processes performed in the various workstations and this is particularly true when the range of the laser power required is widespread.

In point of fact, up to now there are only few industrial applications in production requiring a laser power in excess of 5 kW.

Therefore the number of laser sources with an output power greater than 5 kW installed in an industrial environment is now very limited. As a consequence it is very difficult to give a figure of merit about the "industrial reliability" of such lasers. Moreover, when a single laser unit is used in a multistation system, there is another drawback: in the case of malfunctioning of the laser unit none of the workstations can operate anymore.

Another possible configuration is based on a single laser unit whose beam is split into several beams sent to the various workstations.

In this case a laser source with a power equal to the sum of the powers required in the various workstations is needed.

This configuration allows to work simoultaneously in many workstations, where the laser beam is always available.

However, there are several drawbacks concerning this solution: besides those discussed regarding the latter configuration, in this case the ratios of the amount of laser power in every workstation are fixed. This implies a decrease in the system flexibility because the process performed in a workstation cannot be indipendent, in terms of laser power, from those performed in the other workstations.

In order to overcome the drawbacks related to the previous described laser systems and to achieve the maximum degree of flexibility we propose a new laser manufacturing system configuration in which, instead of using a single high power laser unit, a group of laser sources is put together and their laser beams are suitably composed in order to match the various technological process requirements.

Schematically the system is composed of:
- a group of laser sources;
- an optical group for composing the single laser beams and addressing them to the workstations;
- some optical groups for making the laser beams suitable for different technological processes, placed in
- workstations equipped with specific workpiece handling systems.

At any moment the total available laser power can be delivered to the different workstations, which can operate simultaneously. In particular any workstation can be reached by a laser beam with a power variable between the single laser source power and the sum (partial or total) of the powers of all the laser sources. Moreover in each station it is possible to perform on-line different processes by automatically changing the beam manipulating optics.

3. SYSTEM GENERAL DESCRIPTION

Given a number K of laser sources (with different levels of output power) and a number $N \geq K$ of workstations, the system we have studied can perform the following operations:
- to direct the laser beams emitted by whatever laser source to any workstations;
- to direct simultaneously the laser beams emitted by K laser sources to K workstations no matter how chosen among the N available workstations;
- to direct a laser beam obtained from composition of a number $S = 2,....,$ K of laser beams, no matter how chosen among the K available laser beams, with a power equal to the sum of the powers of the considered beams, to whatever workstation, while the remaining K-S laser beams are directed to K-S workstations no matter how chosen among the N-1 remaining workstations.

The block diagram referring to the generation and the manipulation of the laser beams is shown in fig. 1, where:
- block L represents the K laser sources;
- block D optically decouples the laser sources from the addressing optical system;
- block S represents the laser beam composition unit;
- blok I is the optical group which directs the laser beams to the workstations;
- block C represents the workstations.

The block D is conceived for alignment purposes. In fact in case of misalignment of one laser optical cavity or after a maintenance procedure it is possible to align the optical cavity without changing the following optical groups alignment.

The block S, which is the most original part of the system, composes the input laser beams in order to obtain a single laser beam with a greater power.

The output beam from the block S and, if any, the remaining laser beams enter the block I, where they are properly delivered to the workstations by means of a group of moving optics.

In any workstation, contained in the block C, there are optical system for manipulating the laser beam and workpiece handling systems.

4. DESCRIPTION OF THE ACTUAL CONFIGURATION UNDER DEVELOPMENT

We have chosen a configuration utilizing three carbon-dioxide c.w. laser sources and five workstations. The output laser beams must be of anular cross section.

The system can be equipped with laser sources having a maximum output pow er of 5 kW.

In the system we describe in the following there are three laser units with a power of 2.5 kW.

The maximum laser power obtained in this configuration is equal to 7.5 kW with which it is possible to perform almost all the technological processes of interest for the automotive industry.

A schematic lay-out of the chosen configuration is shown in fig. 2.

For the sake of semplicity in this figure the blocks S, D and I of the figure 1 have been put together in the block T.

The laser beams emitted by the three laser sources L_1, L_2, L_3 travel at a height of about 2 m and have an annular cross section with the same inner and outer diameters. They enter the block T, where they are suitably composed and addressed as described later. Coming out from the block T, the laser beams reach the appropriate workstations $C_1 \div C_5$ in accordance with the particular processes to be performed, and there, specific manipulation optical systems $F_1 \div F_5$ deliver the laser power to the workpieces. The workpieces are conveyed to nearby the workstations by means of a system SC. A group of recognition devices $SV_1 \div SV_5$ identify the workpieces which are positioned in the appropriate workstation by means of the handling systems $R_1 \div R_5$ and unloaded after the end of the laser processing.

All the operations inside the system (i.e. laser beam composition and delivery, workpiece loading and unloading, and so on) are computer controlled (CPU unit). The computer is also used to optimize the performances of the system even in the case of a malfunctioning of one of the components described before

This paper deals only with the problems related to the composition, addressing and focusing of the laser beams.

With reference to fig. 3, which schematically shows the configuration of the blocks D, S and I, the laser beams enter the block D. This block is composed of three groups of two plane copper mirrors facing each other, both adjustable in orientation with respect to the incident beam direction. In this block are also placed, not shown in fig. 3, three power monitor devices, three mechanical shutters for stopping the laser beams and three beam shape visualization units.

The block S is divided into two parts: an addressing unit and a summing unit. The addressing unit is used to direct the beams toward the summing unit and contains three plane copper mirrors M_1, M_2, M_3 mounted on pneumatic slides moving in the direction orthogonal to the plane of the figure. If one mirror is inserted, it intercepts the laser beam and directs it to the summing unit; on the contrary the laser beam travels straight on to the block I. The mirror M_1 has also a movement in the x direction. This is due to the fact that when only two laser beams have to be sent to the summing unit, it is better to use channels 2 and 3, because, as will be discussed later, in this case the resulting beam can be better focused. The summing unit is shown in more details in fig. 4. It has three input channels and one output channel. In channel 3 the laser beam is expanded by a factor M, equal to the ratio between the external and internal beam diameter. The beam expander is made up of a spherical concave copper mirror and a spherical convex one, mounted in a positive confocal telescopic configuration. The laser beam coming out from the beam expander is bent at a right angle by a plane copper mirror, with an internal hole whose diameter is $\sqrt{2}$ times the external diameter of the beam entering the beam expander. In channel 2 the laser beam is simply bent at a right angle by menas of a plane copper mirror with an internal hole whose diameter is equal to $\sqrt{2}$ times the internal diameter of the beam. The laser beam reflected by this mirror is coaxial with the bent expanded beam. In channel 1 the diameter of the beam is reduced by a factor M by means of a two lens collimating unit. The laser beam is then reflected coaxially with the former two bent beams by a plane copper mirror. The beam coming out from the summing unit S is made up of two or three coaxial annular beams and has an external diameter M times the external diameter of the original laser beam. If for example the incoming beams have the external diameter of 45 mm and M=1.8, the outer diameter of the beam coming out from the summing unit S is equal to 81 mm.

The block I is composed of three plane copper mirrors (M_4, M_5, M_6 in fig. 3) to direct the single beams and a larger plane copper mirror (M_7) to direct the composed beam.

The block C is made uf of the workstations in which there are the optical systems to suitably manipulate the laser beams for welding and heat treating applications.

In particular there is an optical system which can assume three different configurations:
a) a focusing head provided with a ZnSe lens used for welding applications accomplished with the single laser beams;
b) a focusing head provided with an off-axis paraboloidal mirror used for welding applications accomplished with a beam given by the sum of two or three single laser beams;
c) a beam manipulating unit provided with a beam integrator used for heat treating applications.

The three types of focusing systems are obtained by means of moving optics and the operator can automatically select the particular optical chain accordingly to the particular technological process to be performed in that particular workstation.

In our opinion a possible critical point of our system is represented by the problem of maintaining the proper alignment of all the optical paths.

There are two kinds of problems, related to:
a) beam point stability of the laser sources;
b) mechanical stability of the optical components.

Up to now we have developed two different diagnostic devices for off-line control of the phenomena described in a) and b).

Depending on the experimental data of the behaviour of the system that we are going to collect, we will decide whether or not to install an on-line control of the alignment.

5. THEORETICAL AND EXPERIMENTAL ANALYSIS

Taking into account the number of optical components encountered in the various optical paths corresponding to the configuration described in paragraph 4., we have calculated the total power losses associated with i) a single laser beam, ii) a beam made up of two indipendent laser beams and iii) a beam made up of three indipendent laser beams.

In these calculations we have assumed a tipical power loss of 2% for every mirror in the optical chain and we have considered a three-mirror manipulating optical device in each workstation.

The calculated total power losses amount respectively, to 9,5%, 15,5% and 14,5% for the cases i), ii), and iii).

We have also calculated the theoretical spatial behaviour of the far field intensity distribution of a beam composed of two and three indipendent laser beams. We made the assumption that there is no interference among the various laser beams: this assumption has been found in agreement with the preliminary experimental results, as described in the following.

When this assumption is made, the e.m. field intensity distribution is simply given by the sum of the three single e.m. field intensities.

Here we consider the laser beams coming out the sources with an annular cross section having an external diameter of 45 mm and an internal diameter of 25 mm (M = 1.8).

For the sake of semplicity both the e.m. field strength and phase are assumed constant over the annular cross section.

As described in paragraph 4, in the summing unit one laser beam diameter is enlarged by a factor M, another one is reduced by the same factor M, while the third laser beam diameter is left unchanged.

These three laser beams are then made coaxial one another by menas of flat mirrors. By assuming a nominal laser power of 2.5 kW coming out each source, the ideal intensity profile in the laser beam cross section at the exit of the summing unit is that shown in fig. 5a, where R is the radial coordinate. Fig. 6 shows the normalized intensity profile I_n in the far field as a function of the normalized radial coordinate r, in the case of the sum of two and three laser beams, respectively.

I_n and r are given by:

$$I_n = \frac{I(r)}{I_{max}} \qquad (1)$$

$$r = \frac{r_f}{\lambda f/D} \qquad (2)$$

where:
r_f = radial coordinate in the far-field;
λ = laser beam wavelength;
f = focal length of the focusing system;
D = external diameter of the laser beam coming out the laser source;
$I(r)$ = far-field intensity;
I_{max} = $P/(\lambda f/D)^2$, being P the power of a single laser beam.

As a comparison in fig. 6 also the intensity profiles obtained with a superposition of two and three equal laser beams are shown. It is evident that it is better to sum a laser beam with its enlarged replica instead of summing it with its reduced replica. This is due to the fact that the former case allows to obtain a reduction in the equivalent f-number of the system, so a greater intensity value is achieved on the optical axis in the far field. Moreover even in the case of the sum of three laser beams there is an improvement of the on-axis intensity profile, in that more energy is concentrated near the optical axis. The analytical form of the intensity profile behaviour in the far field is given by:

$$I_n(r) = \frac{1}{I_{max}} \cdot [M^2 I(r_1) + I(r)], \text{ (sum of two beams)} \qquad (3)$$

$$I_n(r) = \frac{1}{I_{max}} \cdot [M^2 I(r_1) + I(r) + \frac{I(r_2)}{M^2}], \text{ (sum of three beams)} \qquad (4)$$

where:
$I(r)$ = far-field intensity of a single laser beam ($\int_0^\infty I(r) 2\pi r dr = I_{max}$);
M = magnification factor;
r_1 = Mr;
r_2 = r/M.

Fig. 7 shows the comparison between the far field intensity profile from three coaxial and adjacent beams (see fig. 5a) and that from three coaxial beams with 1 mm radial separation (see fig. 5b) which is a more actual case. Because of the absence of interference phenomena, there is pratically no difference in the two cases.

Fig. 8 shows the integral curves of far field intensity in the case of the sum of two and three laser beams. These graphs represent the fraction P_f of the total laser power (5kW and 7.5 kW respectively) contained in a far-field circle of radius r.

For comparison the integral curve in the cases of superposition of two or three equal laser beams is also shown.

In the case of the sum of two laser beams the integral curve is always higher than that of two superposed beams, while in the case of three laser beams there are some crossing of the two curves.

However, in particular, at r=1 (first zero of the Airy function) integral curve in the case of the sum of two laser beams is 35% higher with respect to two superposed laser beams, while in the case of the sum of three laser beams it is 15% higher with respect to three superposed laser beams.

The theoretical analysis and considerations we have made so far point out that the laser beam coming out the proposed summing unit has a better focusability than a single laser beam of equal power.

The results of preliminary feasibility tests on beam composition are shown in fig. 9 and fig. 10. Figure 9 a represents a burn pattern on lucite of an annular cross section laser beam. In fig. 9b there is the burn pattern of a similar laser beam with a different value of M, expanded in such a way that its internal diameter is equal to the external diameter of the former one. Figure 10a shows the sum of these two laser beams, while in fig. 10b is reported the burn pattern of the composed beam after focusing with a long focal lenght mirror in order to display the intensity distribution.

The experimental results obtained show clearly that there is no significant interference phenomena, so the assumptions we have made at the beginning of this paragraph are correct.

These and other experimental results support the feasibility of our scheme for the composition of high power laser beams. As a consequence a patent request has been deposited for the reported configuration.

6. CONCLUSIONS

A study has been presented of a new configuration of a flexible manufacturing system provided with several laser sources, based on the concept of the sum of several laser beams.

This system seems appealing in that it exhibits a great flexibility and an improvement in the achievable laser beam focusability. Moreover it is possible to obtain a large amount of laser power starting from relatively low power and reliable laser sources.

A particular configuration has been described, including three laser sources and five workstations.

The results of theoretical calculations and preliminary experimental tests have been presented and discussed.

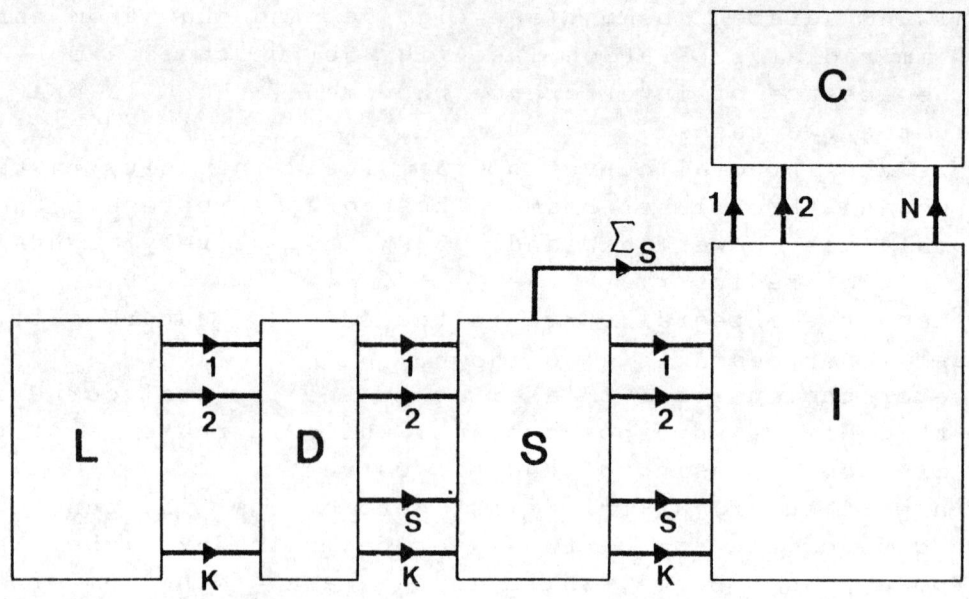

Fig. 1. Block diagram of the proposed flexible laser system

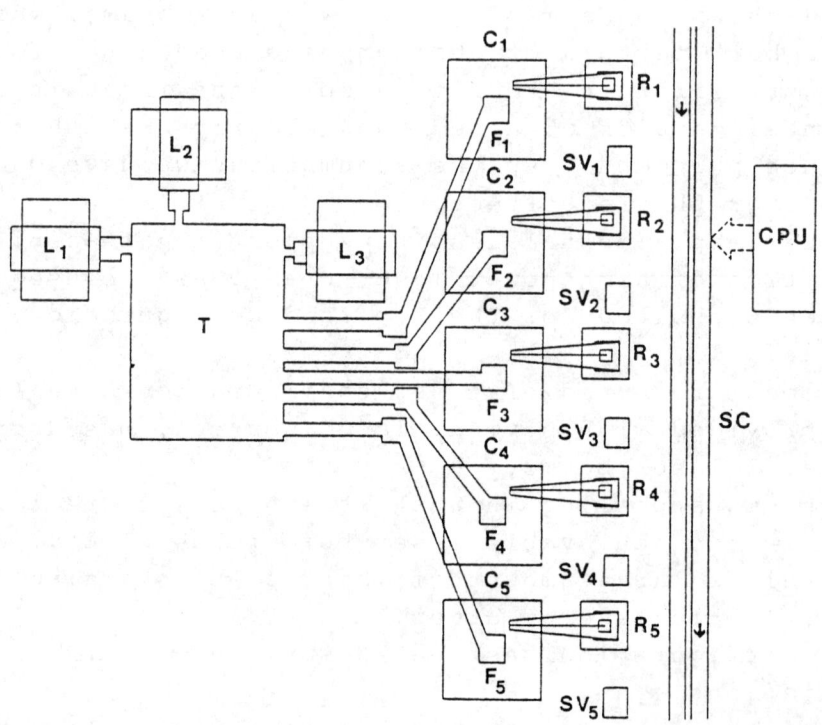

Fig. 2. Schematic lay-out of the actual laser system configuration

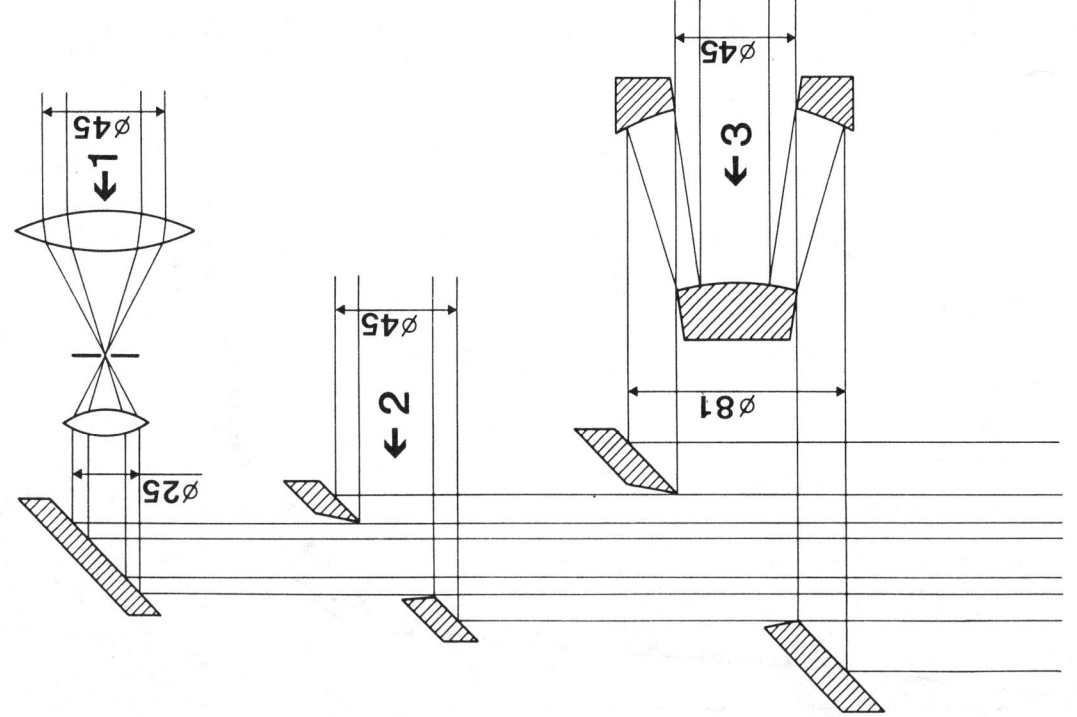

Fig. 4. Scheme of the laser beams summing unit

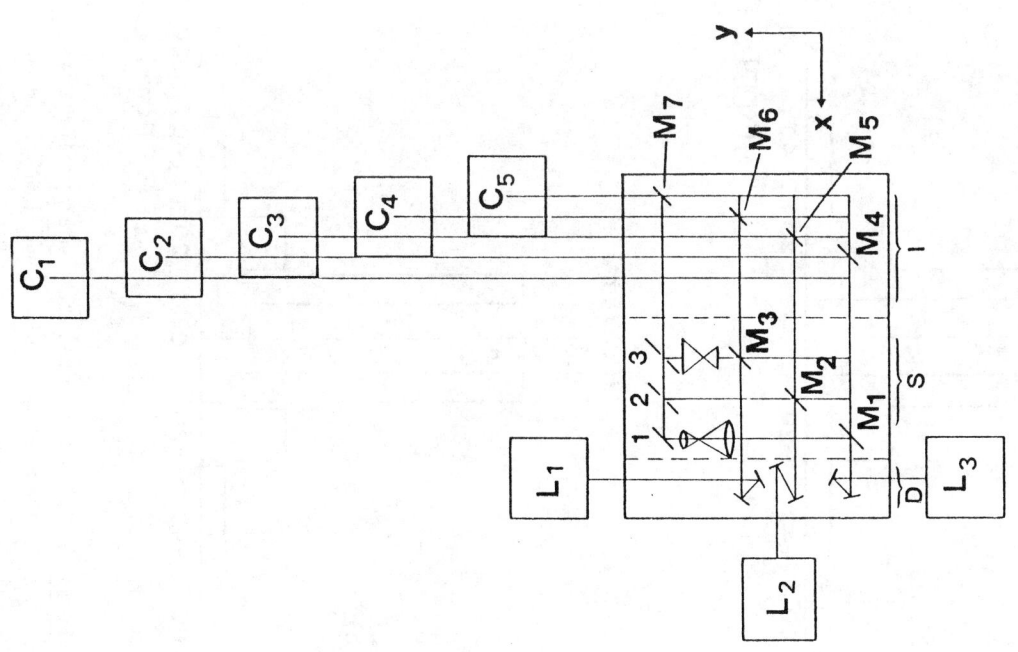

Fig. 3. Scheme of the optical paths in the block T

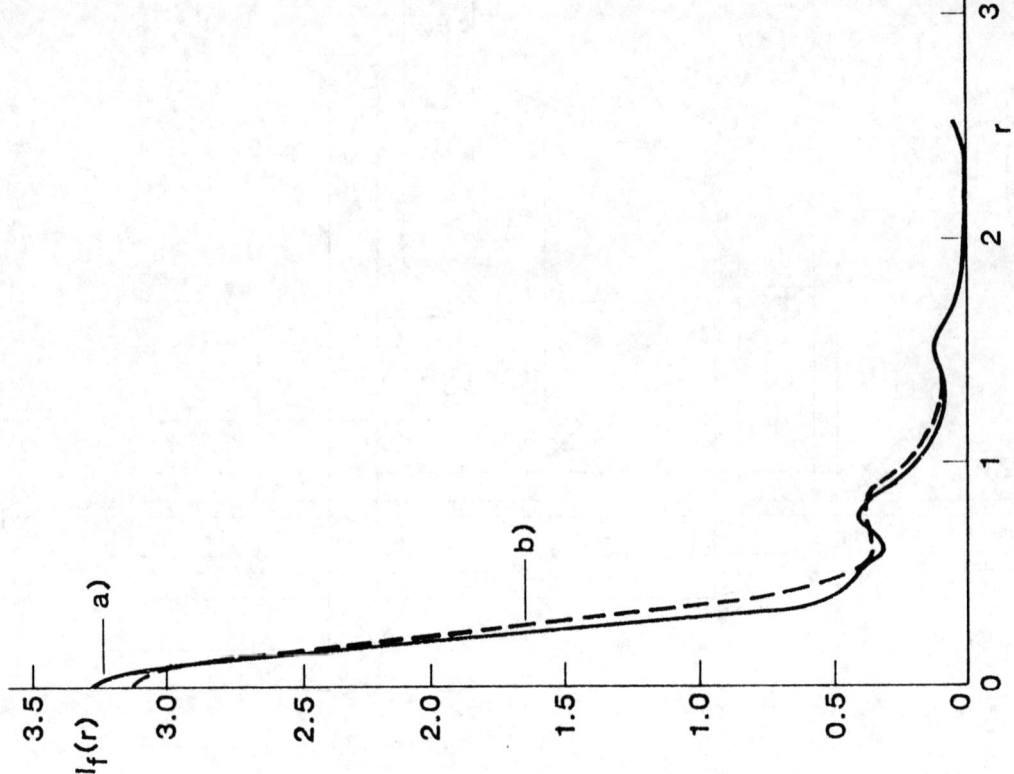

Fig. 6. Laser beam far field intensity distribution as a function of the radial coordinate

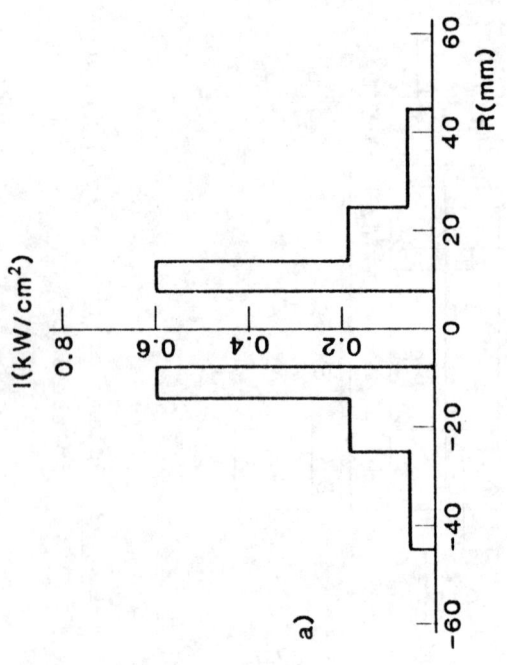
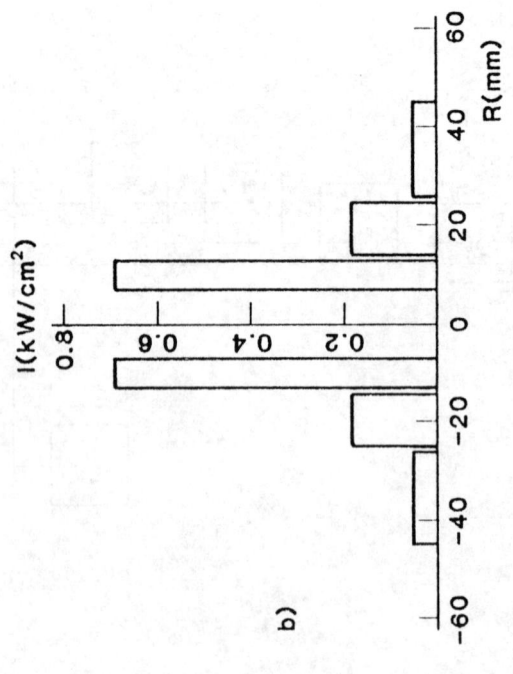

Fig. 5. Ideal intensity distributions of the three component laser beam:
a) adjacent beams
b) 1 mm separated beams

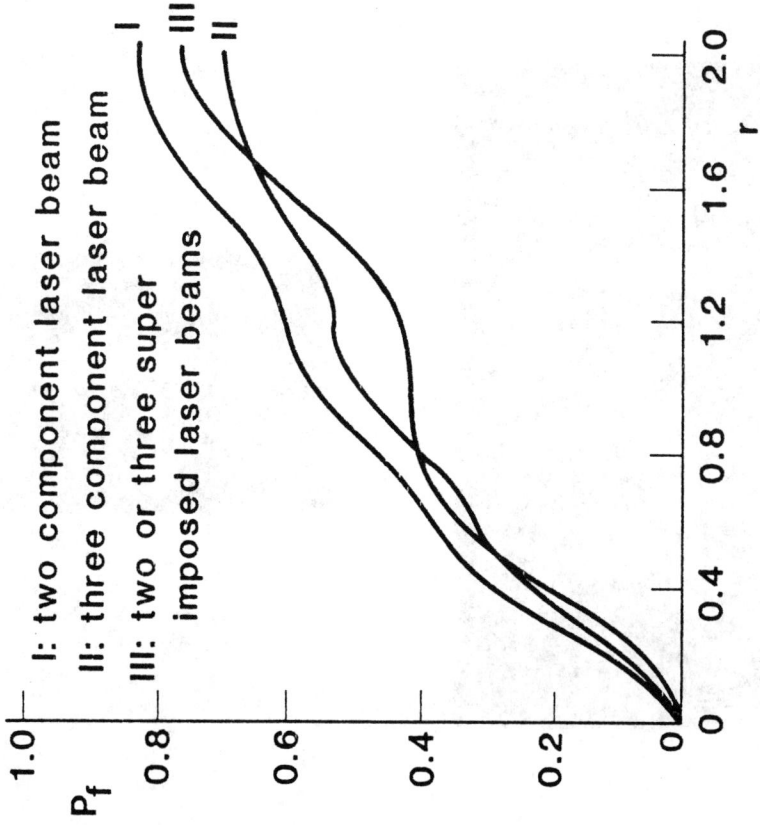

Fig. 7. Three component laser beam far field intensity distribution:
I: expanded beam+unchanged beam
II: two superimposed unchanged beams
III: unchanged beam+reduced beam
IV: expanded beam+unchanged beam+reduced beam
V: three superimposed unchanged beams
a) adjacent beams
b) 1 mm separated beams

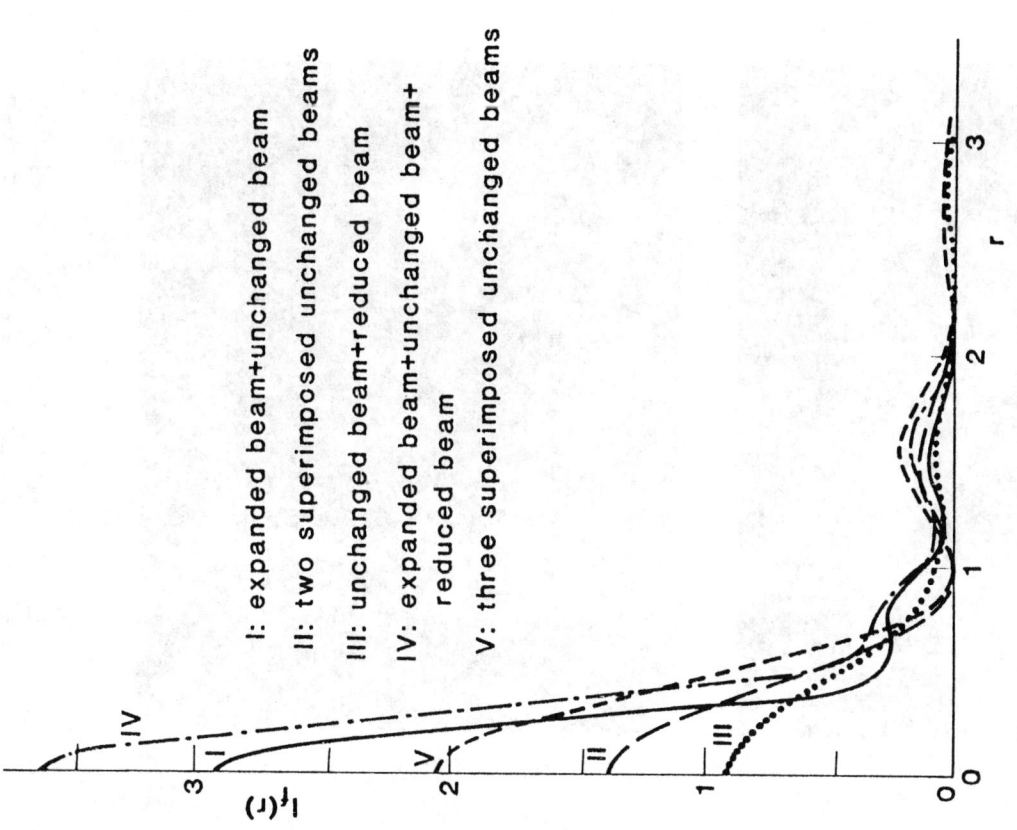

Fig. 8. Integral curves in the far field
I: two component laser beam
II: three component laser beam
III: two or three superimposed laser beams

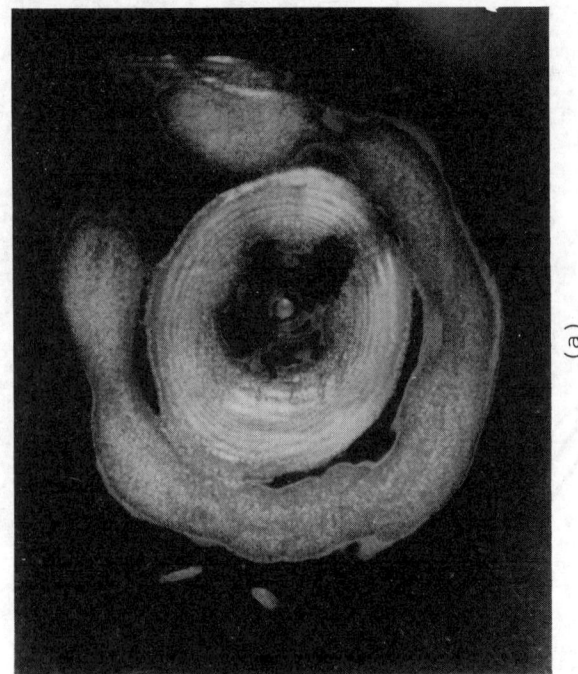

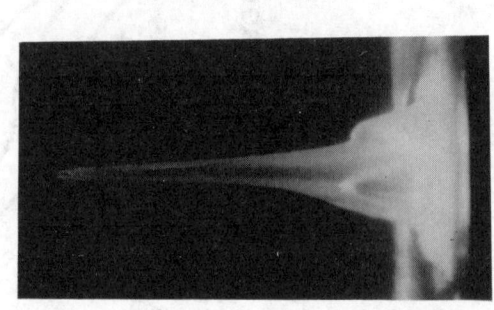

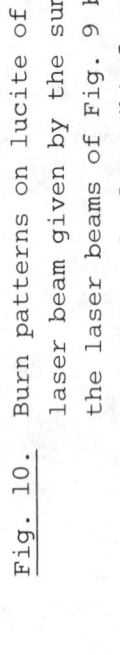

Fig. 10. Burn patterns on lucite of the laser beam given by the sum of the laser beams of Fig. 9 before (a) and after (b) focusing

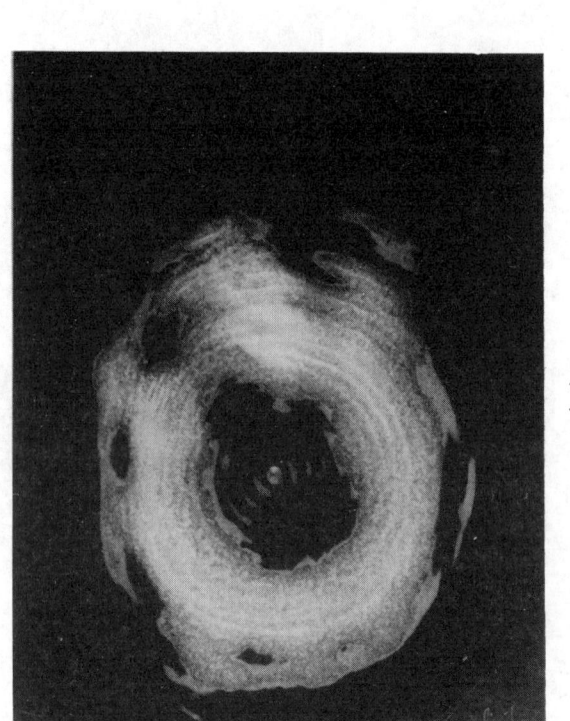

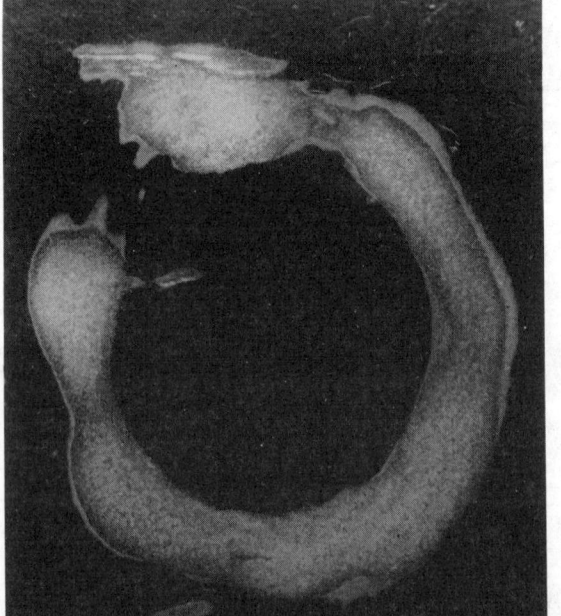

Fig. 9. Burn patterns on lucite of the unchanged (a) and the expanded (b) laser beams

Laser-cutting and welding in a flexible manufacturing system

H. Uetz, G. Hardock
and
H.-J. Warnecke
Fraunhofer-Institute for Manufacturing Engineering and Automation (IPA), West Germany

ABSTRACT

Utilizing traditional machinery, sheet metal parts are usually produced in large series. Decreasing lot sizes and quickly changing demand calls for highly flexible production systems. In spite of their rather high system cost, high power laser production systems could prove profitable on account of their big advantages in terms of flexibility.

A uniquely flexible manufacturing system was planned for a broad spectrum of 3-dimensional sheet metal parts. The 3-dimensional parts are produced from 1,25 mm thick sheet metal by stamping, cutting, welding and finishing. Parts assembly, painting and other operations follow. Medium sized series are to be produced in 3 shifts with the third shift being unattended.

Planning of a Flexible Manufacturing System for Cutting and Welding of Thin-walled Hollow Sheet Metal Parts by Laser

1. Introduction

The application of innovative laser technologies in various fields of material processing is increasing rapidly, especially in the sheet metal sector. Cutting as well as welding procedures are being successfully applied today, whereby the combination of both processes in selected cases can lead to highly favourable production structures and costs.

In contrast to conventional technology laser technology involves a number of steps so structured that only new planning methods lead to good results. Based upon the specific demands of laser processing, the design of the workpieces must be thought over completely anew, as must the placement of the processing stations and the production structure to which these belong. Only then can this promising technology be applied successfully and economically.

Some of these planning steps and intermediate results are presented in this paper. In particular experiences in the planning of laser production systems are treated.

2. General tendencies in sheet metal production

In the field of sheet metal processing, the production of work pieces of especially complex form and small batch size from flat sheet metal cutouts has been already carried out successfully for several years using laser technology. Here the market today offers very interesting laser processing facilities with the necessary peripherals and the accessory know-how. In constrast to this with regard to three-dimensional formed sheet metal workpieces, laser technology has only been able to reach a successful breakthrough in a few isolated cases.

For cost-effective design, the specific advantages of three-dimensional formed workpieces are of increasing importance. The sheet metal parts often attain their stability and stiffness only through three-dimensional forming, e.g. by ribbing or lengthwise flanging. In sheet metal technology, cutting as well as welding is of central importance. Next to forming, these are the most important processes used today in industrial production of sheet metal parts. The trend toward smaller batch sizes is clearly evident in sheet metal part manufacture.

Firms are being forced to reconfigure their production on short notice. They must react quickly to new products or product modifications. Rapidly-changing products require very high flexibility in production. In relation to this the following tendencies may be listed:

- drastic reduction of setup time, e.g. by means of quickly changing diesystems for presses (fig. 1)

- partial replacement of tools with fixed form by other processes, such as laser cutting instead of punching (fig. 2)

- enhanced application of programmable, flexible basic units, such as industrial robots for welding.

One example of the applications being processed today with laser technology is the basic forming of valve lifters in the automobile industry; other workpieces are displayed in the figures.

3. Selected task and workpieces

For frequently reoccuring processing tasks in sheet metal production, a representative workpiece with appropriate elements of form such as linear contours and radii is selected.

After being formed, this representative workpiece is laser trimmed and then welded with two different types of seams. Here a square butt weld and a overlap weld are to be investigated (fig 5). These selected tasks are representative for a very broad workpiece spectrum containing parts with various forms and dimensions. An upper limit of 1000 mm has been placed on workpiece length; maximum width is 600 mm (fig. 6).

In particular, the transition from linear contours into radii and vice versa are to be characterized as problem areas. As such they require special attention. A laser manufacturing system can be planned for these work-pieces, if some fundamental facts are clarified:

- feasibility of processing the material

- ability of the moving system to follow the cutlines and seams

- achieving the required quality

- economic justification

4. Basic examination of material

The material to be used is to be examined with special regard to its welding suitability. The following requirements are to be fulfilled:

- killed steel, ideally with silicon, since the highly reflective properties of aluminum alloy can result in a feedback of a part of the laser beam. This would exert a negative influence upon beam quality and plasma formation at the workpiece.

- consistent material quality with regard to alloy components, since there are no procedures known this far to adjust the laser parameters according to the material to be worked.

- consistent material quality regarding material paramaters and alloy components is a basic requirement for consistent weld quality, since processes to adapt the laser parameters to the corresponding material are not known.

- material suppliers should support the user with results of material analysis parameter values for the respective coil charge. Later on, this information could be documented and used in the corresponding CNC control to avoid time-consuming testing and alignment. It must be pointed out here that the laser processing facilities are located on the shop floor; hence the operating personnel do not have the same basis of experience as e.g. the physicist or technician in the applications laboratory.

- traces of corrosion protection oil, drawing lubricant, or cutting oil should be removed from the material before laser welding.

5. Combination of laser cutting and welding

Besides weldability, the combination of laser cutting with laser welding is of great interest. Here the central question is whether the oxide layer formed at the kerf edges during laser cutting influences the welding results negatively. Experiments have led to the following conclusions:

- The quality of a laser-cut seam depends in large part upon the cutting process applied in material preparation. With mechanical cutting processes such as shearing, the moving cutting tool (i.e. the shear) exerts a drawing effect upon the material as it submerges therein, producing rounded corners (fig. 7). For the focus of the laser beam, the cross section of the material to be melted becomes larger at the workpiece surface and the beam coupling and the introduction of energy into the material become more critical (fig. 13, fig. 14). Due to corner rounding and preexisting cutting burrs, the mating surfaces of the workpiece halves are not flat, resulting in a welding gap which cannot specifically be defined (fig. 4). In production-technical applications, this leads to increased fixturing problems at the local workpiece clamping fixture as well as recognition problems in weld seam sensing systems (fig.4).

- In seam preparation by laser cutting, the problems mentioned above are no longer of concern as the kerf corner is formed at a right angle to the surface and has no burr (fig. 8).

  Additionally, with the application of gas beam techniques in burning cutting processes, an oxide layer is built up at the cut surface by the exothermic reaction between iron and oxygen. This oxide layer was seen by the laser users as a hindrance to further processing by laser welding.

  Due to the much more limited process speed, it appears to be unwise to fall back upon melting cutting processes with inert gas. Only recently, experiments have been carried out in Europe regarding oxide layer thicknesses and their influence upon weld seam quality; for this reason it was necessary to broaden the scope of the feasibility study to include metallurgic examination of the oxide layer.

- the oxide layer thickness of workpieces cut by linear-flow lasers emitting modes near the fundamental mode lies in the region of 1-4 um (fig. 9). For transverse-flow lasers with a higher mode structure, however, the resulting values are clearly higher (fig. 10).

During welding, the molten oxide layer becomes mixed with the melt bath (fig. 11, fig. 12). Due to the high energy density in the focus field, the oxygen content of the oxide may be split off, leading to gas emission and porosity. According to the available results of metallurgic tests, no disturbing conglomerates of oxide or pores have been found in the weld grain structure. Thus it has been proven that laser-cut workpieces of St 12.03 also may be welded again with the laser (fig. 15, fig. 16).

6. Requirements of laser processing under production conditions

Laser processing is today still burdened with high costs. This is due to investment costs as well as operating costs. Although a certain reduction is to be expected, a very strong entry, such as that of electronic parts, will not take place. This situation places special demands upon laser processing systems:

- very high temporal utilization, ideally in three-shift operations.

- production automated to the greatest possible extent in order to reduce the safety risk for the personnel to a large degree.

- very rapid movement and transport systems in order to take full advantage of the short process times of laser processing

- work with multiple-station machines in order to avoid as much as possible idle and empty time of the laser aggregate

- integration of laser processing into a flexible overall system. Here, especially, the periphery of the laser system must have high flexibility comparable to that of the laser system itself.

- environmental conditions suited to laser application: here the laser aggregate and especially the beam guidance are either to be protected from steam, dust particles, and macromolecules such as CCl_4 or to be held below excessive pressure by means of "purified air" or gas mixtures.

A further specification of laser processing lies in the very small focus diameter required of the focused beam. Typical values are approximately 0.05 - 0.10 mm. The demand for very high accuracy of moving systems, only, gives rise to welding tasks, with the same or even considerably smaller deviations than the focus diameter for three-dimensional contours.

Therefore laser welding is quite distinctive from classical welding methods; obviously this must be taken into account during planning.

7. Flexible manufacturing system with laser work stations

The flexible manufacturing system consists of a flexible laser processing system with several laser processing stations, a transport system and an information/production control system as well as sensors and clamping systems. Under the point of view of the production engineer, the requirements of the laser have to be lined out exactly before deciding about a producer preference. In this case dominant criteria are:

- Beam quality e.g. mode, variation of mode

- Power stability e.g. longterm tests

- Divergence

- Beam pointing stability in accordance to power output and resonator temperature

- Power adaptability e.g. ramp up, ramp down, pulsing

- Beam availability e.g. guaranteed MTBF

- Maintainance e.g. service intervals, spares

Because the beam intensity of the laser is reduced only marginally even over great distances, the work stations may be placed quite far apart; one laser may supply several work stations via beam conduits. The processing stations may be configured thereby as single or multiple stations. (Fig. 17)

8. System of movement

For the required beam or workpiece movement systems, several different concepts are conceivable:

- stationary workpiece, mobile laser processing nozzle

- mobile workpiece, stationary laser processing nozzle

- a combination of the concepts mentioned above

In special applications of the workpiece variants pictured, a combination consisting of the work-piece moving about the x-y table with a laser processing head which may be positioned in the Z direction, turned 360o in the A axis, and tilted +90o in the C axis was chosen. All axes are numerically controlled.

To avoid the near/far field problematics often occuring in mobile beam direction systems, the axes with the longest linear movement and the largest feed dimension of the workpiece are positioned into the x-y plane. For the welding of radii in corner regions, high speeds are to be expected of the movement system, so that a high speed of travel can be realized. To minimize the dynamic forces upon the workpiece, the clamping device and the x-y table, rotational movement at the radii regions is effected through the rotary A-axis.

9. Requirements of the movement system

- exactness e.g. to +0.05

- speed of travel e.g. up to 20 m/min (path control)

- high static/dynamic stability

- work space

- teach-in function of control system

10. Work stations

The respective processing station has two workplaces which can be supplied alternately by means of beam switching. Workpiece loading and unloading are thus carried out during the laser processing utilization time. In order to execute the production task in the system and the required production volume, the execution steps are grouped into four areas of partial execution:

- cutting system: drawing flange cutting

- welding system sheet reinforcement

- welding system external seam

- welding system connective supports

In order to maintain emergency operation with reduced capacity in case of disturbance or total malfunction of a laser aggregate, beam switches are available at all processing stations as well as beam guidance tubes inbetween. Because a cutting laser with limited storage capacity would, in case of stoppage, bring the entire system down at the system entrance, plans should specify that the laser type introduced into the welding system also have cutting ability based upon its mode structure.

11. Sensor systems

The realization of the overall system depends to a great degree upon the sensor system applied. The following sensory tasks are to be carried out:

- detection of distance between workpiece and laser processing nozzle

- weld seam tracking system

Technical solutions based on different physical principles exist for both systems. While distance sensing already can be considered industry-proven, weld seam tracking systems are still developing. Owing to the very high demands for accuracy in laser welding (i.e. around 0.05 - 0.1 mm), extensive development work must be carried out to bring about a system suitable for industrial application.

12. Clamping

The clamping device must be adapted to the respective laser processing technique. Different designs for clamping devices are to be applied for laser cutting rather than for laser welding.

Catalog of demands for workpiece clamping:

Two separate systems are required, one for positio-ning and one for clamping. Flexibility requirements regarding workpiece size: e..g.

    11 different variants in length

    10 different variants in width

- different degrees of automation of the workpiece clamping device are:

a) Fixed-type quick change clamping devices with standardized interface to x-y table

b) Clamping device is fixed for respective length. Manual setup takes place for different width variants.

c) Clamping device is fixed for respective length; automatic setup takes place for different width variants.

d) A clamping device with the highest of flexibility can be set up automatically for different length and width workpiece variants.

- <u>Workpiece processing position</u>

a) laser cutting: horizontal

b) laser welding: horizontal

- Workpiece insertion and removal

Before laser processing the workpiece halves are inserted into the clamping device by a handling system and held prepositioned there.

Following processing, the workpiece must be removed automatically from the clamping device without jamming.

Working space for laser processing unit

In order to avoid collisions between movement systems, clamping cylinders, clamping device lifting fixtures and clamping device anchorage points, a free space must be supplied for the movement system of the laser processing optics above the level of workpiece processing.

Distinctions are to be made between

a) clamping device for laser cutting

- clamping device has only limited ability to absorb thermal reaction force from the workpiece

- workpiece loading system must facilitate removal of separated sheet sections

- inner sides of the hollow body to be cut should be protected from the cutting beam (flying sparks, hot metal particles and burning gases) by shield panels or workpiece loading of corresponding shape.

b) clamping device for laser welding

- clamping device is to be laid out in design such that it can absorb large thermal reaction forces from the workpiece

- to protect the weld seam root, the clamping device should allow the hollow body to be flooded with shield gas.

13. Conclusion

Laser cutting and welding of sheet metal parts of St 12.03 could be carried out under industrial conditions. The carried out material experiments show the feasibility for the use of a suitable laser. If laser parameters are suited according to the cutting process at hand the oxid layer produced by cutting is no insurmountable obstacle.
(Layer thickness is from 1 - 5 um.)

A flexible manufacturing system with five independent laser stations was planned regarding the preexamination mentioned above and will be realized in the near future. The main remaining problems are:

- sensors applicable under production conditions

- rapid systems of movement, probably combinations of gantry types with rotation devices

- high flexibility in clamping

- reliability of lasers under tough production conditions

Acknowledgments

The authors are grateful to the Max-Planck-Institut PML for carrying out metalurgical tests. Furthermore we thank our colleagues of the Laser Group.

Fig. 1.   Quick - Change of a pressing tool

Fig. 2. CO_2-Laser Cutting Machine for sheet metal and plate fabrication

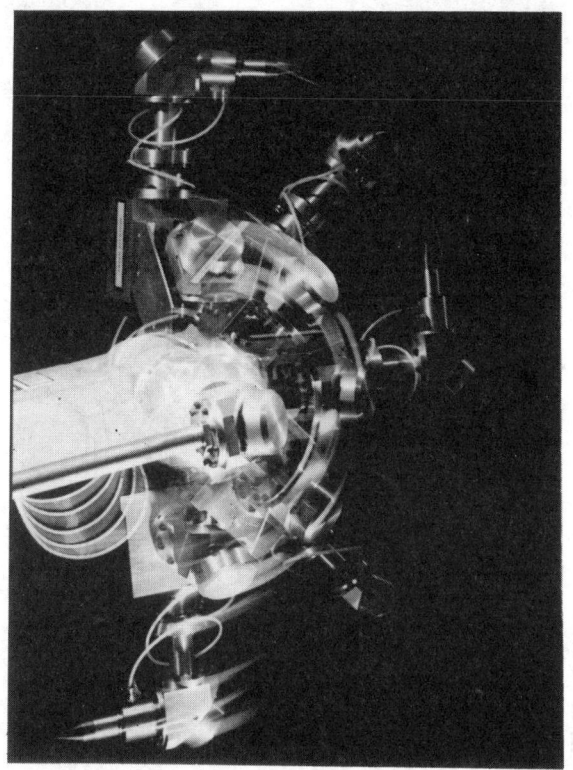

(WERKBILD FERRANTI-FLS)

Fig. 3. Flexible laser beam handling by an industrial robot

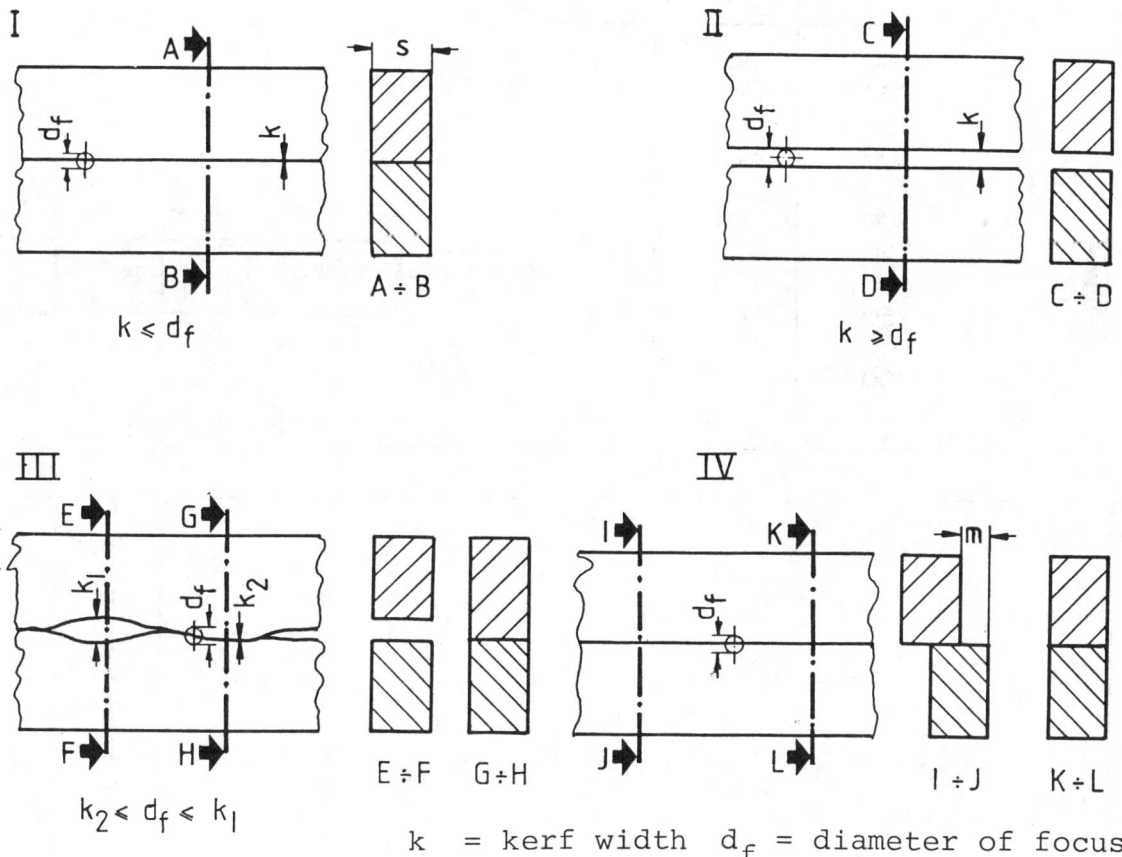

k = kerf width d_f = diameter of focus

Fig. 4. Problems of tolerance and mismatch in laser welding at awelding seam during series production

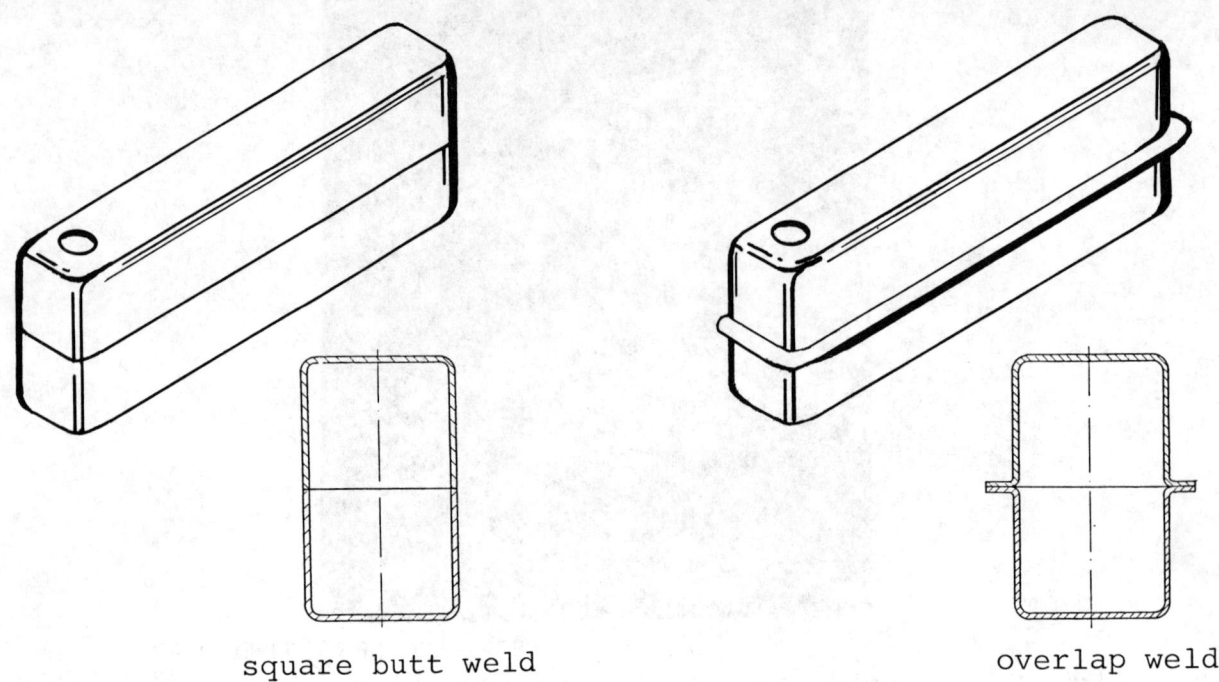

Fig. 5. Types of welds applied for the workpieces

L	W	H
200	60	60
300	120	
350	180	
400	240	
450	300	
500	360	
550	420	
600	480	
750	540	
900	600	
1000		

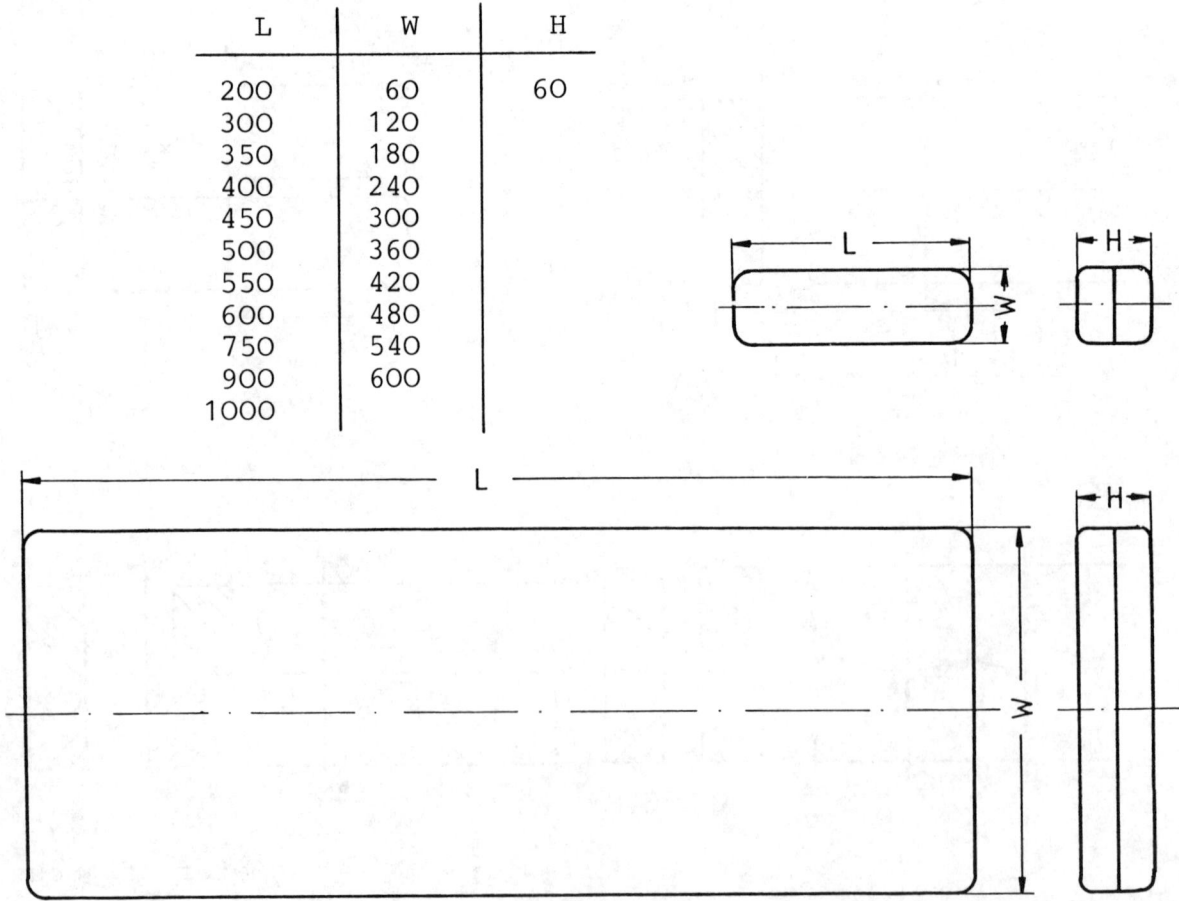

Fig. 6. Range of workpieces produced in the FMS

Fig. 7. Mechanically cut

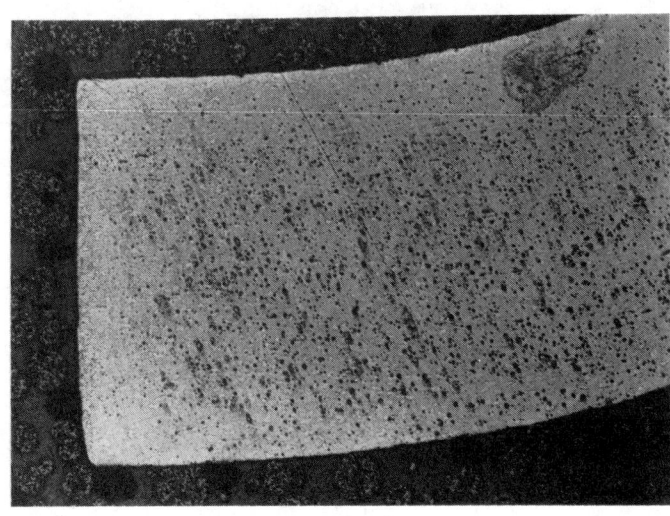
Fig. 8. Laser cut

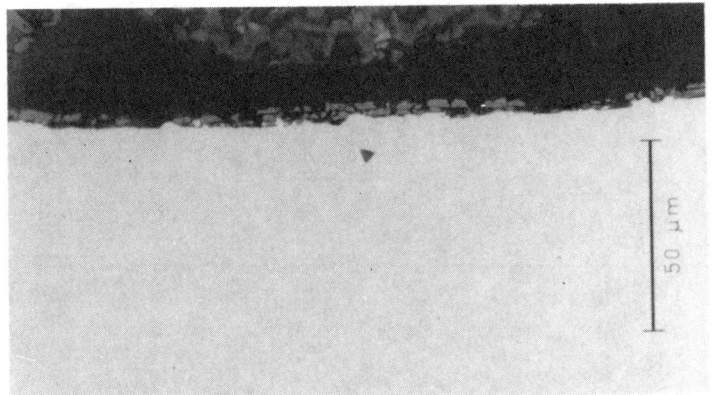
Fig. 9. Oxide layer thickness of laser cut by laser with TEM 01

Fig. 10. Oxide layer thickness of laser cut by laser with high order mode

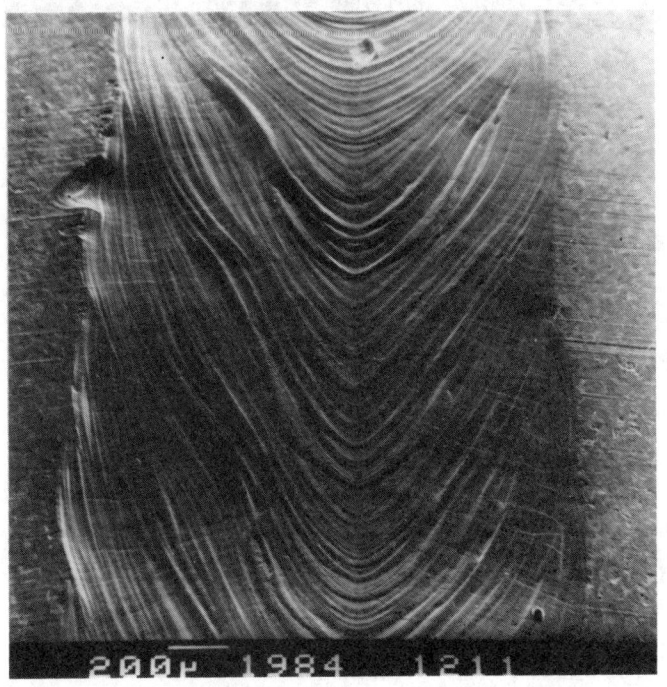

Fig. 11. Surface of laser weld by scanning microscope

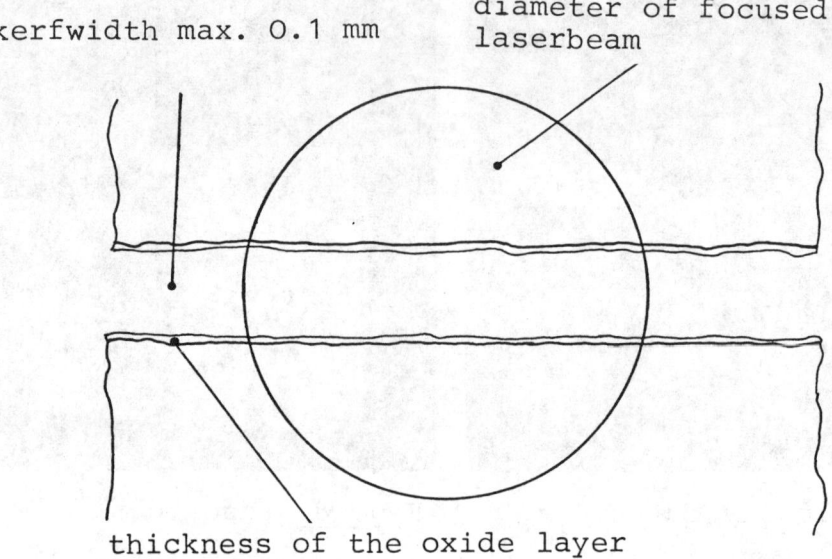

Fig. 12. Diameter of focused laser beam in proportion to oxide layer

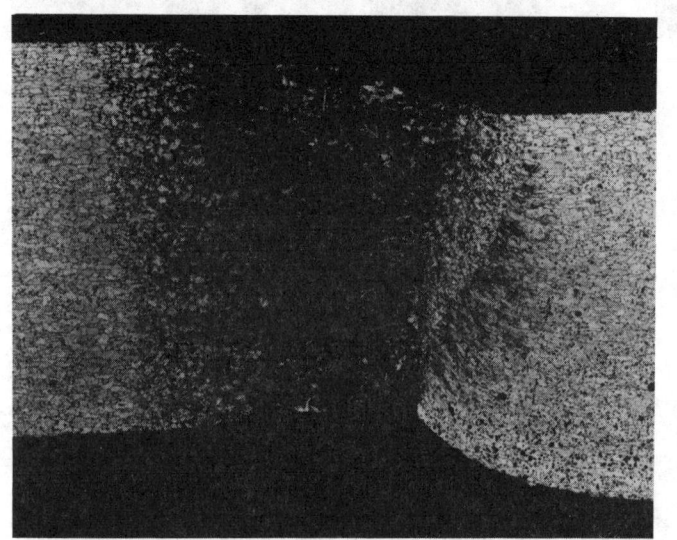

Fig. 13. Mechanically cut and laser welded

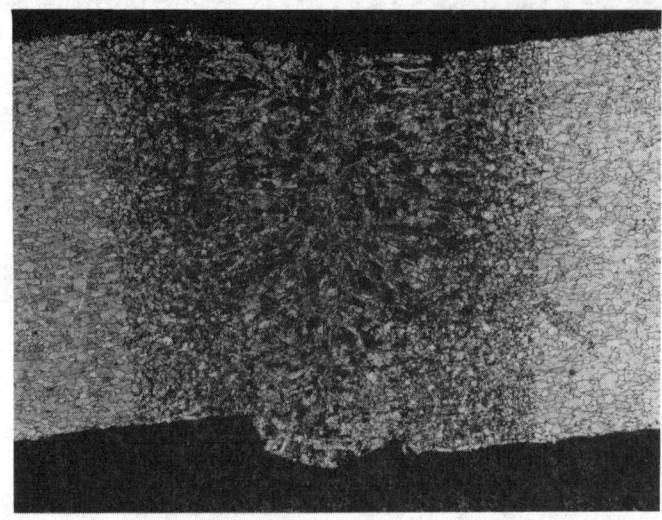

Fig. 14. Laser cut and laser welded

Fig. 15.1. Laser cut and weld by laser with TEM 01 (Square butt weld)

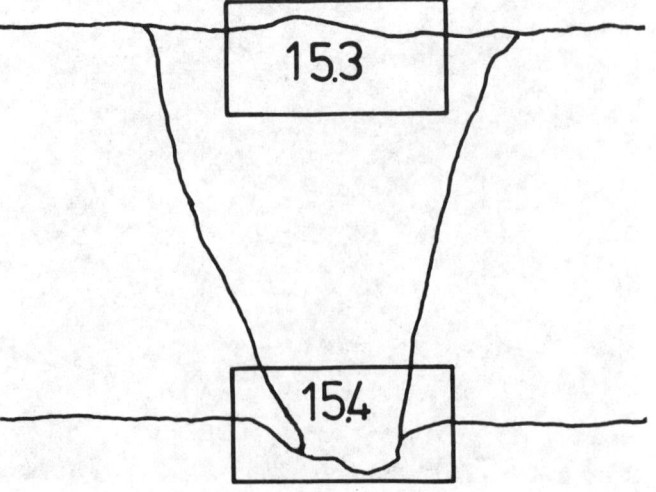

Fig. 15.2. Position of enlarged details

Fig. 15.3. Oxide particles in the weld pool

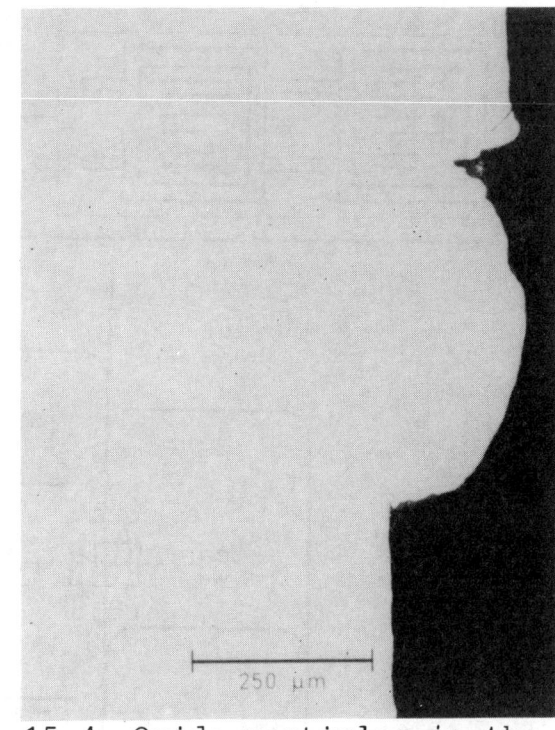

Fig. 15.4. Oxide particles in the weld pool

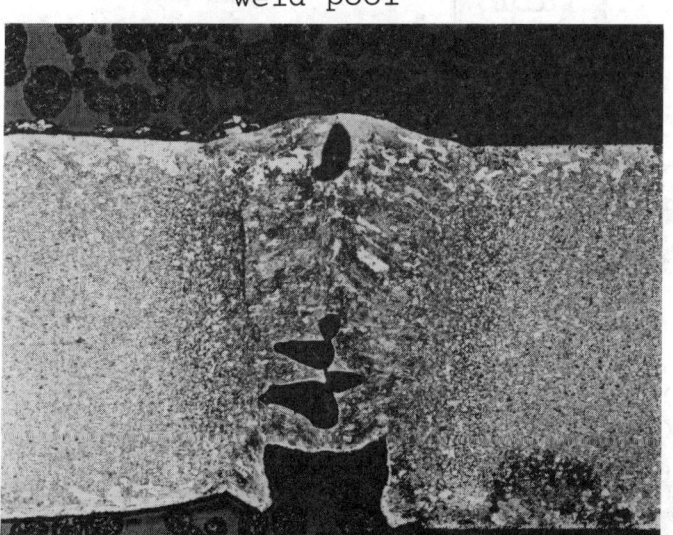

Fig. 16.1. Laser cut and weld by laser with high order mode (Square butt weld)

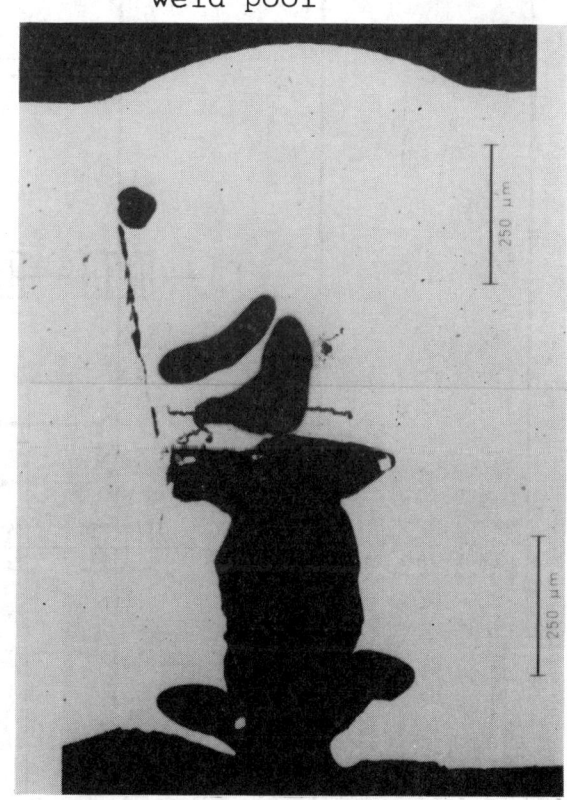

Fig. 16.2. Oxide particles in the weld pool

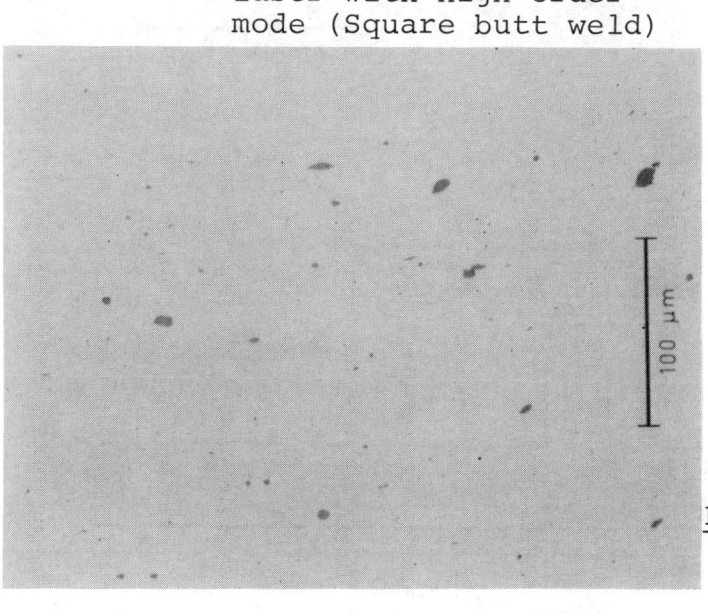

Fig. 16.3. Enlarged section of the weld pool showing oxide particles

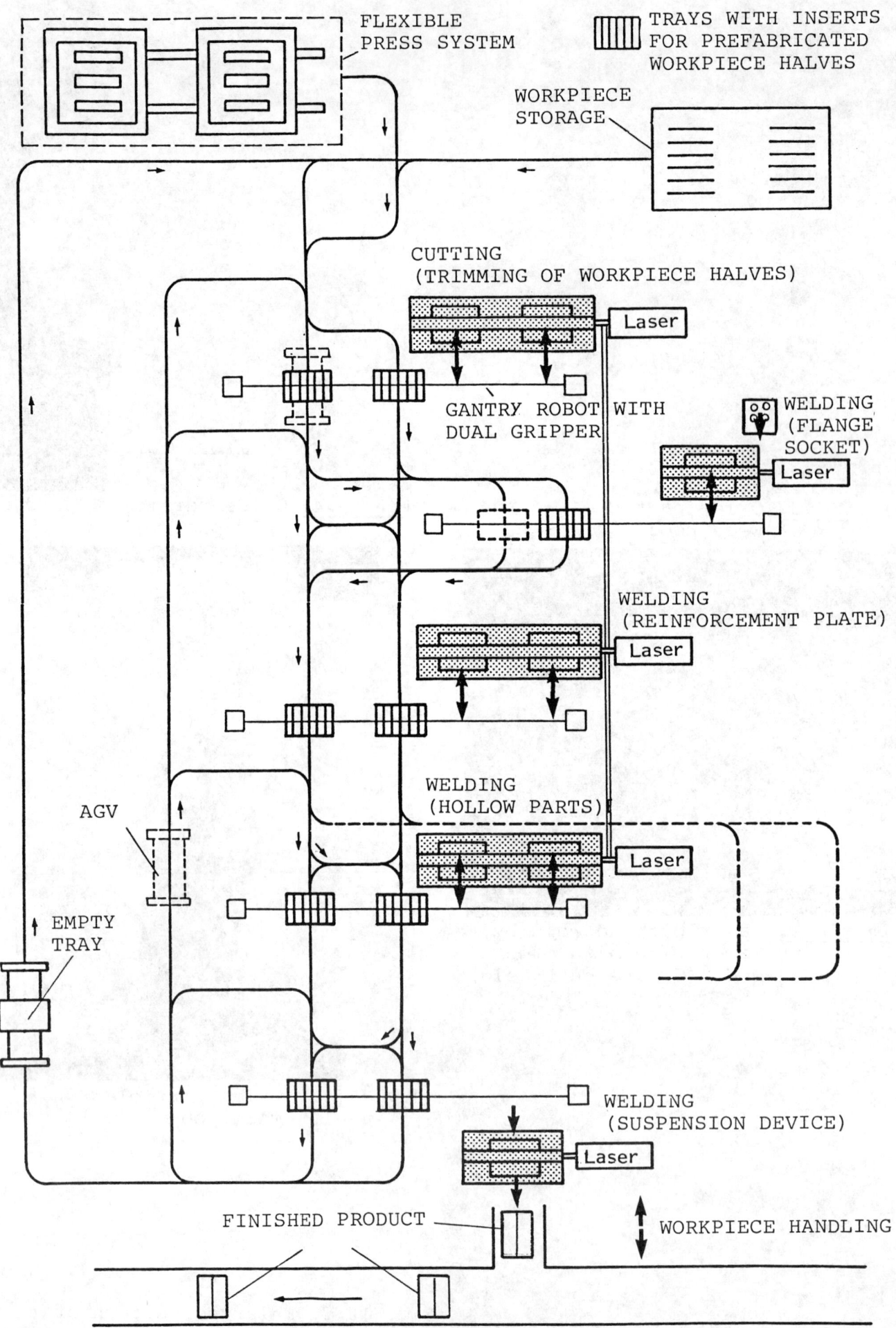

Fig. 17. Layout of the FMS provided by laser

Programmable laser character generation
D. I. Greenwood, A. McNeish
and
J. J. Harris
Isomet Laser Systems Ltd, UK

ABSTRACT

Acousto-optic techniques for the generation of spot matrix alphanumerics on continuously moving target material are described. The systems utilise a high p.r.f. Nd:YAG laser to mark the characters point by point.

Two approaches are considered: a simple system using a single acousto-optic deflector is briefly discussed which relies on the movement of the target material to provide the necessary shift in the orthogonal direction; and a two axis system is described which enables a high utilisation of the available laser pulses, resulting in a high throughput rate of the marked target material.

INTRODUCTION

The subject of this paper is the generation of alphanumerics for the identification by laser marking of continuously fed materials, such as cables and tubing. The system utilises the output from a high p.r.f. Nd:YAG laser to generate 7 x 5 spot matrix characters.

A conventional method of producing the required deflection is by the use of two galvanometer mirrors. Due to the limited speed of these devices, however, the material must be stopped while a section is scanned to produce a string of characters. Alternatively, a polygon mirror could be used to scan the laser pulses orthogonally to the target motion, allowing the material to be continuously fed. In this case, however, synchronisation between the laser pulses and the polygon position can be problematical.

The deflection systems described here are acousto-optic. Of these, the first is a simple system which effects a cross-scan analogous with a polygon scanner. The second is a more complex system giving a high utilisation of the available laser p.r.f., providing a high material throughput rate.

SINGLE AXIS DEFLECTION

This system utilises a single acousto-optic deflector to perform a scan orthogonal to the material transport direction, Figure 1. While analogous to the polygon cross-scan approach, it does not suffer from the severe synchronisation problems encountered where the laser is the master timing reference.

The laser pulses are passed through the cell via appropriate coupling optics and then focussed onto the target material.

The cell can be driven by seven frequencies corresponding to the seven positions of a column of the 7 x 5 matrix. Each frequency is generated in turn, effecting a scan down each column. The alphanumeric is generated by enabling or not enabling the r.f. amplifier at each position. Where the cell is not enabled, the laser pulse is wasted, impinging on a beam dump situated at the zero order position of the deflector.

Tilt of the columns created by target movement can be corrected merely by tilting the deflection plane.

With this system, the time taken to generate a character is equivalent to 35 laser pulses, plus an allowance for the space between characters of 14 pulses. Thus, with a laser p.r.f. of 15 kHz as considered here, the throughput rate is 306 characters per second.

TWO AXIS DEFLECTION

While the single axis deflection system is simple, it is wasteful of the available laser p.r.f., as with other conventional deflection techniques. Since normal alphanumerics are generated by a maximum of 20 pixels, less than 40% of the available laser pulses are utilised.

By introducing a second acousto-optic deflector, it is possible to take advantage of the 20 pixel per character maximum, and to eliminate the dead time corresponding to character spacing. The material throughput rate is then increased to 750 characters per second for a 15 kHz laser.

Principle of Operation

The arrangement is shown in Figure 2. The first cell deflects orthogonal to the material feed (X axis) with seven addressable positions. The second cell deflects in the plane of the feed (Z axis).

The combination of two cells can be addressed to any position within the 7 x 5 matrix, and so, with an allowance of 20 laser pulses per character, any normal alphanumeric can be generated. At the completion of a character, spare laser pulses are dumped.

Target movement is a prime consideration in this system; tilting of the cross axis deflection cannot provide a correction in this case since there is no regular scan pattern. The correction is achieved by tracking the target material using the second deflector. This cell thus has two functions as described below.

System Design

The operation of this dual function Z axis deflector is illustrated in Figure 3. The target material is fed past the marking station at a rate of one character, including spacing, every 20 pulses. During this time, a linear sweep must be added to the five Z axis pixel positions in order to 'freeze' the target during the exposure of a character. Thus the deflection range of the cell must be sufficient to encompass 10 times the pixel to pixel spacing, compared with 4 times the pixel to pixel spacing for a stationary target.

The resolution of both cells must be sufficient to provide well separated dots; values based on the Rayleigh criterion of 13 spots resolution for the first cell, and 23 spots resolution for the second have been taken.

A further consideration is optical power. The laser produces pulses of 3mJ with a 200nS pulse length at a repetition frequency of 15 kHz; a mean power of 45W. This indicates the use of fused silica as the interaction material, together with a substantial aperture. Values for the apertures of the cells of 6mm in the deflection plane, by 2mm to restrict the drive requirements have been chosen.

With a 6mm aperture, the acoustic transit time, τ, with fused silica is 1 µS. Thus the required bandwidth, Δf, of the cells to meet the resolutions required can be calculated from the expression

$$\tau . \Delta f = R-1$$

where R is the number of spots. This gives a bandwidth of 12 MHz for the X-axis deflector and 22 MHz for the Z-axis. Using a centre frequency of 70 MHz makes the fractional bandwidths small, enabling the use of a single transducer electrode. For the Z-axis, an electrode length, L, of 26.5mm can be used, giving a 1dB roll-off in interaction efficiency at the bandwidth edges. For convenience, a similar geometry can be used for the lower resolution X-axis cell. With an electrode width, H, of 2.5mm the required power, P, given by

$$P = \frac{\lambda_0^2 H}{2LM_2}$$

is 35W. This high power is due to the low figure of merit, M_2, of fused silica, 1.5×10^{-15} s^3 Kg^{-1}. However, the low duty cycle in this application, approximately 2 µs every 66 µs, reduces the average drive power to approximately 1W. The efficiency of each cell is greater than 85% over the bands.

This cell design is well within the current capabilities of acousto-optic technology.

The optical system is shown in Figure 4. The optical train begins with a Galilean beam expanding telescope to increase the beam size to 6mm diameter. The final lens assembly is a telephoto lens which focusses the deflected beam into the target plane with an effective focal length set to give the required character height. For a 1mm high character a focal length of 500mm is required. The remaining elements have the function of compressing the beam in the non-deflection planes while maintaining a 6mm beam size in the planes of deflection.

To drive the system, incoming ASCII characters are processed to give a series of coordinate voltages (z,x) over the range (1,1) to (5,7) which are fed to the a-o cell drive circuits, synchronised with the laser pulses, Figure 5. The X-axis deflector has seven addressable frequencies generated by a VCO.

The five voltage inputs to the z-axis driver are summed with a linear ramp voltage before feeding into a linear VCO, thus generating the required sweep to track the target. The length of this ramp is adjusted to maintain synchronisation with the laser p.r.f. By feedback from the target material transport system the voltage swing of the ramp is adjusted to maintain a frozen character despite target speed variations.

CONCLUSIONS

The two systems described enable permanent marking of 7 x 5 dot matrix alphanumerics onto continuously fed material at rates of 306 and 750 characters per second, using a 15 kHz Q-switched Nd:YAG laser. The higher writing rate is obtained on the basis of each character only requiring a maximum of 20 pixels within the 7 x 5 dot matrix. This better utilisation of laser p.r.f. could enable the use of a lower performance laser if a lower writing speed is sufficient for the application.

Finally, the two axis system described could readily operate with an 8 x 5 dot matrix with no speed penalty.

REFERENCES

1. Young, E.H. and Yao, S.K., "Design considerations for Acousto-Optic Devices", Proc. I.E.E.E., Vol 69, pp. 54-64, 1981
2. Gottlieb, M, , Ireland, C.L.M. and Ley, J.M., Electro-Optic and Acousto-Optic Scanning and Deflection, Dekker 1983.
3. Chang, I.C., "Acousto Optic Devices and Applications", I.E.E.E. Transactions on Sonics and Ultrasonics, Vol. SU-23, pp. 2-22, 1976

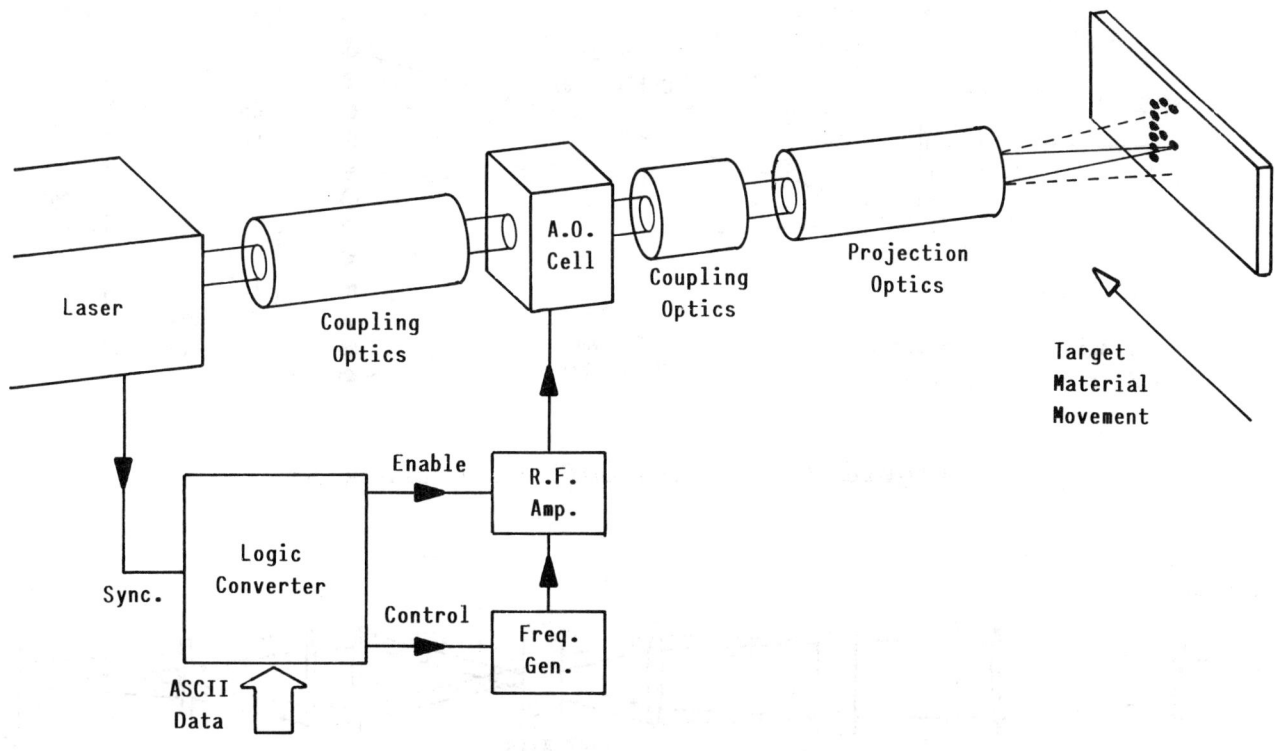

Figure 1. Character generation by single axis deflection.

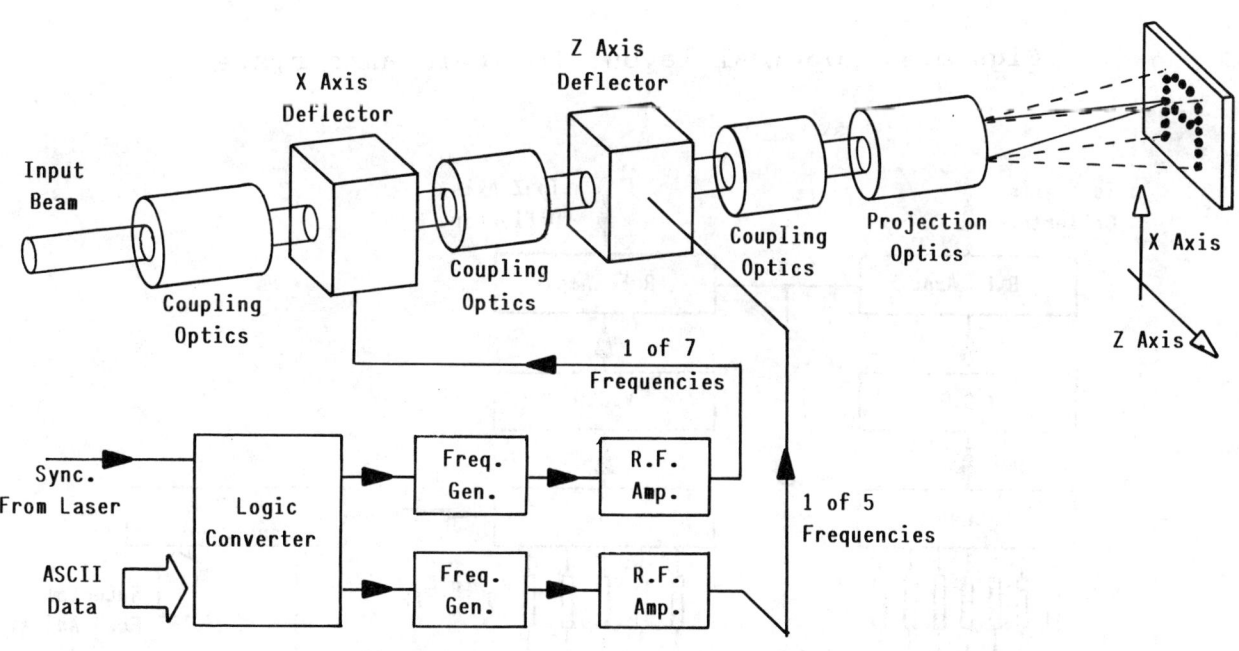

Figure 2. Character generation by twin axis deflection.

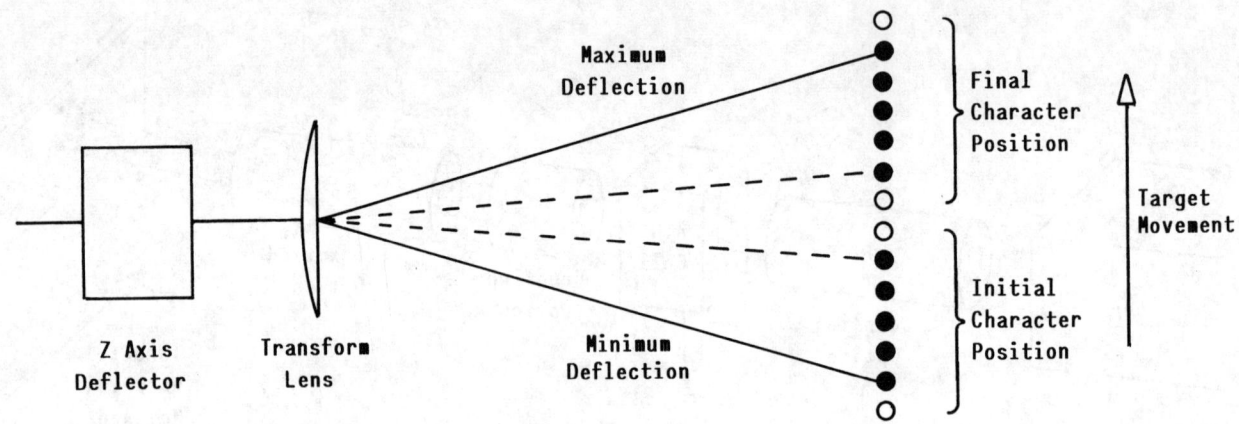

Figure 3. Z axis deflector tracking.

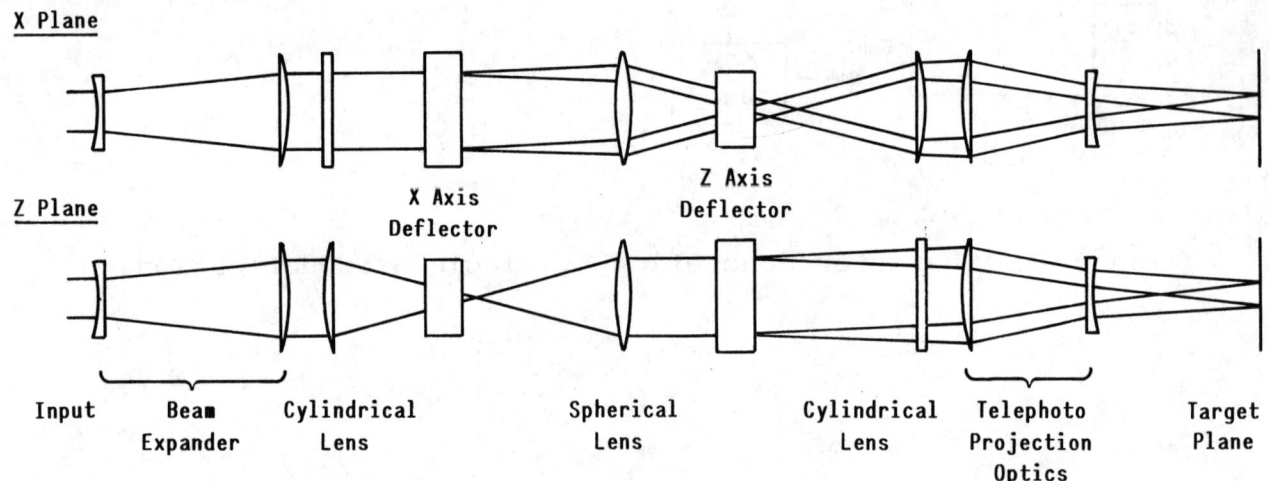

Figure 4. Optical layout for twin axis system.

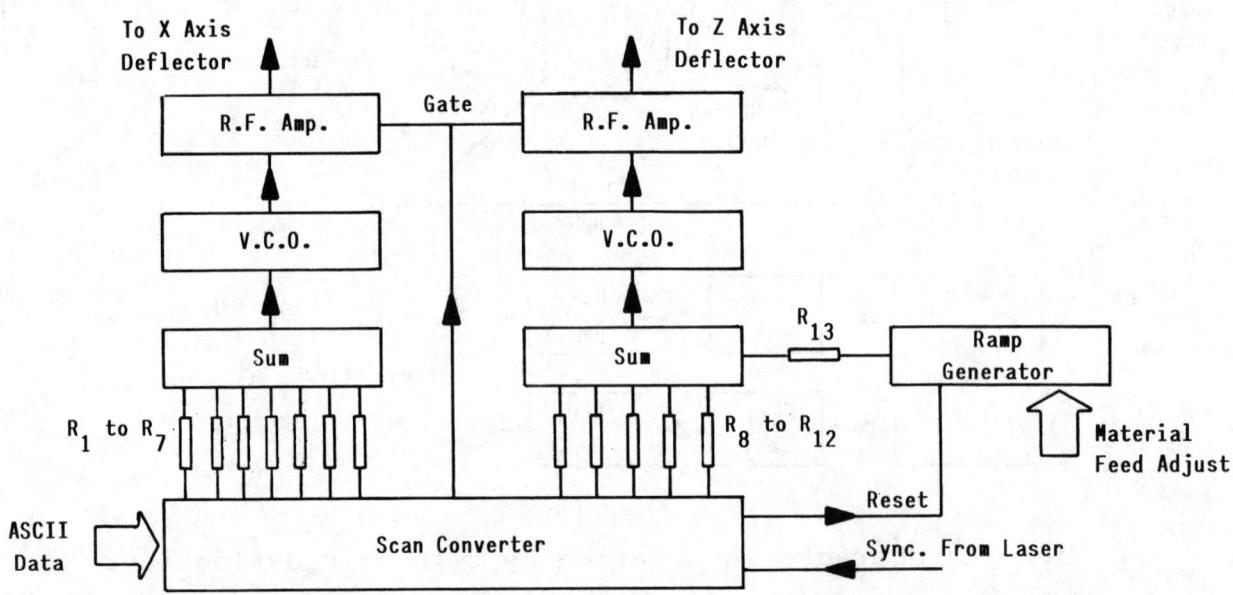

Figure 5. Twin axis system electronics.

The management of industrial lasers
B. G. Green
and
M. J. Bragg
Laser Scientific Services Ltd, UK

This paper, of particular interest to industrialists and commercial laser operators, covers the full spectrum of subjects relevant to the Management of Industrial Lasers in 1985. Equipment aspects covered are identification of requirement; how to procure a laser system; the advantages of purchasing from a supplier with sub-contract facilities; pitfalls; staff training; day-to-day running and quality assurance. A financial appraisal shows that an industrial laser can be operated profitably without difficulty, particularly where high utilisation is achieved.

INTRODUCTION

This paper gives an account of experience in managing industrial lasers. It is based on lessons learned by Laser Scientific Services which has more laser systems carrying out a wider variety of laser applications than any other in Europe. The major emphasis is on the problems of purchasing a laser system, spending the money wisely, and generating real profits as a return on investment.

IDENTIFYING THE REQUIREMENT

In U.K. Industrialists and Production Engineers are only now beginning to become aware of the role that an industrial laser plays in the production process. What then are the considerations that should lead a Production Engineer to decide that an industrial laser might be the answer to his problem? We suggest that there are several possible areas:-

a. It may be that out-dated machinery, in need of replacement, has carried out a process that will continue to be required for the foreseeable future. Perhaps the equipment is subject to excessive down-time due to its age or is, by today's standards, unacceptably slow.

b. Similarly current equipment may be too labour intensive, perhaps due to the amount of corrective or fettling work which needs to be done on completion of a standard machining operation.

c. A new material may have been introduced which is not amenable to processing by the existing machinery.

d. Increasing demands for accuracy for high technology applications may take a process beyond that which is within the capacity of existing equipment.

e. There may be a specific requirement to keep the heat-affected zone as narrow as possible to avoid adverse mechanical effects.

f. The tooling costs involved in setting up a new operation using traditional machinery may be unacceptably high.

g. Distortion in thin metals may be a major problem and the possibility exists that the use of a laser may overcome this difficulty.

h. A high quality edge finish may be a feature which is of a particular advantage.

i. It is possible that a machine is required which will process only part of a component without adversely affecting the remainder.

j. In welding applications, it may be a requirement that there is no additional weld material.

k. The intricate profile cutting that can be achieved by a laser may be a particular advantage.

HOW TO PROCURE A LASER

In advising on this matter we assume that the Company concerned does not have in-house laser expertise and the first question is, therefore, how does a company obtain that expertise at a minimum cost. There are in the U.K. several companies who market industrial lasers and who have the expertise and experience to advise a client company. There is no doubt that carefully planned approaches to these laser companies can be a cost effective way of moving forward. Most will carry out a limited feasibility study at little or no cost and will advise the potential customer of the likely requirement. Some companies exist solely to market industrial lasers produced elsewhere; one or two companies manufacture and market lasers in U.K.; other companies, such as Laser Scientific Services, market and service industrial lasers manufactured in U.K. or abroad, design and assemble the work handler/CNC to provide a turnkey system; and also run sub-contract facilities. For the reasons given below we recommend the latter type of company.

The Laser Company should provide a formal report as a result of its application study. This will recommend the type of system that should be used; the power of laser that is likely to be required; and give budgetary estimate of the likely cost. An estimate will be provided of the cost of tooling to produce samples or it may be possible to produce samples without a tooling cost. There are then two ways of proceeding towards procuring a laser. The customer can either use a sub-contract facility first or go straight to a laser purchase.

Use of Sub-Contract

If the cost of a system is more than had been forecast or expected, or is too much to commit without further detailed study, then the use of sub-contract facilities is a prudent course of action. Placing sub-contract work with a laser supplier will bring with it many advantages:-

a. The laser supplier will demonstrate the work being done.

b. Expertise will be gained in actually doing the work, which will benefit the customer.

c. The tooling will be tested or perhaps, by using prototype tooling, it will be possible to develop an improved system.

d. Software difficulties can be ironed out.

e. Achievable accuracy can be demonstrated.

f. Large sums of money are not committed unnecessarily.

g. The customer firm wins time to study the matter in depth and can carry out an appraisal of a range of possible laser applications.

Direct Purchase

If it is decided to move direct to a laser purchase there are several pointers towards achieving success:-

a. It is risky to purchase a laser on the premise of the receipt of a few samples. The wise purchaser will insist on a small production run being done by the laser supplier before he accepts the equipment. Inevitably there will be tooling costs to enable this work to be done but if designed carefully this tooling can be used by the customer for production so the money is not wasted. Time and trouble can be saved by doing it properly first time.

b. The laser supplier must specify the environment that is required for the laser to operate satisfactorily.

c. To achieve maximum utilisation, a flexible system should be sought where this is possible. The financial benefits of flexibility are detailed later in this paper.

d. The laser supplier should specify the area that is required for his installation but the customer must consider other factors such as material handling and the proximity of other machinery before deciding the location of the laser in his works.

e. The laser supplier must specify the power supply required.

f. The requirement for laser gas and oxygen should be specified and an estimate made of the likely usage.

g. The need for the air compressor must be established and procurement action taken.

h. The need for the water cooler and its size must be established and procurement action taken.

i. The likely total running cost should be established bearing in mind the cost of a service contract on expiration of warranty and the cost of spares and replacement parts. The latter figure is likely to be higher for a solid state laser than for a standard CO_2 laser. Further comment on this aspect is given under Running Costs below.

j. Safety requirements must be worked out bearing in mind the working environment of the laser installation.

Choosing a Company

In choosing a laser company a wise customer will check on the following:-

a. Whether the laser company has achieved success in providing installations to their customers.

b. The depth of expertise and speed of response of the service back up.

c. Whether the laser supplier's location is convenient from a geographical point of view.

d. Where the laser is manufactured.

e. Spare parts availability.

f. The warranty terms and, in particular, the date from which the warranty runs. Purchaser and supplier should also agree the date of delivery, the date of installation, and the date of commissioning, the latter being the time at which the complete system is handed over to the customer and, normally, the final payment is made.

g. Provision of operator training is an important matter. Some companies offer training at the customer's site on the customer's equipment doing the work which the customer intends for the laser that he has procured. Other laser suppliers provide training at their own works or premises which, while at no cost to the customer, will involve him in travelling or accommodation expenses for his staff. The training facilities offered by some suppliers are orientated towards the laser installation, while other training is geared to the carrying out of the work. This is most likely to be the case where the laser firm runs a subcontract facility.

Workhandling

An important factor in the selection of a laser supplier and a laser system is the matter of the associated workhandling equipment and control system. One or two companies offer complete workhandling packages, other Companies offer a range of options. It is important to define clearly the responsibilities of the laser supplier for this equipment at an early stage and, in particular, to lay down responsibility for the almost inevitable interface problems that will occur with a custom-made system. It is sometimes worthwhile to produce a schematic drawing actually delineating areas of responsibility but even then difficulties can occur. There is much to be said for appointing one contractor as the prime contractor who will be responsible for handing over the complete installation to the customer. Alternatively, depending on the size of the customer company, it may be perfectly satisfactory to purchase the laser off-the-shelf and carry out the workhandling and control system package work in-house.

PITFALLS

This list is based on our considerable experience gained in running well managed laser installations. It is not intended to alarm the reader by placing undue weight on operating problems. Rather the aim is, by advising him in advance, to enable any adverse effects to be mitigated.

a. The user is advised that the use of laser gas and oxygen may well be in excess of his initial estimates. However, the supply situation is good in U.K. and price competition is strong. It is, therefore, worthwhile to keep a watch on this particular aspect of the operating cost, negotiating the best deal as consumption increases.

b. Optics - lenses and the like - are expensive and should be carefully stored and used. They must be kept perfectly clean and it is very important to ensure that any gas supplies are clean and dry. The installation of an air drying facility for example may well pay for itself in a reduction in the number of lenses used.

c. It is a false economy to use poor quality tools and ancillary equipment with a laser and attention paid at the procurement stage to the ancillary equipment will be time well spent.

d. The customer should also assess various options where a workhandler is being purchased at the same time as the laser. All too often we see a customer devote a disproportionate amount of time to selecting the laser but seemingly to be disinterested in the task of selecting the best workhandler. The reason for this is unclear, for the workhandler and associated equipment may well cost as much as the laser and it is not unlikely that the customer will have in-house expertise on these or similar production engineering facilities enabling him to make judgements based on experience.

e. As part of the procurement process the customer should assure himself as to the throughput that the system will achieve. If he is unsure on this point the economics of the whole installation will be based on unsound foundations. It is for this reason that, as stated above, we recommend subcontracting as part of a system procurement process.

f. Particular note should be taken of the beam delivery system used by the laser. Much time may be lost in production if adjustment of beam delivery is a cumbersome and lengthy operation.

g. Maintainability is important. The customer should look for a system where parts are readily accessible; where cable connections are readily made; where instruments can be used with ease and where system performance can be monitored using the built-in instrumentation or test equipment provided.

h. Speed of back-up service is another important factor. For the economic reasons given below, loss of productive time is, as with all machine tools, a serious matter and the provision of an on-call service is, therefore, worth paying for.

i. The final subject under the heading of pitfalls concerns safety. The customer is advised to get full information from the supplier on the safety requirements of the laser and he should gauge whether the introduction of the equipment into his premises is likely to provoke an adverse reaction from his workforce. Time spent in educating the workforce as to the real nature of the safety hazard is always well spent, for it is only too easy for mischief-making rumours to be circulated with adverse effects.

STAFF AND TRAINING

A properly designed workhandler with clear instructions to the operator should enable work to be carried out perfectly satisfactorily by the normal machine operative. There is a tendency to try to make laser operation into a black art and this should be resisted. There are two areas, however, where training is necessary and will be rewarding in the long run.

a. Instruction for management in how to make best use of a laser and keep it working.

b. Instruction for operators in programming tasks and use of CNC controls.

It should be appreciated that, as with other machine tools in a repetitive production environment, operating a laser can be a boring and lacklustre form of employment. It is important, therefore, to select personnel for this repetitive work with care and the highly intelligent ambitious employee may not be the best person to choose.

DAY-TO-DAY RUNNING

The laser manufacturer or supplier will recommend a routine maintenance schedule specifying what is to be done and when. It really is important that this schedule is strictly followed as we have found that a proportion of our laser breakdowns have been directly attributable to failure to carry out routine maintenance.

In our experience some 80% of laser breakdowns or other unserviceabilities occur during the switching-on period. Often the reasons are not directly attributable to the laser, but the customer's own technical expertise should be available on call when a laser is switched on and, as a general rule, lasers should not be switched on and off unnecessarily.

Following from the above it is important to ensure that the flow of material to the laser, instructions to the operator, programs and in-production inspection arrangements is maintained on a continuous basis. This is, of course, a basic production engineering principle but it applies to a laser installation as much as to any other high technology, high productivity machine.

Care taken to ensure cleanliness of the workhandler and other mechanical moving parts will be amply repaid. It is essential to keep swarf and other metal particles or alumina dust out of worktable drives, spindles, etc.

A record of laser operation should be maintained giving details of causes of unserviceability, spares used, settings, maintenance activities. Mention has been made of the need to ensure that a laser's beam delivery system can be readily adjusted although in a well designed modern system frequent adjustment should not be necessary. Supervision of this aspect of a laser operator's work is a particularly important feature for the quality of the output will depend directly on the accuracy of the beam setting.

QUALITY ASSURANCE

There is no reason why a laser should not produce regular, repeatable, high quality products. Points worthy of mention are:-

a. The setter or supervisor must check, not only the first off a production run, but also the settings which achieved that first-off.

b. For any production inspection requirements it is essential that the actual dimensions to be measured and the tolerances permitted are specified.

c. If work is required to a high standard of accuracy then the inspection equipment required to ensure that it is maintained should be considered as part of the laser procurement arrangements.

d. Keeping the surface flat is of importance and the quality will be adversely affected if the workpiece is allowed to flex in any way.

e. Attention should be paid to deburring as, if this work is done to a low standard, the advantages of high accuracy and superior edge finish of the laser operation will be negated.

FINANCIAL APPRAISAL

In this section of the paper we study financial aspects of procuring, installing and running an industrial laser. We have taken as our example a typical medium power CO_2 laser with associated worktable and CNC equipment. We consider the capital costs, the running costs and the total costs. Hence we derive the break-even point above which the installation runs at a profit. The effect of the contribution which might be made by Inland Revenue is explained and comments made on utilisation.

Capital Cost

A realistic budgetary estimate of the capital cost of the laser and associated equipment is:-

a. Medium power CO2 laser   £ 70,000
b. Worktable and CNC facility   £ 40,000
c. Installation including compressor
   and water cooler   £ 10,000
                                    £120,000

Experience in the United States supported by evidence in U.K. indicates that, even allowing for obsolescence considerations, it is reasonable to expect that the life of a standard CO2 laser will be at least eight years. We therefore assess, an an initial approach, that three years should be allowed for repayment and a further five years allowed for building-up a reserve against the purchase of a replacement. Therefore, for three years the annual repayment of the capital cost will be:-

$$\frac{120,000}{3} \text{ or } £40,000 \text{ per year}$$

without allowance being made for interest or for the effect of contribution by Inland Revenue.

Running Costs

For this section we have not used estimates but actual annual costs experienced on one of our own laser installations at a subsidiary company.

Wages of operator	£ 4,160
Electricity: power, heat & light	£ 1,200
Gas	£ 800
Petty cash/sundry expenses	£ 300
Insurance	£ 1,000
Maintenance/repairs	£ 1,200
Spares and components	£ 1,000
Telephone	£ 350
Rent/rates	£ 1,800
Audit/accountancy fees	£ 500
	£ 12,310

Notes on the above figures are:-

a. No account is taken of marketing or management overheads.

b. The figure of rent and rates is high as the installation is housed in its own separate building. A more typical rental would be 400 square feet at £2.50 per square foot = £1,000.

c. Our actual maintenance/repair cost of £1,200 is higher than our normal figure as the laser in question is situated in an isolated and distant location. However, this figure is a fair approximation to the charge which a customer might expect to pay for a service contract for the installation under study.

d. The wages of the operator are based on a single shift of eight hours.

e. Significant items in the expenditure on spares and components are lenses which cost £300-£400 each, which supports our earlier comment that lenses should be stored, handled and used with care.

f. If the installation is based on a Solid State laser rather than a standard CO2 system the likely maintenance and repair costs will be somewhat higher, based on past experience. Significant items are replacement flash tubes, which are normally only guaranteed for 200 hours and which cost £80 - £100, and the laser rods which inevitably deteriorate and cost approximately £1,000 to replace.

f. (continued)
   Refurbishing facilities for laser rods are available but the time taken for this work is unlikely to be less than three weeks.

Total Costs

From the above figures, but using the more typical rental quoted, the total annual costs (capital cost repayment plus running cost) is £51,510 with one operator. Accordingly, on single shift working, the maximum annual hours of operation that can be achieved, assuming that maintenance is done out of hours, is 2,000. If we make an allowance of 100 hours to cover downtime, the laser must earn (or save) £27.11 per hour to break even - or approximately £1 every 2 minutes, although management overheads and material must, of course, be added. Any money earned in excess of this is profit and such profits are readily achievable. We know because we are operating lasers profitably now.

Tax Situation

The above figures take no account of the effects of the contribution by Inland Revenue towards the cost of purchase of capital equipment or of alternative methods of making the purchase.

It is inappropriate here to attempt to explain the intricacies of Corporation Tax but the implications of the tax arrangements are so significant that the question cannot be ignored.

From the 1984 Budget the normal rate of Corporation Tax has been fixed as follows:-

Year to 31st March	Normal rate
1984	50%
1985	45%
1986	40%
1987	35%

For small companies and medium sized companies with taxable profits below £500,000, marginal rates apply. However, for this part of the study we assume that the purchasing company is paying the full rate of Corporation Tax.

Expenditure on plant qualifies for a 75% first year allowance provided the expenditure is incurred prior to 1st April 1985. Between April 1985 and 31st March 1986 the first year allowance is 50%. Thereafter only writing down allowances of 25% on the reducing balances will be given. It will be appreciated therefore that the figures change for a laser purchased after 1st April 1985, shown as Case 2 in the tables below.

REVENUE CONTRIBUTION

Case 1			
Year	Tax Calculation	Contribution by Revenue	Residual value
Purchase 84/85	75% of £120,000 @ 45%	£40,500	£30,000
85/86	25% of £ 30,000 @ 40%	£ 3,000	£22,500
86/87	25% of £ 22,500 @ 35%	£ 1,969	£16,875
Total contribution by Revenue:- £45,469			

Case 2			
Year	Tax Calculation	Contribution by Revenue	Residual value
Purchase 85/86 86/87 87/88	50% of £120,000 @ 40% 25% of £ 60,000 @ 35% 25% of £ 45,000 @ 30% (est)	£24,000 £ 5,250 £ 3,375	£60,000 £45,000 £33,750
Total contribution by Revenue:- £32,625			

Thus, from the above calculations, a large profitable organisation paying £120,000 for a laser installation will, over a three-year period, have paid only £120,000 - £45,469 = £74,531 nett if the laser is purchased before 1st April 1985 and £120,000 - £32,625 = £87,375 nett if the laser is purchased after that date. Also, the writing down allowances continue beyond the third year giving further benefit, and if profits are sufficient to absorb the writing down allowances then these can be carried forward to offset against future profits.

A company with reduced liability to corporation Tax receives less by way of contribution by Inland Revenue and is effectively paying more for its laser. This, in turn, is counter-balanced by the possibility, not quantified here, of a company in such a situation receiving some form of grant assistance, for example under the FMS or Small Business Expansion Scheme.

From the above, the nett capital cost, spread equally over three years, is either £24,843 per year if the laser is purchased before 1st April 1985 or £29,125 per year if purchased after that date.

Leasing

An attractive alternative is to consider leasing. Until the 1984 changes it has normally been beneficial for a company to buy a laser rather than lease it provided the company had, as described above, sufficient taxable profits to absorb the capital allowances. However, those companies not paying the full rate of Corporation Tax should now give the leasing option serious consideration. Lasers are perfectly acceptable installations for a leasing arrangement and we know that several leading Companies are active in this field. However, the decision as to whether to lease or buy is one that can only be made by a company in the light of its own financial situation.

New Total Cost

Returning to the case where a laser is purchased outright, but taking account of the contribution by the Revenue, it will be apparent that, taking the less favourable case of a laser purchased after 1st April 1985, the total of capital cost repayment and running cost is £29,125 + £11,510 = £40,635. Using the same basis as before, the laser must earn (or save) £21.39 per hour or approximately £1 every 3 minutes to break even.

Utilisation

All the above calculations have been based on single shift operation. If we work on a two-shift system the situation is even more favourable. The additional annual running costs will be:-

Wages	£4,160
Electricity, heat and light	£1,200
Gas	£ 800
Maintenance/repairs	£ 600
Spares	£1,000
	£7,760

The total annual cost taking the possible Revenue contribution into account is thus £48,395 for 3,800 hours of operation or only £12.73 per hour. Comparison of this figure with that derived earlier demonstrates the benefit of a high utilisation of the laser. So the message is that, at the procurement stage, it is essential to consider as wide a range of applications for the equipment as possible. The laser purchased should, where practicable, have the flexibility to carry out a variety of applications. The firm's designers must be taught to think laser so that benefits are maximised. We quote, by way of example a laser installation which has been sold to a major customer. Already that customer has approached us to provide tooling for other applications. The tooling would be used by us, the laser supplier, to carry out the work on a sub-contract basis. When the laser is installed, the tooling will be transferred to the customer. Wisely, our customer is seeing that the use of his laser is maximised right from the start.

Finally, we make the point that, on the basis of the above calculations, with the capital cost repaid at the end of three years, the business generated needs only to support the running costs and build up funds for purchase of new equipment. On a two-shift system, the running cost of the laser is £11,510 plus £7,760 = £19,270 per annum or a mere £5.07 per hour. The potential for profitable operation is then beyond question.

CONCLUSIONS

This paper has argued a convincing case for the introduction of an industrial laser into the facilities of manufacturing companies. We have given some guidelines on the identification of the requirement and suggested the best way to go about procuring a system. We believe strongly that it is wise to use a company which has its own sub-contract facilities and thus has depth of experience in the managing of a laser operation. Pitfalls and problems of day-to-day running have been described and comments made on quality assurance. Finance is fundamental to the subject and, using the example of a medium powered CO_2 laser, we have shown that profitable operation is readily achieved, particularly if full advantage is taken of contributions which can be obtained from the Revenue. Finally we have demonstrated the dramatic advantage which can be obtained by increasing the utilisation of the laser. If a laser can be run on a two-shift or even three-shift basis and its management is capable of supporting that level of activity, then a substantial return on the investment will be a certainty.

SUPPLEMENTARY PAPER

Manufacturing WO3 and Fe2O3 films by laser chemical vapour deposition

M. S. Chiu, C. C. Chou
and
G. P. Shen
Shanghai Institute of Laser Technology, People's Republic of China

The amorphous WO_3 and Fe_2O_3 films were deposited on quartz substrates by photolysis with an excimer laser for the first time. The compositions, the resistivities, the deposition rates and the constructions etc. of these two films are given in this article.

Laser chemical vapour deposition (LCVD) is a new technology based on the principle that the metal atoms, dissociated from organometallic molecules by photolysis, deposite on the substrate surface. Recently, LCVD has been applied not only to the manufacture of the metal film but also to that of metal compounds, such as Al_2O_3, SiO_2 and ZnO, due to the reaction between the dissociated metal atoms and the dissociated oxygen atoms. Compared with the vacuum coating, LCVD have the advantages of higher deposition rate and higher film purity at room temperature operation in one-step way, so that it is very useful in the investigation of microelectronics and surface physics etc.. It is reported in this article for the first time that the amorphous WO_3 film and the amorphous Fe_2O_3 film were deposited by means of LCVD with an excimer laser as the light source. Being available for display, the amorphous WO_3 film is an electrochromism material[1], the absorption spectrum of which, i.e. the colour of which, can change either with the strength of the electric field exerted on it or with that of the electric current flowing through it. Unlike other electrochromisms only suitable for high temperature, it has been proved to be a good one at room temperature. The amorphous Fe_2O_3 film is one of the magnetic materials, which is profitalbe for data recording medium in microelectronics. The device array of this medium can be obtained by scanning the laser beam on the substrate surface during the deposition process. The compositions, the resistivities, the deposition rates and the constructions etc. of the WO_3 and the Fe_2O_3 film were measured in the experiments.

$W(CO)_6$ and NO_2 or $Fe(CO)_5$ and NO_2 were used as the operation gases in our experiments. Because $W(CO)_6$ and $Fe(CO)_5$ are multi-atom molecules, there are many vibration degrees of freedom and rotation levels which overlap each other forming several quite wide absorption bands. Figure 1 shows the absorption spectra of $W(CO)_6$ and $Fe(CO)_5$ with some obvious absorption peaks in UV region, observed with a Spectrometer UV 190. From figure 1, the absorption cross-sections are $1.36 \times 10^{-17} cm^2$ for $W(CO)_6$ and $5.98 \times 10^{-19} cm^2$ for $Fe(CO)_5$ at 308nm. After absorbing photons, the excited molecules generally relax to the ground state by fluoresence emission or by the intramolecular energy transference. Excited by photons with the energy higher than the bond energy of them, the molecules may happen to dissociate. The dissociation probability of a weak bond in a molecule is much larger than that of the strong one.

The bond energy of ON-O is 3.12ev[2], equivalent to the energy of a photon at 397.7nm. When the light wavelength is shorter than 360nm, the photodissociation quantum yield of NO_2 is about 1, and when the light wavelength is longer than 360nm, the yield decreases and the fluoresence quantum yield increases rapidly. The latter reaches about 1 at the light wavelength of 390nm, and keeps constant for the light wavelength longer than that. Under the irradiation of UV light, the NO_2 molecules dissociate into a NO free radical and an oxygen atom:

$$NO_2 + h\nu \longrightarrow NO + O \qquad (1)$$

The bonds between carbon and metal atoms, C-M, are often weakest in most organometallic molecules with the dissociation energy about 1~3ev. The average bond energy of W-C in $W(CO)_6$ molecule is 1.64ev[3] and that of Fe-C in $Fe(CO)_5$ 1.2ev[4]. The energies of bonds are much less than that of a UV photon, 4~6ev, so that the bonds between carbon atoms and metal atoms can be broken up after absorbing UV photons as following:

$$MCB \xrightarrow{h\nu} M + CB \qquad (2)$$

The light source is a XeCl excimer laser at 308nm with the output over 50mJ and with the pulse duration of 15ns. The cylindric gas cell, 45mm in length and 22mm in diameter, has two quartz windows at both ends and a shealth in the wall, through which the temperature controlled water flows to keep the cell at a desired temperature in order to control the vapour pressure of the $W(CO)_6$.

In depositing tungsten oxide film, the $W(CO)_6$ powder was set in the cell which was evacuated to a pressure of 10^{-3} torr and then was filled with NO_2 gas of 10 torr and Ne buffer gas of 700 torr. The temperature of the cell was raised to 50°C to evaporate the $W(CO)_6$ inside the cell to the saturated pressure of it, 0.35 torr. In depositing iron oxide film, the cell was filled with $Fe(CO)_5$ of 25 torr, instead of $W(CO)_6$ powder, as well as NO_2 of 60 torr and He of 500 torr. Converged by a quartz cylindric lens with a focus length of 45mm, the UV light beam was perpendicular to the quartz substrate with a rectanglar spot of 4mm X 25mm where the light yellow tungsten oxide film and black-red iron oxide film were deposited on the substrates respectively.

No obvious peak can be observed in the X-ray diffraction spectra of the deposited films, as demenstrated in Fig.2a and Fig.2b. It can be said that the films are amorphous.

The composition of the deposited iron oxide film was examined with a X-ray Photoelectron Spectroscope. The binding energy of $4f_{7/2}$ valence electron in tungsten atom of the film was determined to be 35.6ev, which was close to that of WO_3 specimen with spectroscopic purity, 35.7ev. Thus, we concluded that the tungsten oxide film is composed of WO_3.

The similar examination was made for the iron oxide film with a Laser Raman Spectroscope. In the Raman spectra of the iron oxide films, there existed three peaks at 450, 800 and 1100 cm^{-1} respectively, which were exactly those of Fe_2O_3 specimen with spectroscopic purity. Therefore, the deposited iron oxide film is Fe_2O_3. The thickness of WO_3 and Fe_2O_3 film, measured with an Alpha-Step Thin Film and Surface Profile Measurement, were 3.0μm and 1.5μm respectively, thus the average deposition rates were 41Å/pulse for WO_3 and 30Å/pulse for Fe_2O_3. The evaporation rate of WO_3 film by means of the vacuum vapour coating, 30Å/sec, is much lower than that by LCVD[5]. Observed with a Normarski Microscope 8Q-2, these films appeared quite uniform. These films are so adhered to the substrates that scraped with a knife, they could hardly been taken off. Fig.3 shows the construction of Fe_2O_3 film taken with a Scanning Electronic Microscope. The size of the Fe_2O_3 particle in the film is about $5\mu m$.

The resistance of WO_3 film of 1.72mm long and 1.16mm wide was measured to be in the order of $10^9 \Omega$, which corresponds to a resistivity of $2 \times 10^5 cm \cdot \Omega$, similar to the result obtained by Faughnan [6]. The same parameter was measured for Fe_2O_3 film with an area of $3 \times 10 mm^2$, too. Its resistance is $2 \times 10^7 \Omega$ and the resistivity $9 \times 10^2 cm \cdot \Omega$. Four days after the exertion of 3000 Gauss magnetic field on the Fe_2O_3 film, the remanent magnetism near the film surface was about 1.7×10^{-3} Gauss.

The photolysis of $W(CO)_6$ and $Fe(CO)_5$ molecules may be multi-step processes under the irradiation of the XeCl excimer laser at 308nm. The dissociation energies of W atom in $W(CO)_6$ and of Fe atom in $Fe(CO)_5$ are about 9.85 ev and about 6.05ev, respectively, estimated from the formation heat of $W(CO)_6$, 227.2kcal/mole and $Fe(CO)_5$, 140kcal/mole [4]. Because the energy of a photon at 308nm is 4.03ev, at least three photons and two photons are needed for the dissociation of $W(CO)_6$ and $Fe(CO)_5$ respectively. Generally, this is multi-step processes with the single photon excitations rather than the multi-photon ones. During the multi-step processes, the molecule dissociation rate has a linear dependence on the laser intensity because of the long lifetimes of the intermediate products. The experiment method reported in this article is rather practical and convenient to manufacture WO_3 and Fe_2O_3 films with one-step-process. The LCDV method can also be adopted for the deposition of other oxide films. Recently, with KrCl excimer laser, we have been depositing the fluoresence film of PbO and photoconductive film of PbS by means of LCDV and have obtained exciting results.

Reference

1. Kanui F., Kurita, S., Sugioka, S., Li, M., and Mita, Y. "Optical characteristics of WO_3 electrochromic cells under heavy Li". Journal of the Electrochemical Society, Vol.129, No.11, pp.2633-2635(November 1982).
2. Hideo Okabe. Photochemistry of Small Molecules. John Whiley and Sons. Inc., The United States of America, 1978.
3. Wender, I. and Pine, P.Organic Syntheses Via Metal Carbonlys. Interscience, New York, 1968.
4. Yardley, J.T.,Gitlin, B., Nathanson, G.and Rosan, A.M.The Journal of Chemical Physics, Vol.74, No.1, pp.370-378 (January 1981)
5. Fang, Y."Study of cathodic coloration in the function-optical films based on α-WO_3". Acta Optica Sinica, Vol.4, No.2, pp.188-192 (February 1984).

6. Faughnan, B.S., Crandall, R.S. and Heyman, "Electrochromism in WO_3 amorphous films". RCA Review, Vol.36, No.1, pp.177-197 (March 1975).

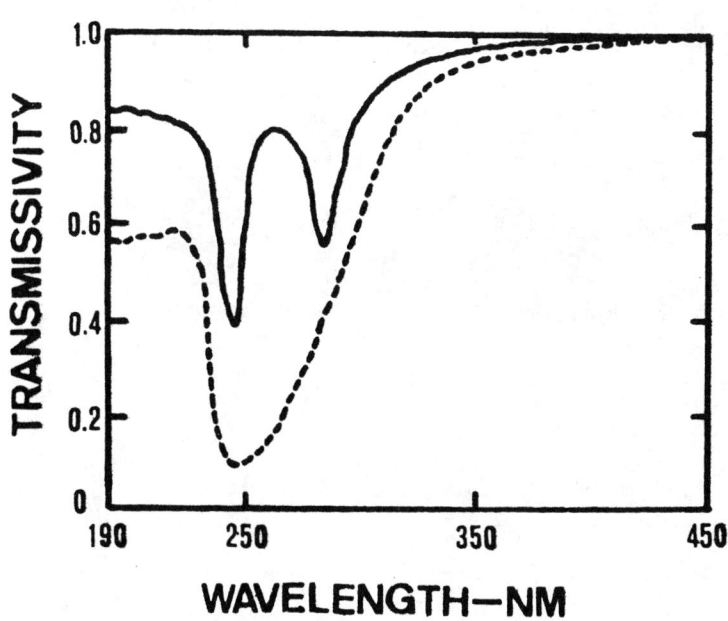

Fig.1. Absorption spectra of $W(CO)_6$ (full curve) and $Fe(CO)_5$ (dash curve).

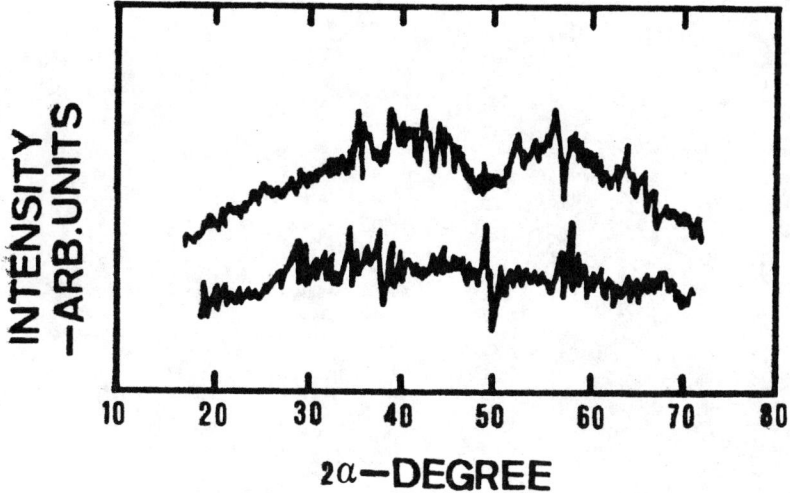

Fig.2. (a) X-ray diffraction spectrum of WO_3.
(b) X-ray diffraction spectrum of Fe_2O_3.

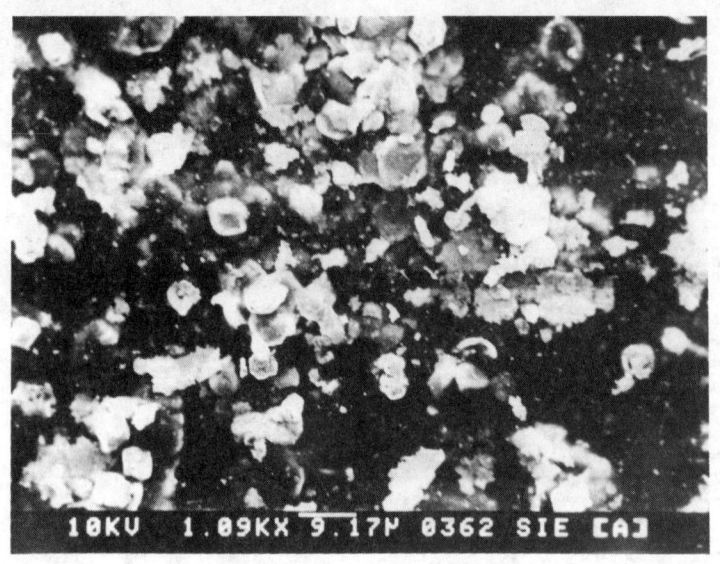

Fig.3. SEM photograph of Fe_2O_3 film

LATE PAPERS

Heat treatment of steels using different high power CO2 laser beam intensity distributions

W. Cerri and A. Vendramini
CISE SpA, Italy
and
E. Ramous
Università di Padova, Italy

ABSTRACT

High power lasers have been used since several years in heat treatment, mainly for surface hardening of steels.

To determine the best operative conditions a set of surface hardening tests has been carried out on a constructional steel using laser beams generated by two different optical cavities:

- a stable multimode optical resonator and
- a stable-unstable optical resonator.

A comparison of the obtained results has been made.

1. EXPERIMENTAL DEVICE

The technological results of laser surface hardening process depend on the incident laser beam intensity distribution on the workpiece. To evidence this influence hardening tests have been carried out using two laser beam distributions generated by different optical cavities. These hardening tests have been performed by a transverse flow self-sustained discharge C.W. CO_2 laser designed and realized by CISE (1).
The following optical cavity configurations have been adopted:
a) stable-unstable resonator

b) stable-multimode resonator.

The stable-unstable resonator has been developed by CISE in the frame of the Italian National Research Council Special project on High Power Lasers (2). This resonator behaves like on off-axis unstable confocal resonator in one trasverse dimension and like an on-axis concave-convex stable resonator in the other orthogonal direction.

Such an optical cavity produces a rectangular cross-section laser beam. The laser beam intensity distribution is shown in fig. 1a (intensity distribution in the stable plane) and in fig. 1b (the same in the unstable plane).

The stable resonator adopted in the second set of tests is a common multimode concave-flat cavity with a high Fresnel-number. The intensity distribution of the laser beam produced by this resonator is reported in figure 2.

In the two sets of hardening tests the beam has been delivered to the target simply by focussing it by a 10" focal length ZnSe meniscus lens and placing the sample at a fixed distance (90 mm) from the focal plane. It was not used any beam integrator optical system.

Construction steel UNI 38 NiCrMo4, which corresponds to AISI 9840, has been used. The range of application for this steel is: axles, axle arms, connecting rods, arbors and so on.

The composition and the mechanical properties of the steel are reported in Tab. 1.

An absorptive coating (colloidal graphite in buthil acetate) has been used to enhance the infrared absorbitivity of the steel. A helium flow of 20 l/m has been utilized as shielding gas. The following experimental conditions (power and interaction time) have been adopted:
- laser power 2.3 kW on the workpiece,
- processing speed ranging from 0.1 m/min to 1 m/min.

2. HARDENING RESULTS

The treated samples have been examined by means of macro and micrographs and microhardness tests.

The hardening results are satisfactory: a uniform, homogeneous, fully hardened structure has been obtained. The hardened case depth is ranging from about 0.5 mm to about 2 mm. The hardness values achieved correspond to a completely transformed martensitic structure (HV 700). Furthermore no quenching cracks of surface microcracks have been observed.

The hardened zone and structure is shown in fig. 3.

The hardened layer case depth has been measured by means of microhardness tests. The results (case depth in different experimental conditions) are summarized in fig. 4.

The microhardness profiles obtained for a couple of tests performed by the stable-unstable resonator (fig. 5) and by the stable multimode resonator (fig. 6) are reported.

3. HARDENING DEPENDENCE ON THE OPTICAL CAVITY TYPE

The results obtained by the two cavities show a few significative differences. The presence of a hot spot in the intensity distribution generated by the stable-unstable resonator (see the central peak in fig. 1) determines the onset of surface melting effects at low processing speed by the stable multimode resonator (see fig. 7).

Furthermore the hardness profiles in the transverse section of the hardened layer are more uniform (see fig. 5 and fig. 6) in samples treated by the stable multimode laser beam.

4. CONCLUSIONS

The hardening tests have proved the technical feasibility of a CO_2 laser surface hardening treatment on a widely used constructional steel using a defocused beam produced by a multimode stable resonator and by a stable-unstable resonator.

The best results have been obtained in tests performed by the extremely flat intensity distribution generated by a stable multimode cavity. The beam produced by a stable-unstable resonator, although its intensity distribution is not optimized for hardening application has given positive technological results in a wide range of hardened layer depths. On the other hand the beam generated by the stable-multimode resonator cannot be efficiently focussed to achieve power density values suitable for laser penetration welding. On the contrary welding processes can be performed by the same laser changing the optical resonator from stable multi mode to stable-unstable which, as well known, is more suitable for welding applications. Some results of penetration tests carried out by the stable-unstable beam are reported in fig. 8, while the bead-on-plate penetration speed versus the material thickness is shown at a power level of 2 kW. These results favorably compare with those reported for commercially available lasers of this power range.

In conclusion these tests indicate that the stable-unstable resonator can be considered a flexible optical cavity able to generate a beam suitable to be used in different technological applications.

(1) V. Fantini, G. Incerti, W. Cerri, V. Donati, L. Garifo
    "A 5 kW CW CO_2 laser for industrial applications"
    Fifth International Symposium on Gas Flow and Chemical Lasers,
    Oxford (U.K.) 20-24/8/1984.

(2) A. Borghese, R. Canevari, V. Donati, L. Garifo
    "Unstable-stable resonators with toroidal minors"
    Applied Optics, Vol. 20, pp 3547-3552 (1981)

TABLE I

Chemical composition (%)

C	Mn	Si	P	S	Ni	Cr	Mo
0.40	0.74	0.25	0.011	0.007	0.99	0.95	0.23

Mechanical properties
- Tensile strenght : 1005 N/mm^2
- Yield point : 898 N/mm^2
- Elongation : 16 %
- Reduction of area: 64 %

Heat treatment : quenched and tempered.

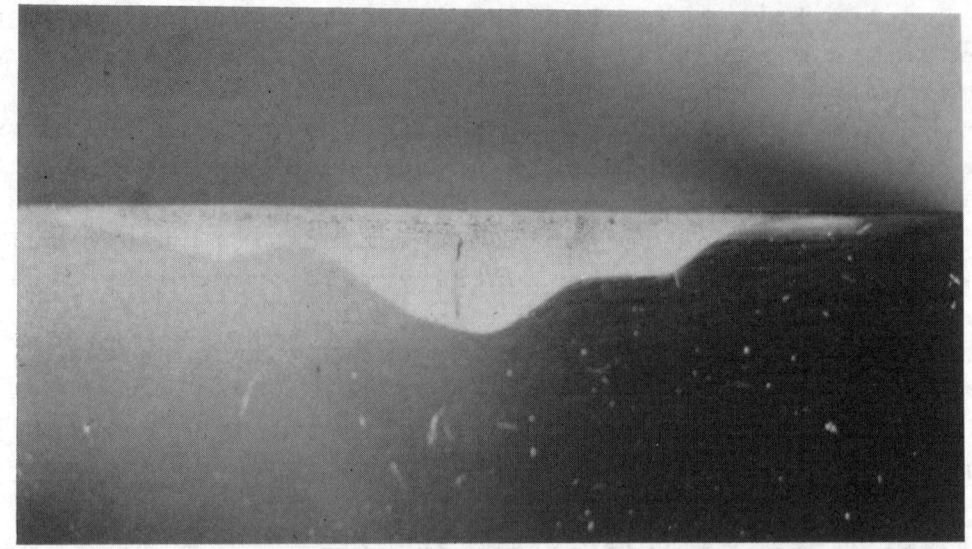

A

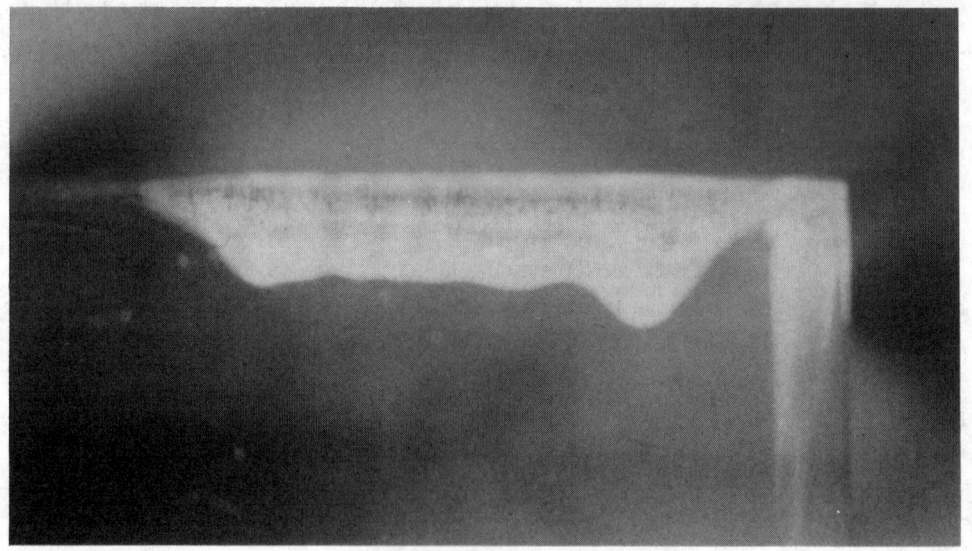

B

Fig. 1 - Intensity distribution of the laser beam by the stable-unstable resonator:
(A) in the stable direction,
(B) in the unstable direction.

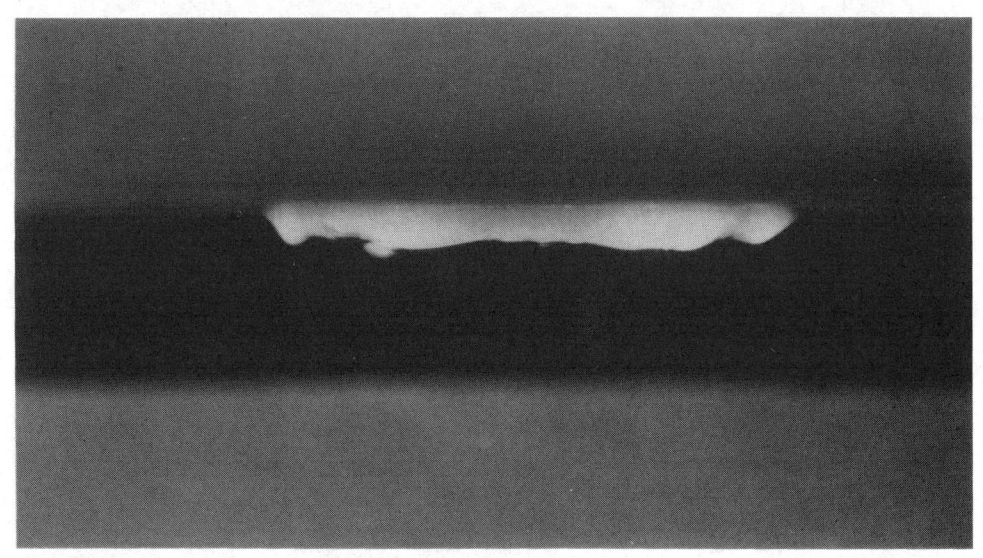

Fig. 2 - Intensity distribution of the lser beam generated by the stable multimode resonator.

Fig. 3 - Stable multimode resonator: microstructure of the hardened zone (25x).
Processin speed: 0.6 m/min.

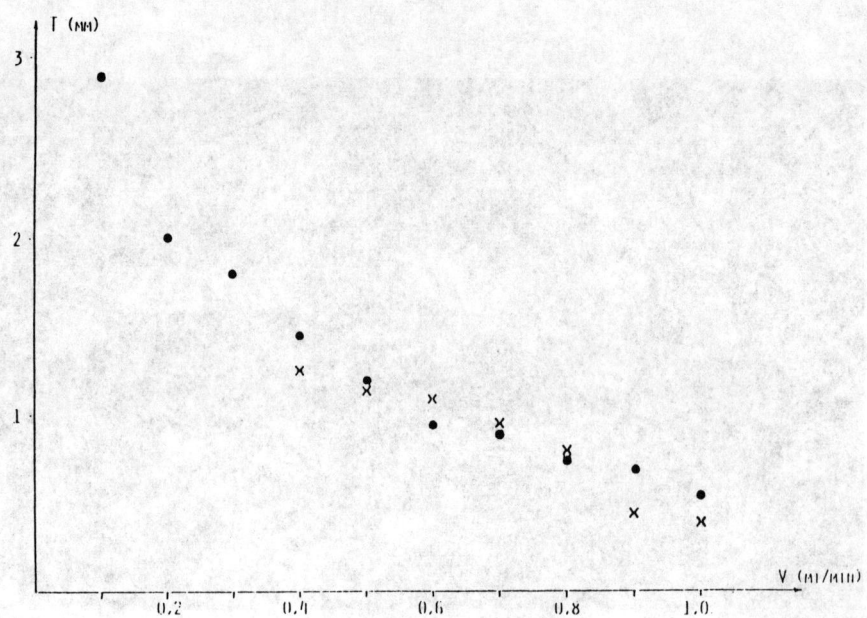

Fig. 4 - Hardened layer case depth (T) vs. processing speed (V): comparision between stable-unstable (x) and stable multimode (●) resonator.

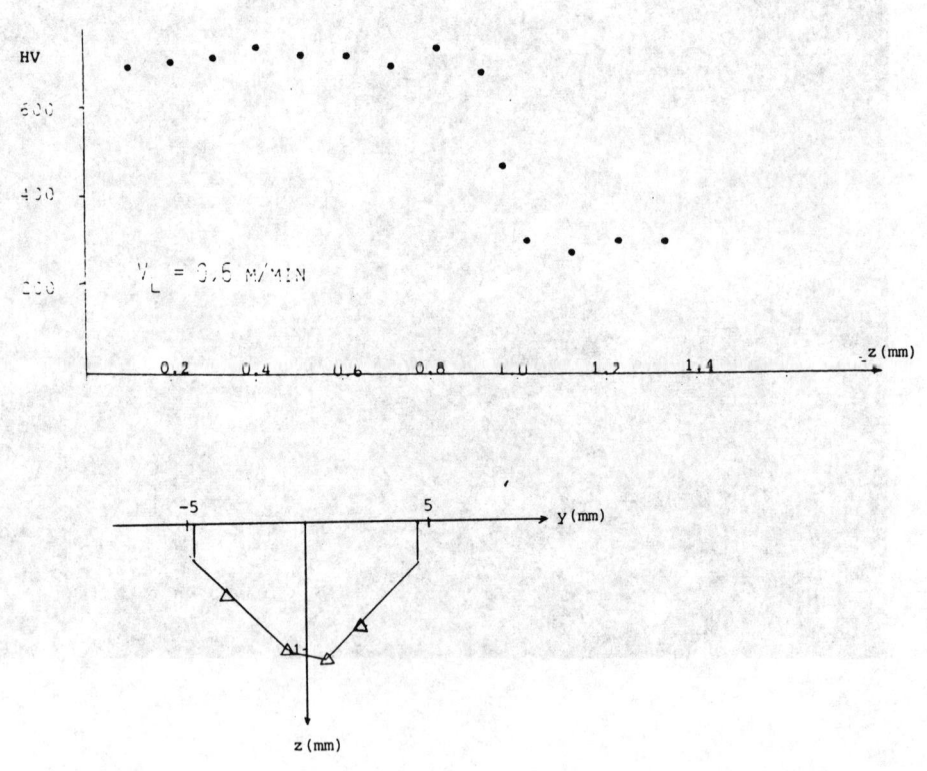

Fig. 5 - Hardened zone case depth an geometry: stable unstable resonator.

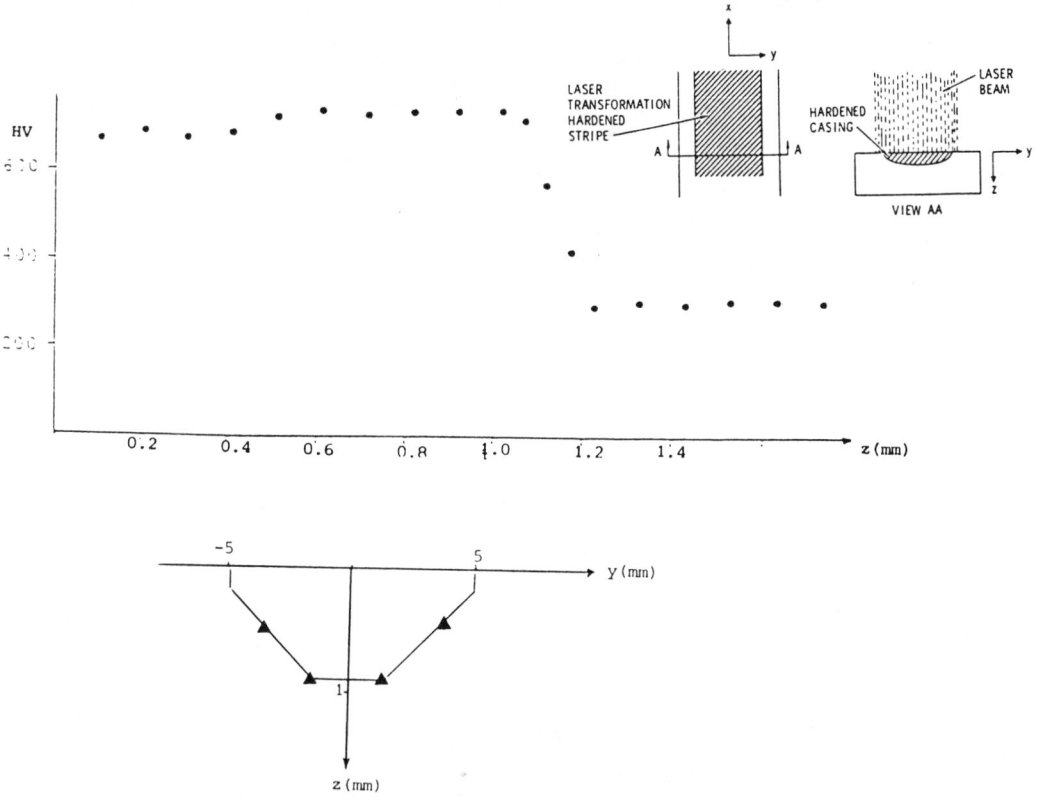

Fig. 6 - Hardened case depth and geometry: stable multimode resonator.

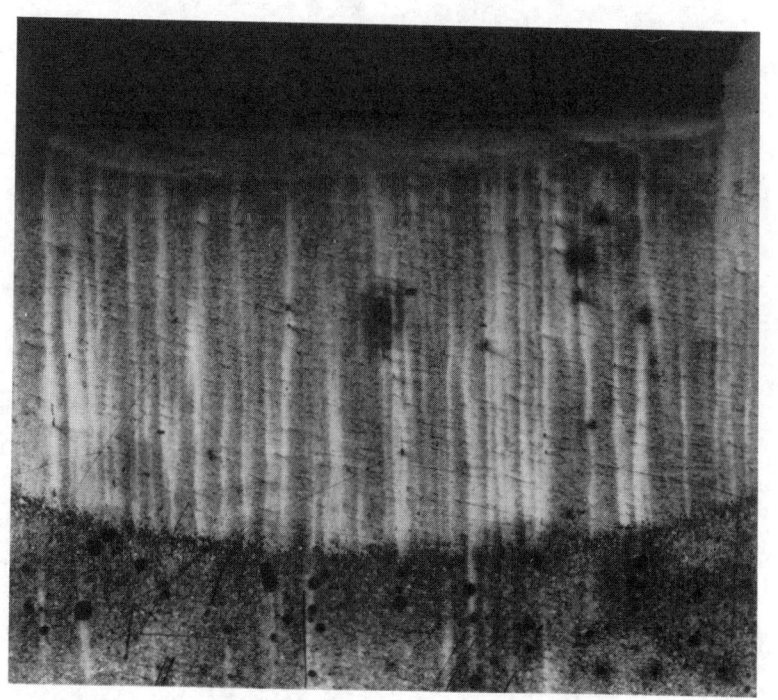

Fig. 7 - Hardened zone structure with a melted region on the surface: stable--unstable resonator.

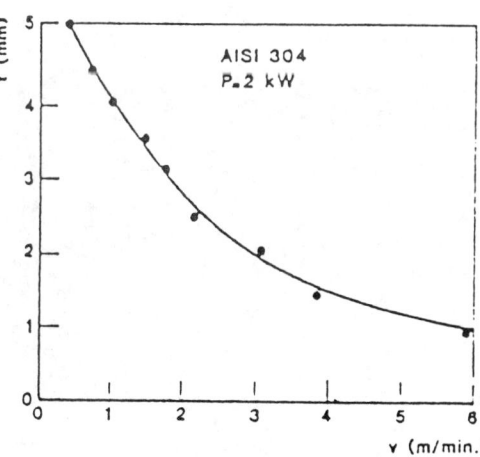

Fig. 8 - Bead-on-plate penetration thickness vs welding speed: stable-unstable resonator.

Optical problems of beam delivery
N. Forbes
Ferranti Industrial Electronics Ltd, UK

Aspects of optical beam delivery will be considered. Typical components used in standard beam delivery systems will be discussed and material and processing problems associated with the manufacture of such components briefly mentioned. Modern methods of beam delivery will be described, together with projections for future systems. Novel forms of beam delivery are emerging and will be of growing importance to articulated arms and fibre optic delivery systems in future years. The use of mirrors to shape the beam will be described as well as a production example of a more advanced optical system for cigarette manufacture.

INTRODUCTION

Lasers are used in a wide variety of applications, ranging from the processing of solid materials to interaction with liquids or gases. Almost all applications use optics to deliver the beam to the interaction region. This paper has been restricted to some considerations affecting the beam delivery from high power lasers above one watt since this covers an area of growing importance to industry.

RESONATOR CONSIDERATIONS

The simplest form of laser resonator is formed by two spherical mirrors of suitable radii of curvature. At one extreme the radii are infinite and at the other the radii are equal to the separation of the mirrors. These configurations allow free space modes to propagate and, by careful arrangement of the gain element and these mirrors, the laser output can be constrained to the TEM_{00} mode giving a gaussian intensity profile (Fig 1).

More complex resonators are used in solid state lasers, such as neodymium in YAG, but the complexity arises from other requirements. Waveguide lasers also employ a more complex resonator structure of which the walls of the gain element form part. The output of these devices is usually in the form of an EH_{11} mode (Fig 2), 98% of which can be converted into the TEM_{00} free space mode for subsequent use. Another resonator configuration is also used where the walls of the gain element do not contribute a guiding effect or a constraint. This type is known as an unstable resonator and has at least one of the mirrors with negative radius of curvature. The output takes the form of an annulus which leaks past one of the mirrors (Fig 3a). When such an output is brought to a focus by a lens or mirror the resultant intensity profile differs from gaussian but the shape is such as to allow highly effective interaction with materials of interest (Fig 3b).

Another factor inherent in the behaviour of laser resonators which must be borne in mind is the polarisation state of the output. This affects the design of subsequent optics and also the interaction of the output with the material being processed.

BEAM OPTICS

The final factor which gives rise to problems is the power of the laser beam itself. Most machining operations involve at least one mirror and one lens, as shown in Fig 4. The usual situation is more complex, especially when the laser optics are also considered. Fig 5 shows a typical set of optics associated with a high-power CO_2 Laser. The end mirror and folding mirrors are high-reflectivity dielectric mirrors on silicon or germanium substrates. These, together with the output mirror, control the performance of the laser. The output mirror is a partial reflector on a transparent substrate such as gallium arsenide or zinc selenide. The outer surface of the output mirror is anti-reflection coated to enhance the power available in the output beam and to minimise the feedback of power from this surface into the laser cavity, thus causing instability in the output.

Depending on the fold configuration employed in the laser, the polarisation may be fixed or varying in time. In general, lasers employing "roof edge" folds have a fixed polarisation, while those with "Z" folding or with no folds have a polarisation which can vary. To fix the polarisation of the latter two a polarising element is employed. This element, known as a "Brewster window", is set at such an angle to the laser beam that one plane of polarisation dominates. Zinc selenide is the preferred material for CO_2 lasers since it has a low absorption. Great care must be taken when polarisers of this type are used and they should only be employed when absolutely necessary.

External optics, though critical to manufacture, have a little more latitude associated with them. In recent times it was realised that the state of polarisation of the laser beam affected the quality of the cut in metal. To overcome this problem phase shifting mirrors were introduced to render the beam circularly polarised.

In some applications a marker beam from a helium-neon laser is combined with the main laser beam. This is achieved by producing a coating which is transparent to the main beam while reflecting the helium-neon laser beam. The 45° angle of incidence and the polarisation of the beams make this a difficult component to manufacture.

The beam-bending mirrors are used to change the direction of the beam and these are the least critical optics in the system. These are formed either by dielectric high reflectivity coatings on silicon, or diamond machined copper with a gold coating.

An enlarged beam diameter allows a smaller spot size at the focus which is more effective for cutting processes. The increase in diameter is produced using beam expander lenses which must be of low absorption with highly efficient anti-reflective coatings.

In some systems the laser beam is split into two or more beams so that several work stations can be employed. The beamsplitters are normally at 45° angles of incidence and have the same difficulties as the beam combiners since the polarisation state of the laser beam has a significant effect on the design of the coating.

The final component in the system is the focusing lens which produces the high power density required for machining operations. The component must be of low absorption with highly efficient anti-reflective coatings which are easily cleanable, since debris and other contamination from the workpiece can reach the lower surface of the lens, causing damage due to burning at this surface.

MATERIAL CONSIDERATIONS

Several conditions have to be met, particularly for the transmissive optics in such a beam delivery system. Some of these also apply to the less critical components. Absorption arising from the bulk substrate, the state of surface polish of the substrate and the type of coating employed all affect the choice of material used for these components. Germanium, zinc selenide, gallium arsenide and cadmium telluride have proved the most successful materials in CO_2 machining lasers. Other materials such as potassium chloride and sodium chloride are used in high power short pulse systems but these materials suffer disadvantages due to their hygroscopic nature which limit their use to very specialised applications. For low power applications, below 50 watts, germanium is a suitable choice of material. However, this suffers from an effect known as 'thermal runaway', where the absorption increases exponentially with temperature as shown in Fig 6. Germanium experiences thermal runaway slightly above room temperature and careful cooling is therefore required when using this material. In practice, the choice of substrate comes down to gallium arsenide or zinc selenide and where visible transparency is required, such as when a helium-neon marker beam is being used, then the focusing lens is restricted to zinc selenide.

The problems experienced by these components during operation arise from excessive heating, due to poor-quality substrate material (bulk absorption), surface absorption, coating absorption or dirt on the surface of the coating. In addition to thermal runaway, optical distortion can also occur. This latter effect, known as thermal lensing, is due to the change of refractive index with temperature which changes the focal length of the component of interest. Thermal runaway can be experienced in the output window where debris from the laser discharge, electrodes etc, can stick inside the reflecting surface giving rise to heating, and in the focusing lens where debris from the material interaction can form on the lower surface of the component. It would appear that, under ideal conditions, zinc selenide is the preferred material to use, ie when there are no coatings or low absorption coatings, zinc selenide is the best performer. If coating absorption is rather high, then gallium arsenide behaves better. In a practical situation, therefore, if dirt can form on the surface of the coating, gallium arsenide is a more tolerant substrate. If ideal conditions can be maintained, then zinc selenide is the clear choice. As a zinc selenide component degrades due to surface contamination, the distortion of the component is much faster and an operator will notice the degradation in performance and carry out maintenance operations, thus maintaining the zinc selenide in good condition. However, it must be pointed out that this is an ideal operator and my own experience has been that the distortion in the lenses has reached severe proportions before maintenance is carried out and, in this situation, gallium arsenide is the more robust material.

All these components require a great deal of care on the part of the optical manufacturer. The substrates must be measured for absorption prior to coating. In this way the bulk absorption and the surface absorption due to polishing can be determined. Table 1 shows the variation in absorption due to surface polishing processes for a list of materials and various suppliers. The coatings themselves are a compromise between low absorption and robustness and, again, must be measured using laser calorimetry.

TABLE 1: ABSORPTION IN COMMERCIALLY POLISHED SUBSTRATES

Supplier		10.6 Micron Absorption (%)			
		Ge	GaAs	ZnSe	ZnS
U.K.	A	1.46 ± 0.65	0.36 ± 0.08		5.33 ± 1.61
	B	0.66 ± 0.12	0.30 ± 0.04	0.23 ± 0.05	
	C	2.38 ± 0.79		0.52 ± 0.38	5.61 ± 1.24
	D	0.99 ± 0.17		0.29 ± 0.07	5.04 ± 0.76
	E	0.74 ± 0.08			
U.S.A.	F	0.75 ± 0.14		0.23 ± 0.03	
	G				4.69 ± 0.62

Published figures show the batch mean with standard deviation.

When satisfactory components are supplied by the optical vendor care must be taken by the user not to introduce more contamination to the components by handling them with bare hands, allowing grease and acid to pass on to the components and to minimise exposure to the environment around the cutting machine. High-humidity environments can also lead to changes in the performances of optics which are usually water cooled. Water condensing on the surfaces can leave absorption residues and care must be taken to adjust the temperature of cooling water, particularly when the laser is shut down, so that condensation cannot occur.

ALTERNATIVE BEAM DELIVERY SYSTEMS

In the early 60's an articulated arm delivery system was demonstrated by AWRE. Little progress in employing this device was made until the mid 70's when the Culham Laboratories produced a variation capable of delivering 400 watts. In the early 1980's this type was coupled to an industrial robot which held the cutting nozzle. The position of the robot-held nozzle is controlled either by numerical control or by a playback routine. The mirrors used in the articulated arm are plane diamond turned copper mirrors with gold coatings. Such systems are used in a controlled access area since they do provide a flexible machining facility which will work equally well on the workpiece or on an intruder in the area. This type of beam delivery is being used with robots in automobile manufacture and there is a significant and growing interest in the United States in such machines.

Another form of beam delivery which is coming into operation is that of fibre optics. Fibre optic beam delivery systems have been employed for several years in medical applications. These systems used silica fibres of the "Nath" type and were used in conjunction with continuous wave neodymium in YAG lasers. Their power handling capability was a few watts. Development work in Japan on their flexible manufacturing system has led to fibre optic delivery systems giving 300 watts to the work piece. These delivery systems are again used with CW neodymium in YAG lasers. The power delivered in this way is used for chip breaking in turning applications or for deburring processes in the machining of gear teeth. The Japanese have also developed infrared transmission fibres but progress in their use in CO_2 laser machining has been restricted due to their high cost. Recent announcements by Gallileo in the States that infrared optical fibres are available should lead to their use by the industry. It is quite possible to envisage a properly engineered and cooled fibre bundle capable of delivering 500 watts of power to the workpiece. Specialised optics will be required to adapt the laser output to a two dimensional array of inputs to match the fibre optic bundle. A specialised focusing system will also be required to recombine the power at the workpiece.

COMPLEX DELIVERY SYSTEMS

Many applications call for complex optical techniques to achieve the required result. Most systems presently deployed rely on lenses to transmit and focus the laser power. After early work on mirrors these have fallen into disuse except for very high power devices. This has arisen because of a lack of understanding as to the aberrations introduced when badly positioned off-axis parabolas are employed. The angular position of an off-axis parabola is as critical as standard centring tolerances are for lenses. A tilt of one minute of arc from its correct position with respect to the laser beam can result in a doubling of the spot size at the focus. Advances in diamond turning technology now allow the machining of suitable off-axis parabolas and toric mirrors which enable the laser beam to be adapted to suit particular requirements.

The output from the lasers, as discussed earlier, tends to give rise to gaussian or near-gaussian energy distribution, as shown in Fig 1. However, for applications such as heat treatment, the wings of the gaussian cause some annealing of the already hardened material as the beam passes over the workpiece in a raster pattern. A more desirable state of affairs would be to have a top-hat distribution of intensity as shown in Fig 7. Preliminary investigations by an optical designer show a solution which may achieve this end with the minimum spot size which can be produced. A possible configuration is shown in Fig 8 and comprises two toric mirrors, designed for a 10mm diameter 1 kw laser beam which is gaussian on its input to the first of the two mirrors, as shown in Fig 9. Fig 10 shows the output beam cross-section in front of and behind the focus of the two-mirror system. It can be seen that at -0.4mm from focus the desired energy distribution has been obtained. It should be emphasised that the spot diagrams in Figs 9 and 10 essentially show the light flux and arise from a geometric investigation. The distribution shown at -0.4 mm in Fig 10 will be modified when diffraction is taken into account and work is in hand to develop a wavefront analysis to allow this to be carried out. The result demonstrated for the pair of mirrors can also be obtained for a lens which has aberrations introduced into it to allow a top-hat power intensity distribution to be achieved at focus. However, it is felt that as the power in lasers rises, the toric mirror approach will provide a valuable method of concentrating the energy into a desirable profile.

By way of a final example of the problems of optical beam delivery, the system used for the perforation of cigarette filter paper will be briefly described. The cigarette industry has long desired to reduce the tar content of cigarettes by perforating the filter paper in an on-line process. A system has been devised which allows this to be done in an on-line process where the pattern of holes can be varied, virtually at will, to meet the cigarette manufacturers' requirements. The types of pattern required are shown in Fig 11 and the unfolded optical arrangement is shown in Fig 12. An acousto-optic beam deflector is used to vary the power and spot size of an array of spots on the paper. A cylindrical lens is used in the system to shape a laser beam profile to match the elliptical aperture of the beam deflector. After deflection a further cylindrical lens restores the beam profile to normal to allow focusing by the final lens. Such a system is now operating on an industrial basis allowing paper to be perforated while moving at 1.8 m per second.

CONCLUSION

Standard means of transmitting power to the workpiece are reasonably well understood and suitable components can be manufactured. The careful analysis of mirror systems can give rise to novel sources of intensity distribution and these will receive further exploitation in years to come. Novel systems using active beam deflection are now operational in industrial environments.

ACKNOWLEDGEMENTS

Some of the contents of this paper have been freely adapted from the references cited below. In particular, I am grateful to Glenn Sherman of Laser Power Corporation for some thoughts in a private communication and to Dr Heshmaty of the Optical Components Group of Ferranti Industrial Electronics Ltd. for his work on toric mirrors.

REFERENCES

Sherman & Frazier, "Transmissive Optics for High Power CO_2 Lasers, Practical Considerations". Optical Engineering.

Reid, Ramsay, Dyson & Ross, "Commercial Laser Beam Perforation of Cigarette Tipping Paper Using Acousto-optic Beam Deflector". SPIE Meeting, San Diego, August 1984.

"Advances in Laser Metal Working". American Machinist (January 1985, 79-81).

Spalding, I J, "Laser Applications". Phys Bull, Vol 35, 1984.

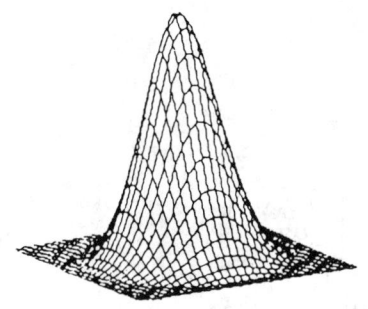

FIG. 1

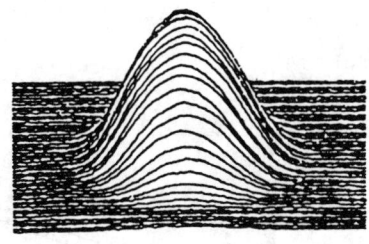

FIG. 2

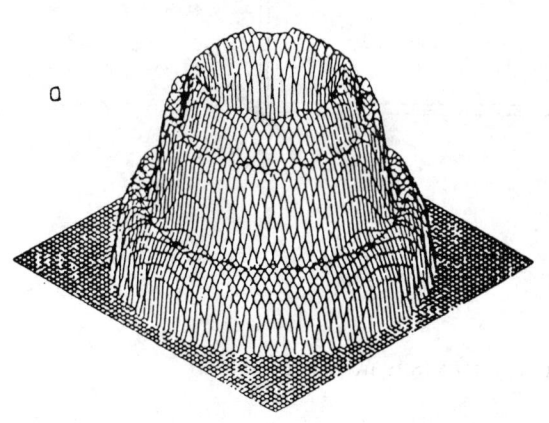

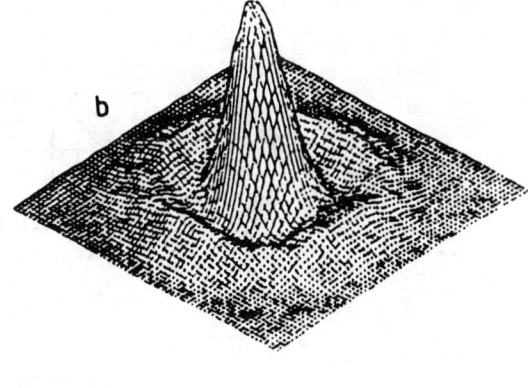

FIG. 3

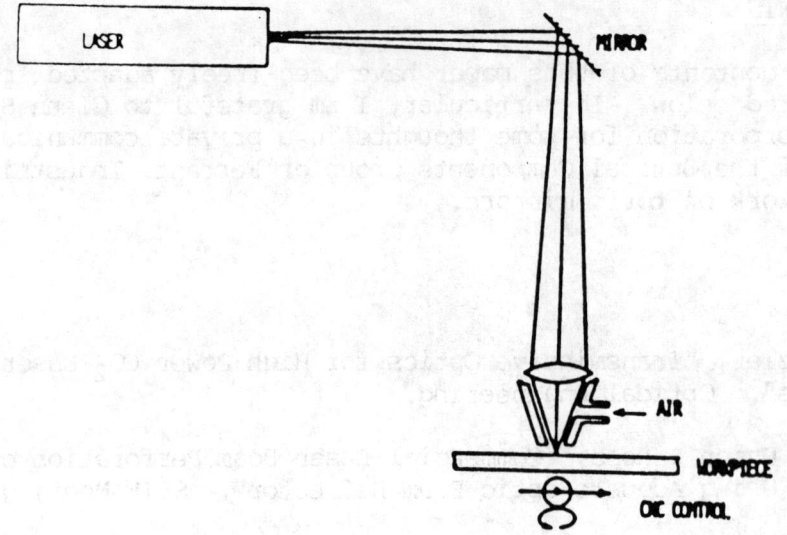

FIG. 4

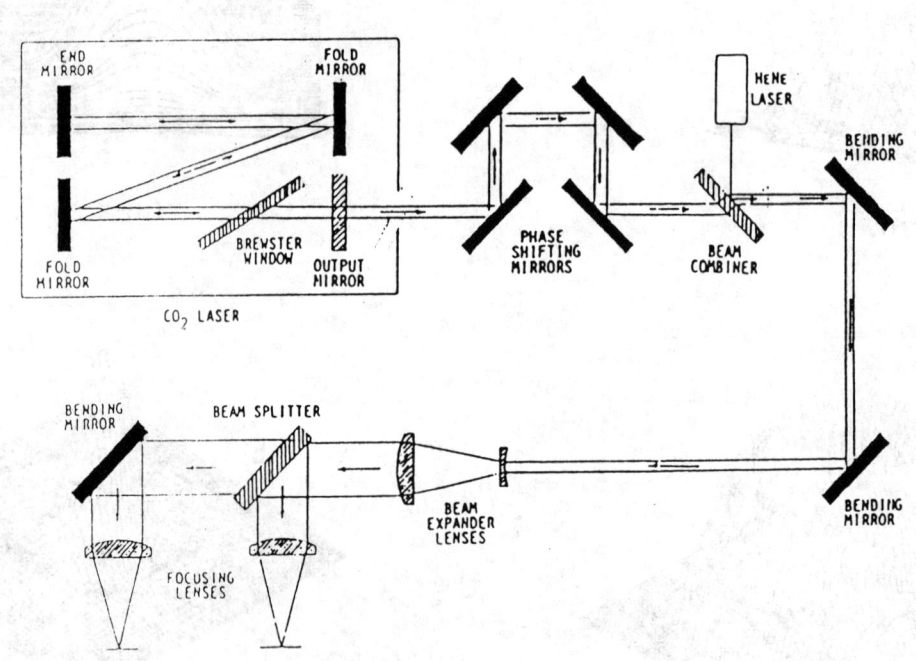

FIG. SCHEMATIC DIAGRAM OF TYPICAL INDUSTRIAL OR MEDICAL CO_2 LASER BEAM DELIVERY SYSTEM

FIG. 5

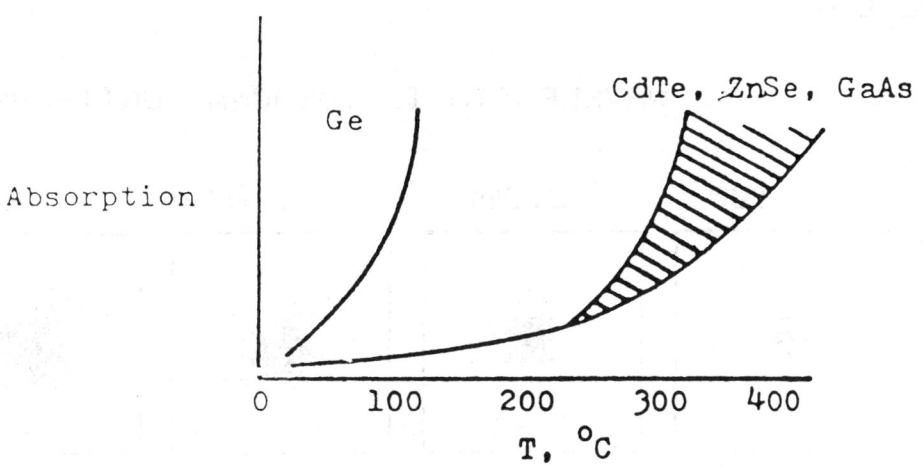

Thermal runaway in semiconductor laser optics.

FIG. 6

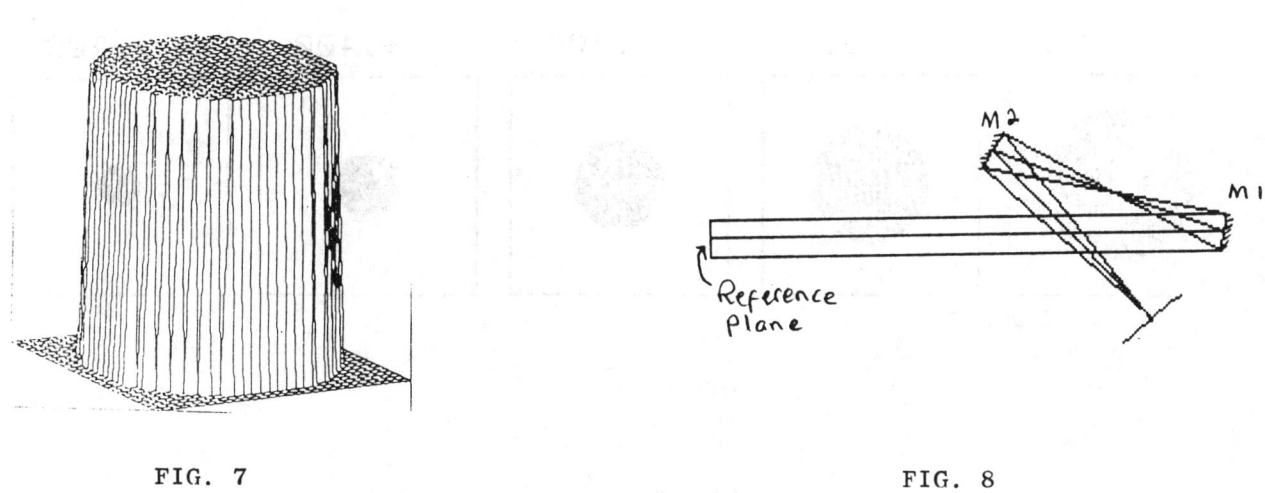

FIG. 7   FIG. 8

INPUT BEAM CROSS SECTION (1000 RAYS)

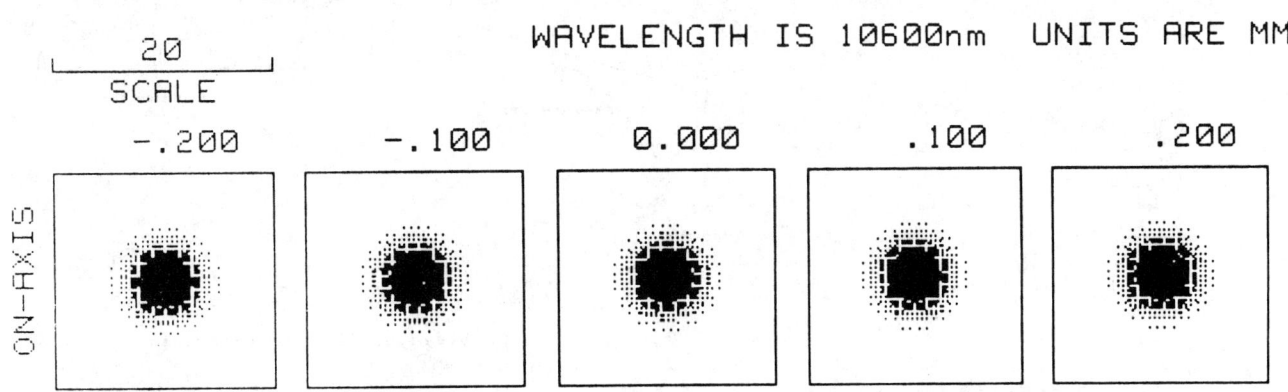

FIG. 9

OUTPUT BEAM CROSS SECTIONS

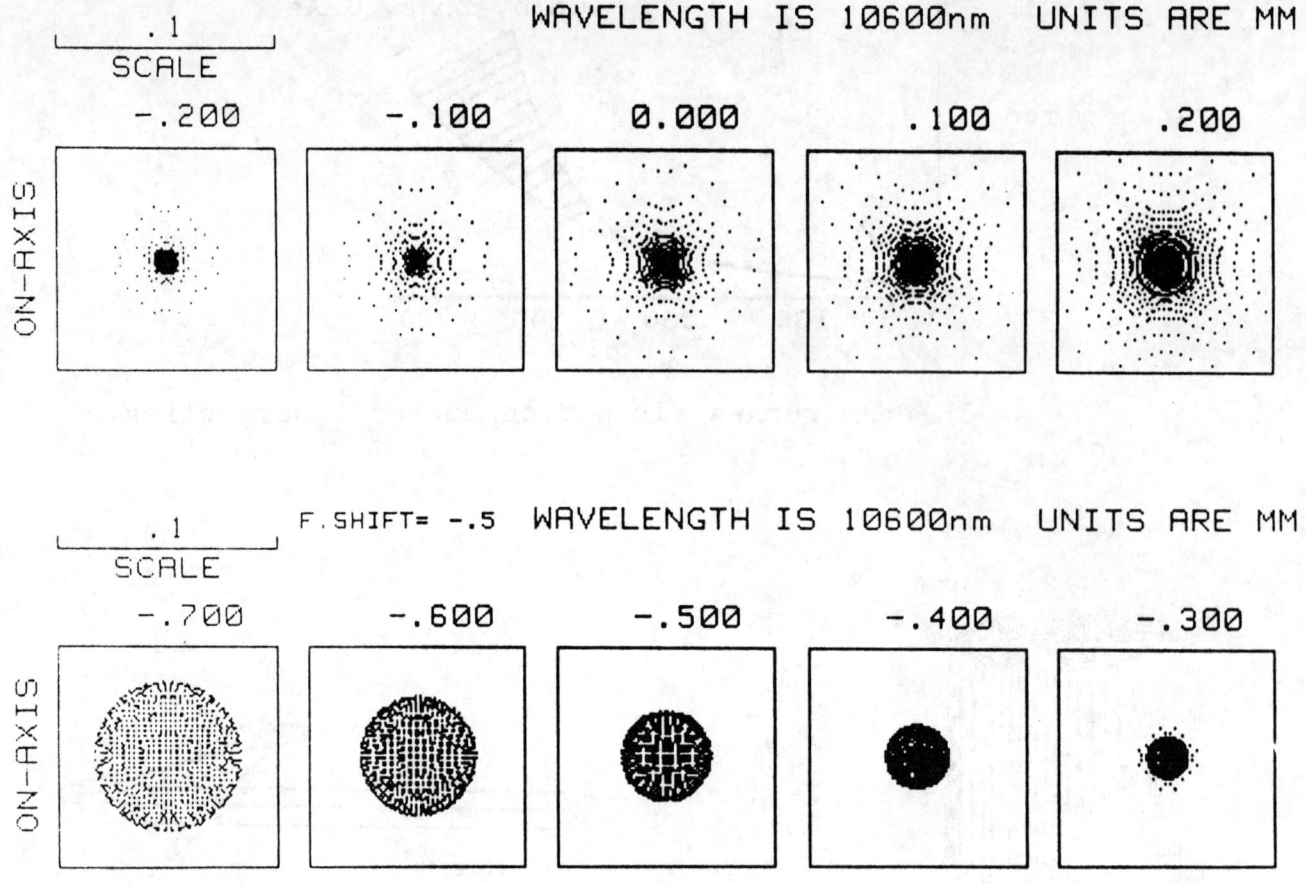

FIG. 10

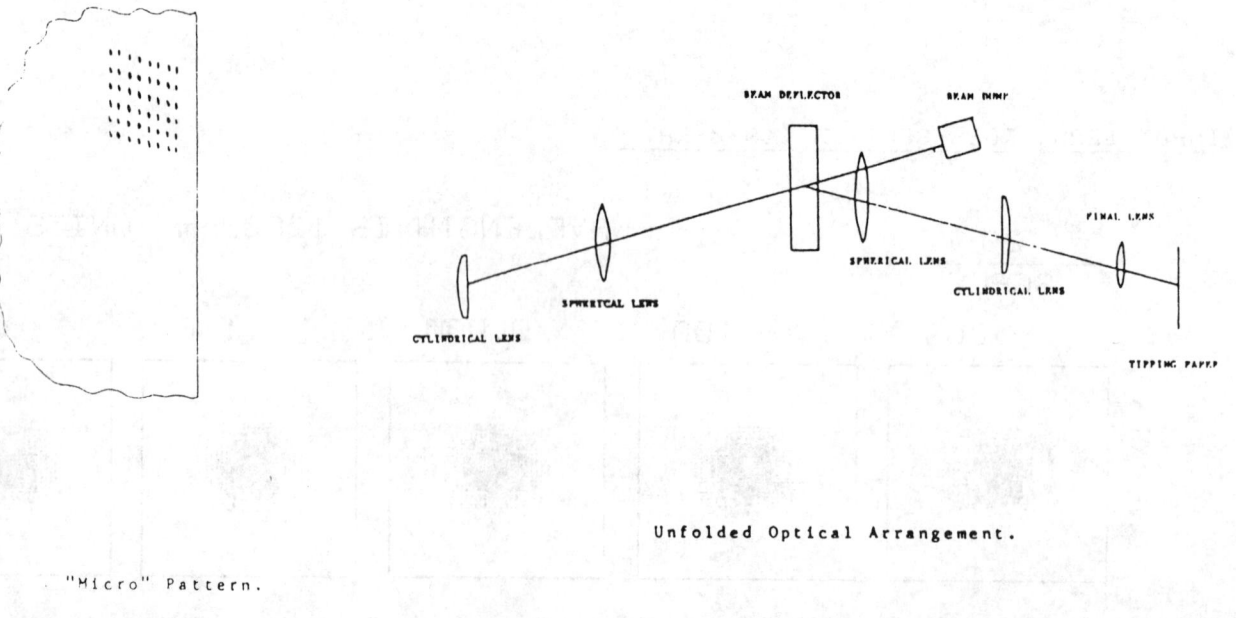

"Micro" Pattern.

FIG. 11

Unfolded Optical Arrangement.

FIG. 12

RAYMOND H. FOGLER LIBRARY
DATE DUE

BOOKS ARE SUBJECT TO
RECALL AFTER TWO WEEKS

NOV 13

79.75
70C
110491933
130
EISEN km
MB